Plastercine.

Ap

Applied Mathematics

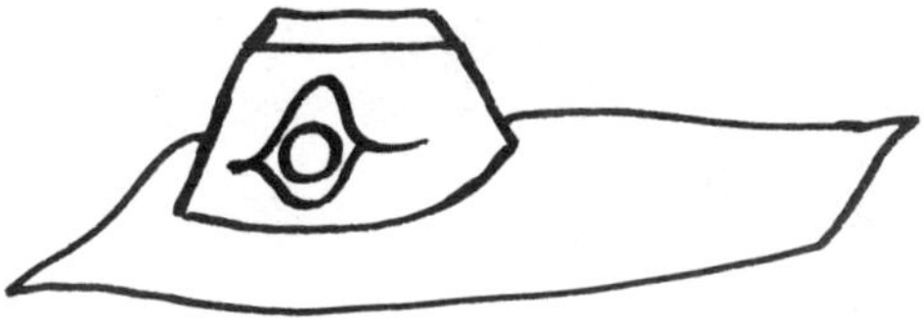

Applied Mathematics

By **E. D. HODGE,** B.SC. *and* **B. G. J. WOOD,** M.A.
Manchester Grammar School

SECOND EDITION
IN SI UNITS

1809
Lucem Libris Disseminamus

BLACKIE AND SON LIMITED

LONDON AND GLASGOW

BLACKIE & SON LIMITED
5 FITZHARDINGE STREET
PORTMAN SQUARE
LONDON · W.1
BISHOPBRIGGS, GLASGOW

SI unit edition 1970

Reprinted 1971 (twice) 1972 (twice)

ISBN 0·216 87403·3 (limp)
0·216 87678·8 (board)
Dewey 531·01

PRINTED IN GREAT BRITAIN BY R. & R. CLARK LTD., EDINBURGH

PREFACE

THIS BOOK PROVIDES a course in applied mathematics covering the requirements of the various examining bodies for the General Certificate of Education, as set out in their A-level syllabuses in mathematics and physics. We assume that the reader is acquainted with elementary algebra and trigonometry, and also with simple differentiation and integration. Such differential equations as occur are, however, fully explained.

Chapters 1–3 deal purely with motion. Variable accelerations, including circular and simple harmonic motion, are introduced at this early stage in the hope that students will be convinced that all acceleration is not uniform—a very common misconception. Chapters 4–7 deal with statics, and 8–13 with dynamics of a particle. In the remaining chapters we consider rigid dynamics, which, though not required by all examining bodies in their mathematics syllabuses, is a necessity for any physicist, and we also apply differential equations to physical and other problems resembling those of dynamics. Those who prefer to do so may start the book at Chapter 4 with statics, and defer the reading of the first three chapters until ready to begin dynamics. It is also possible to leave the chapters on statics until after reading the dynamics of a particle, if that course is preferred.

Our aim has been to provide sound bookwork, and illustrative examples which will show how to approach a problem, and how to set out its solution. The exercises have mostly been arranged in parallel sets marked A and B, an arrangement which is generally found useful. Revision examples on each chapter are also provided.

We gratefully acknowledge the permission given us by the Joint Matriculation Board of the Northern Universities to use questions from its examination papers.

E. D. H.
B. G. J. W.

June, 1960

PREFACE TO REVISED EDITION

This revised edition uses throughout the SI or International System of units, the modern form of the metric system, which is superseding both the F.P.S. and C.G.S. systems formerly used. It has consequently been necessary to recast many of the exercises. The only exception is that in examples concerned with sea and air navigation the nautical mile and knot have been retained, as they are likely to continue in practical use. The abbreviations employed are those officially recommended.

Full information about all SI units is available, for example in the booklet *Changing to the Metric System* (H.M.S.O.).

B.G.J.W.

March 1970

CONTENTS

1

Vectors. Velocity

1. Vectors

The displacement of a moving point has a magnitude and a direction: for example, it might be 2 m horizontally to the north. Such a quantity is called a *vector*; a quantity which has magnitude, but no direction, such as a volume, is called a *scalar*.

It is important to notice that any description of a displacement implies some frame of reference; that is, some background over which the displacement occurs. The frame of reference which we most commonly use is that part of the Earth's surface in the neighbourhood of the points considered, but other frames must sometimes be used. For example, the movement of a person in a railway carriage may be considered with reference to the train, or to the Earth. The motion of the Moon may be considered with reference to the Earth, or to the Sun.

2. Combination of vectors

The combined effect of successive displacements is equivalent to a single displacement. In the figure, ABCD is a parallelogram. The displacements from A to D, and from B to C, are equal in magnitude and direction, and we write $\overline{AD} = \overline{BC}$, the bar denoting a vector. The order of the letters must be observed.

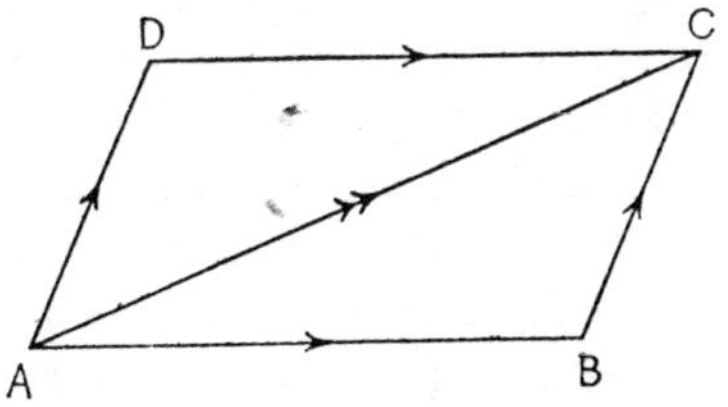

The combination of vectors is shown thus: $\overline{AD}+\overline{DC} = \overline{AC}$, or $\overline{AB}+\overline{BC} = \overline{AC}$. The sign + is used because the operation resembles addition in algebra; in fact, if vectors are in the same direction, vector addition is algebraic addition.

Two given vectors may be added by drawing to scale either of the triangles in the figure. In some cases it is convenient to show the whole parallelogram. The resultant AC may alternatively be calculated by trigonometry.

Displacements are only one kind of vector quantity, and we shall meet other types, all of which are specified by a magnitude and direction, and combined by the method just shown.

3. Resolution of vectors

As two or more vectors can be combined into a single resultant, so a single vector can be split into components. Most commonly we are working in two dimensions, and require two components. The problem is then that of ' solving ' the vector triangle, given one side and sufficient information about the other two. Frequently we require components in two perpendicular directions.

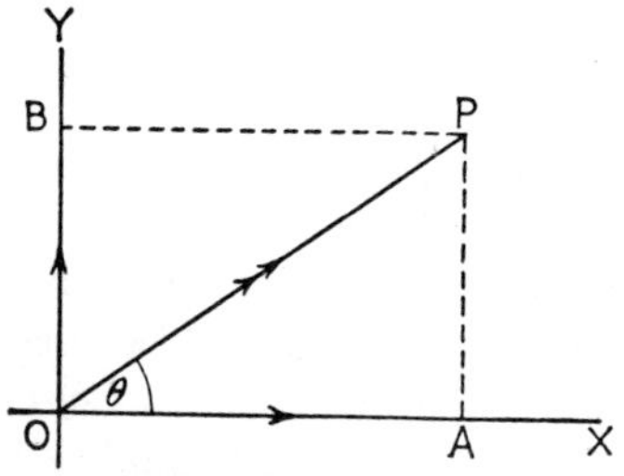

If OP has magnitude r, and direction given by the angle θ in the figure, the magnitudes of the components $\overline{OA}$, $\overline{OB}$ along the perpendicular axes OX, OY are $r \cos \theta$ and $r \sin \theta$ respectively.

One method of calculating the resultant of several coplanar vectors is to resolve each into components along two perpendicular axes, and re-combine the algebraic sums of the two sets of components.

4. Velocity

A vector is said to be multiplied or divided by a scalar when its magnitude is multiplied or divided, the direction remaining unchanged. In particular, multiplying a vector by -1 reverses it, so that $-\overline{PQ} = \overline{QP}$.

If a point moves from P to Q in a time interval t units, the vector $\overline{PQ}/t$ is called the average velocity of the point during that interval, and if the average velocity of a point is the same for all time intervals, its value is called the constant velocity of the point. Note that for a point to have a constant velocity, its direction of motion must be unchanged; that is, it must move in a straight line.

In the general case, let a moving point be at P at time t, and move from P to P′ in the time interval δt. Then $\overline{PP'}/\delta t$ is the average velocity during the interval. If this average velocity approaches a limit as $\delta t \to 0$, the limiting value is called the velocity at the point P, or at the time t. It will in general change in both magnitude and direction as t changes.

If the point is moving in a straight line, let s be its distance at time t from some fixed point on the line. The velocity is then along the same line, and its magnitude is $\frac{ds}{dt}$, which is called the *speed* of the moving point. Thus the speed is given by the gradient of a graph showing distance against time.

5. Relative velocity

Velocity is a vector, and velocities, like displacements, are related to a frame of reference. Consider a passenger in a railway carriage, who moves northward at 2 m/s relative to the train, while the train travels eastward at 20 m/s relative to the Earth's surface. The velocity of the passenger relative to the Earth is the vector sum of these two velocities. In general:

Velocity of A relative to B + velocity of B relative to C
= velocity of A relative to C

The principle is frequently used in another form:

Velocity of A relative to B
= velocity of A relative to C − velocity of B relative to C
= velocity of A relative to C + REVERSED velocity of B relative to C

In considering the motion of ships and aircraft it must be remembered that their engines propel them by acting on the surrounding water or air, and if there is a current or a wind their motion relative to the Earth's surface is different from what it would be in still water or still air.

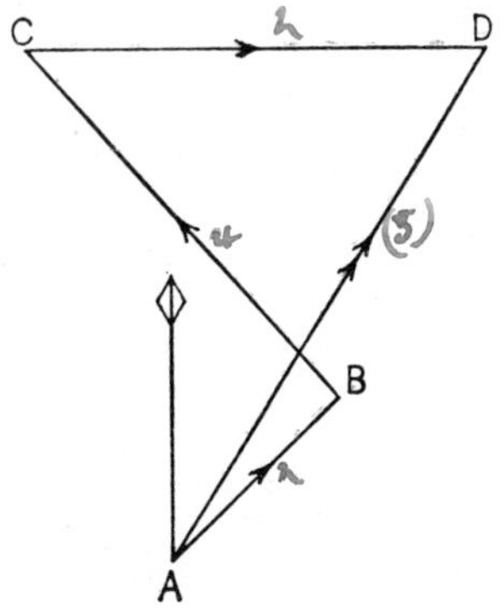

The principle stated above is used in such cases, A being the ship or aircraft, B the surrounding water or air, and C the Earth.

Example 1.—Find the resultant of vectors 2 units NE, 4 units NW, and 4 units E.

(*a*) Graphically.

$\overline{AB}$, $\overline{BC}$, $\overline{CD}$ are the given vectors and $\overline{AD}$ is the resultant, measured as 5 units in direction 031°.

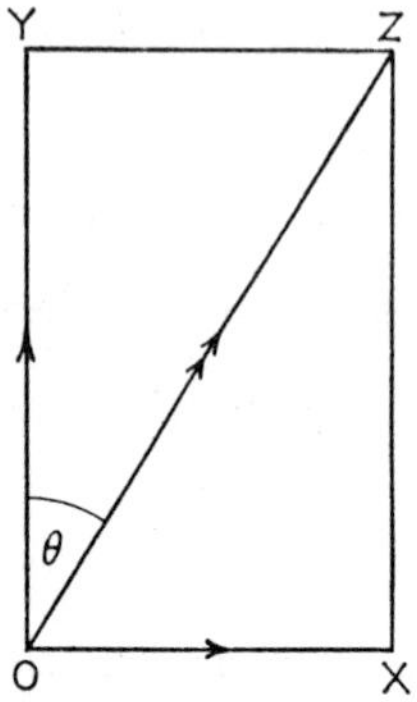

(*b*) By calculation.

The easterly components are $\sqrt{2}$, $-2\sqrt{2}$, 4 units.
The northerly components are $\sqrt{2}$, $2\sqrt{2}$, 0 units.
We combine $\overline{OX}$, $4 - \sqrt{2}$ units easterly, with $\overline{OY}$, $3\sqrt{2}$ units northerly.

$$OZ^2 = OX^2 + OY^2$$
$$= 18 - 8\sqrt{2} + 18$$
$$\simeq 24{\cdot}69$$
$$\therefore \underline{OZ \simeq 4{\cdot}969}$$
$$\tan\theta = OX/OY$$
$$\underline{\theta = 31^\circ\,21'}$$

The resultant is 4·97 units in direction 031° 21′.

Example 2.—Resolve a vector of magnitude 10 units into components at 30° and 50° to its direction, on opposite sides.

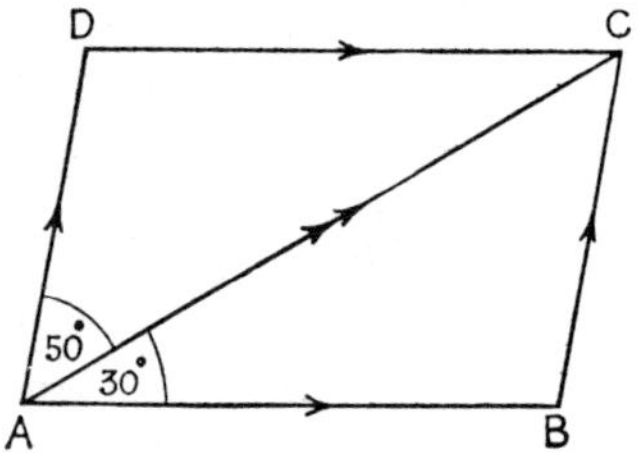

(*a*) Graphically.

Given the diagonal AC and the angles 30°, 50°, the parallelogram can be constructed, and the components AB, AD, measured as 7·8 and 5·1 units respectively.

(*b*) Using the Sine Rule to solve triangle ABC, we find AB = 7·78 units, BC = 5·08 units.

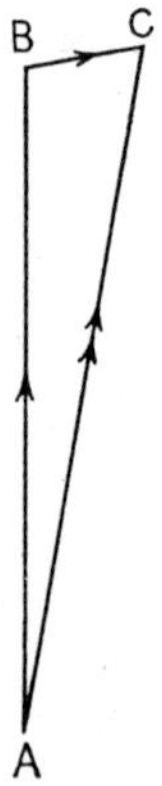

Example 3.—An aircraft whose speed in still air is 200 knots. is headed north. It actually travels over the ground at 210 knots in the direction 010°. Find the velocity of the wind.

In the figure,

$\overline{AB}$ represents the velocity of the aircraft relative to the air.
$\overline{AC}$ represents the velocity of the aircraft relative to the Earth.

Now velocity of aircraft relative to air + velocity of air relative to Earth = velocity of aircraft relative to Earth.

$\therefore$ $\overline{BC}$ represents the wind velocity, and it is found to be 37 knots from the direction 260°.

Note that if the question is solved graphically, a large scale must be used, or the wind vector will be too small for accurate measurement.

Example 4.—A ship B is 20 nautical miles NE of a ship A. B is going south at 10 knots, and A is going east at 16 knots. After what time will A be nearest to B, and what will then be their relative positions?

Velocity of B relative to A = velocity of B relative to Earth + REVERSED velocity of A relative to Earth.

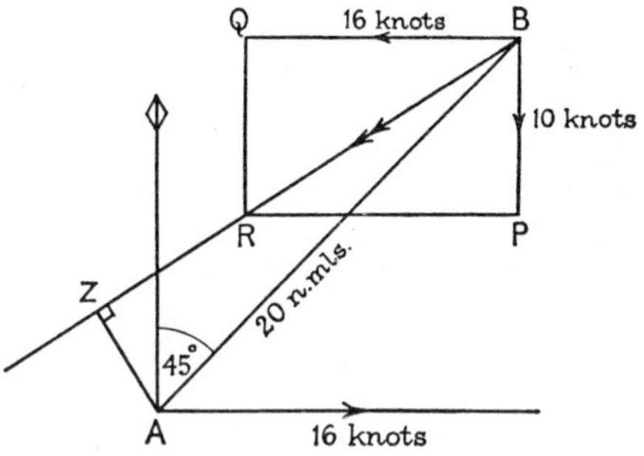

In the figure,

$\overline{BP}$ represents the velocity of B relative to Earth

$\overline{BQ}$ represents the reversed velocity of A relative to Earth

∴ $\overline{BR}$ represents the velocity of B relative to A, and BR produced shows the track of ship B as it would appear to an observer on ship A.

[This construction is sometimes described as 'reducing A to rest', by imposing on both ships the reversed velocity of A.]

If AZ is perpendicular to BR produced, A and Z represent the relative positions of the ships when they are nearest to each other.

If the question is done graphically, the distance scale for AB, AZ, and the velocity scale for the parallelogram BPRQ, must be clearly stated.

By calculation, the steps are:

angle QBR = 32°

∴ angle RBA = 13°

∴ AZ = AB sin RBA ≃ 4·5 nautical miles.

Angle ZAB = 77°, so that when B is nearest to A, its direction from A is 328°.

BR = 18·87 knots, the relative speed.

BZ = 19·49 nautical miles, the relative distance to be covered.

∴ Time required ≃ 1·03 hours.

EXERCISE 1A

1. Find the resultant of vectors 2 units in direction 330° and 3 units in direction 290°.

2. Find the resultant of vectors 3 units in direction 320° and 4 units in direction 070°.

3. The directions of two vectors of magnitudes 3 and 5 units are at an angle of 80°. Find the magnitude of their resultant, and the angle it makes with the 5-unit vector.

4. A vector of magnitude 10 units is to be resolved into components making angles of 30° and 40° with it, on opposite sides. Find their magnitudes.

5. The components of a vector of magnitude 6 units have magnitudes 4 and 5 units. Find the angles they make with the given vector.

6. One component of a vector of magnitude 8 units has magnitude 3 units and makes an angle of 70° with it. Find the magnitude and direction of the other component.
7. Three vectors are given as: 5 units, bearing 100°; 6 units, bearing 170°; 2 units, bearing 140°. Find their resultant. (Bearings are measured from north, clockwise.)
8. Find the resultant of the following vectors: 3 units, bearing 070°; 4 units, bearing 120°; 5 units, bearing 220°.
9. Vectors of 6 units, bearing 030°, and 4 units, bearing 080°, are combined with a third vector. The resultant is a vector of 12 units, bearing 060°. Find the third vector.
10. Find the magnitude and direction of the least vector which can be combined with vectors 10 units, bearing 020° and 12 units, bearing 140°, to give a resultant bearing 090°. What will be the size of the resultant?

EXERCISE 1B

1. Find the resultant of vectors 4 units, bearing 040°, and 6 units, bearing 340°.
2. Find the resultant of vectors $3x$ units and $4x$ units, at an angle of 60°.
3. The directions of two vectors, of magnitudes 2 and 3, are along the sides AB and AC respectively of an equilateral triangle ABC. The line of the resultant is from A to a point X on BC. Find the ratio BX:XC.
4. A unit vector, bearing 090°, is resolved into components bearing 060° and 180°. Find their magnitudes.
5. Two vectors have magnitudes, 5 units and 7 units, and bearings 070° and 150° respectively. What third vector will combine with them to give zero resultant?
6. A vector of magnitude 10 units is resolved into two components. One of them is at an angle of 20° with it, and the magnitude of the other is 5 units. Find the possible directions of the second component.
7. Three vectors are given as: 7 units, bearing 010°; 5 units, bearing 280°; 6 units, bearing 130°. What fourth vector will combine with them to give zero resultant?
8. Two horizontal vectors, 21 units north and 20 units east, are combined with a 10-unit vector vertically upwards. Find the angle between the resultant of the three and the vertical.
9. ABCD is a square. Find the resultant of vectors completely represented by $\overline{AB}$, $\overline{AC}$, $\overline{AD}$.
10. ABCDEF is a regular hexagon. Find the resultant of vectors completely represented by $\overline{AB}$, $\overline{AC}$, $\overline{AD}$, $\overline{AE}$, $\overline{AF}$.

EXERCISE 2A

1. An aircraft is headed due south, and its speed relative to the air is 200 knots. When it has been flying for 30 min, it is over a point 96 n miles from its starting point in the direction 170°. Find the speed and direction of the wind.
2. An aircraft is required to travel due east. Its air speed is 220 knots. The wind is from SW at 40 knots. What course should be set, and what ground speed should be expected?

3. A stream which flows east at 2 m/s is 150 m wide. A boat whose speed in still water is 6 m/s is headed directly across the stream. In what time will it reach the opposite bank, and how far downstream from its starting point? In what direction should the boat be headed to travel directly across the stream, and how long will the crossing take in this case?
4. Drift is the name given to the angle between an aircraft's course (the direction in which it is headed) and its track over the Earth's surface. It is named starboard or port according as the track is to right or left of the course, from the pilot's point of view. Prove that when wind and air speed remain constant the drift when going from A to B is equal and opposite to that when returning from B to A.
5. A ship is steaming NE at 15 knots, and the wind is from south at 20 knots. Find the direction in which a flag flies from the mast head, and the direction of the smoke trail from the funnel.
6. If $\overline{OP}$ represents the reversed course and air speed of an aircraft, and $\overline{OQ}$ the wind velocity, show that $\overline{PQ}$ represents the track and ground speed. Hence find the wind velocity when an aircraft whose air speed is 200 knots experiences a drift of 5° starboard on a course of 060°, and a drift of 10° starboard on a course of 300°.
7. A motor boat whose speed is 30 knots is to intercept a ship whose speed is 12 knots. The ship is 15 n miles north of the motor boat, and travelling NW. What course should the motor boat steer, and how long will it take to intercept the ship?
8. A ship whose speed in still water is u knots is observed to cover a distance of one n mile in t_1 sec when a current is flowing at v knots in the same direction. The time for the same distance against the same current is t_2 sec. Find u and v in terms of t_1, t_2.
9. An aircraft is to fly from A to B, 400 n miles due east. The wind is estimated at 50 knots from NW, and the speed of the aircraft in still air is 200 knots. Find the course to be set, and the estimated position after 1 hour's flying. If the actual position after 1 hour proves to be 20 n miles due north of the estimated position, find the actual wind, and the course to be set to fly directly to B.
10. An aircraft is to fly a distance of x n miles in a straight line, and return to its base along the same line. Its air speed is u knots, and the wind is at w knots in a direction making an angle θ with the intended track. Find α, the angle of drift, in terms of u, w, θ, and prove that the time for the whole journey is
$$\frac{2xu \cos \alpha}{u^2 \cos^2 \alpha - w^2 \cos^2 \theta} \text{ hours.}$$

EXERCISE 2B

1. An aircraft's course is 245°, and its speed through the air is 300 knots. The direction of its track over the ground is 250°, and its ground speed is 310 knots. Find the speed and direction of the wind.
2. The air speed of an aircraft is 280 knots. When the wind is at 50 knots from 225° the aircraft is required to travel on a track of 300°. Find the course to be steered, and the ground speed.
3. When the wind is from 240°, at 40 knots an aircraft is required to make good a ground speed of 200 knots on a track of 175°. Find the air speed and course required.

4. A boat travelling upstream in a river has a breakdown which reduces its speed through the water to half the speed of the stream. At what angle to the up-stream direction should the boat be steered to reach the bank as quickly as possible, and at what angle to the upstream direction will it then travel?

5. An aircraft A, speed 400 knots, is to intercept another, B, whose speed is 300 knots. When A starts B is 100 n miles SE of A, travelling due west. Find A's course, and estimate the time required to intercept B. (There is no wind.)

6. From a point P a ship is observed at distance 10 n miles, bearing 015°. One hour later the distance is 11 n miles, and the bearing 320°. At this time a boat sets out at 20 knots to intercept the ship. Find the course which it must steer.

7. Prove that if any other distance (less than 24 n miles) is substituted for 11 n miles in question 6, the answer will be unaltered.

8. To a cyclist riding NE at 6 m/s the wind seems to be from the north. When he turns a corner and rides north at 5 m/s, the wind seems to be from NW. Find the direction of the wind.

9. The navigator of an aircraft whose course is 050° and air speed 180 knots observes that the drift is 10° to starboard. By observing smoke trails he knows that the wind is from NW. Find the speed of the wind, and the ground speed of the aircraft.

10. An airman has to fly from a point A to a point B due east of A, and at a distance d from A. A wind of constant velocity u is blowing in a direction which makes an acute angle α with AB on its northern side. If the velocity of the aeroplane relative to the air is $v(>u)$, show that the airman should steer in a direction making an angle β with AB on its southern side, where $\sin\beta = (u\sin\alpha)/v$. If the same wind is blowing on the return journey and the velocity of the aero-plane relative to the air is the same, show that the course steered should make the same angle with BA. Show also that the difference in the times taken for the outward and return journeys is $2du\cos\alpha/(v^2 - u^2)$. (No account is taken of the curvature of the Earth.)

2

Acceleration

1. Acceleration

If O is a point in the frame of reference, $\overline{OP}$ is called the position vector of the point P. When the point moves from P to P′, $\overline{PP'}$ is the change in the position vector. Thus the velocity of a point may be described as the rate of change of its position vector.

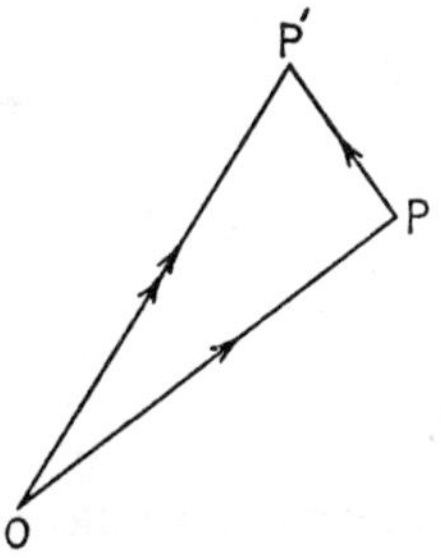

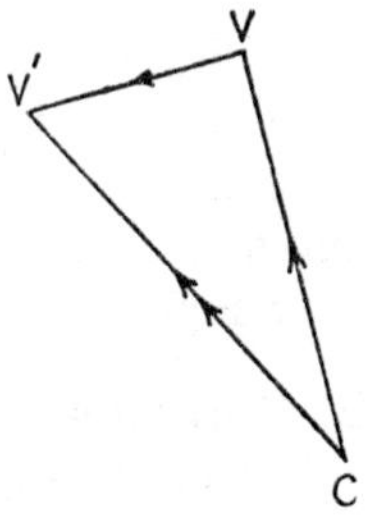

Changes in the velocity of a moving point may be represented by drawing vectors representing the velocity from a point C. If $\overline{CV}$, $\overline{CV'}$ represent the velocities at times t, $t+\delta t$, $\overline{VV'}$ is the change of velocity, and $\overline{VV'}/\delta t$ the rate of change, in the interval δt. This is called the *average acceleration* of the moving point during the interval. If it is the same in magnitude and direction for all intervals it is called a *constant acceleration.* If it varies with the time, the limiting value of $\overline{VV'}/\delta t$ as $\delta t \to 0$ is called the acceleration at time t. Note that acceleration is a vector.

2. Straight-line motion

In the case where the moving point travels along a straight line, the directions of both velocity and acceleration are along that line, and we are concerned only with their magnitudes. We have already shown that if s measures the distance from some fixed point, the speed v is given by $\frac{ds}{dt}$. Similarly, the magnitude of the acceleration is given by $\frac{dv}{dt}$, which may also be written as $\frac{d^2s}{dt^2}$. The notation $\dot{s}$, $\ddot{s}$, $\dot{v}$ for $\frac{ds}{dt}$, $\frac{d^2s}{dt^2}$, $\frac{dv}{dt}$ is often used.

It has been pointed out that in straight-line motion the speed is given by the gradient of the distance-time graph. Similarly the acceleration is given by the gradient of the speed-time graph. The speed-time graph can also be used to find the distance travelled, for since $v = \frac{ds}{dt}$ it follows that $s = \int v\, dt$, the integral being taken between proper limits.

When the connection between v and t is expressed by an equation, the distance can be found by direct integration; but when a number of corresponding values of v and t are supplied, and no law is obvious, the distance will be found by estimating the area between the graph and the axis of t. The most convenient methods are the *Trapezoidal Rule* and *Simpson's Rule.*

If the ordinates $y_1, y_2, \ldots, y_n$ divide the area into strips of equal width h, the Trapezoidal Rule gives the area as

$$h[\tfrac{1}{2}(y_1+y_n)+y_2+y_3+\ldots+y_{n-1}]$$

For an even number of strips of width h, Simpson's Rule gives the area as $\frac{1}{3}h$ [sum of end ordinates $+2\times$ sum of other odd-numbered ordinates $+4\times$ sum of even-numbered ordinates].

Simpson's Rule usually gives the more accurate result.

3. Constant acceleration

The case of constant acceleration frequently occurs and the following standard formulæ must be remembered. [They must not, of course, be used for any but constant acceleration.]

Let f be the constant acceleration, v the speed, and s the distance from the origin, at time t.

$$\frac{dv}{dt} = f$$

$$\therefore \quad \underline{v = u+ft} \text{ where } u \text{ is the initial speed}$$

$$\frac{ds}{dt} = u+ft$$

$$\therefore \quad \underline{s = ut+\tfrac{1}{2}ft^2}$$

there being no constant of integration provided that s is measured from the position of the moving point when $t = 0$.

From these formulæ, by simple algebra, we derive

$$\underline{s = \tfrac{1}{2}(u+v)t}$$

and

$$\underline{v^2 = u^2 + 2fs}$$

Each of these formulæ involves four of the five quantities s, u, v, f, t, and the appropriate formula for any case is most easily chosen by noting which one of the five quantities does not occur in the problem.

4. Units

When these formulæ are used, the units for speed and acceleration must be consistent with those for distance and time. For example, if distance is measured in metres, and time in seconds, speed must be given in metres per second (written m/s). Acceleration, being the rate of change of speed, will be given in (m per sec) per second (written m/s^2).

In drawing speed-time graphs it is advisable to use consistent units on the two axes, or trouble may ensue in interpreting gradient and area. If, for example, the speed axis were graduated in m/s, and the time axis in minutes, the gradient would give the acceleration in m/s per minute, and unit area would represent the distance travelled in one minute at 1 m/s, that is 60 m.

Example 1.—The speed of a car during 8 sec is given in the following table:

Time in sec:	0	1	2	3	4	5	6	7	8
Speed in m/s:	0	1	4	9	14	15	12	6	0

Find (i) the distance covered in the first 4 sec,
(ii) the distance covered in the whole time,
(iii) the greatest positive acceleration,
(iv) the greatest negative acceleration.

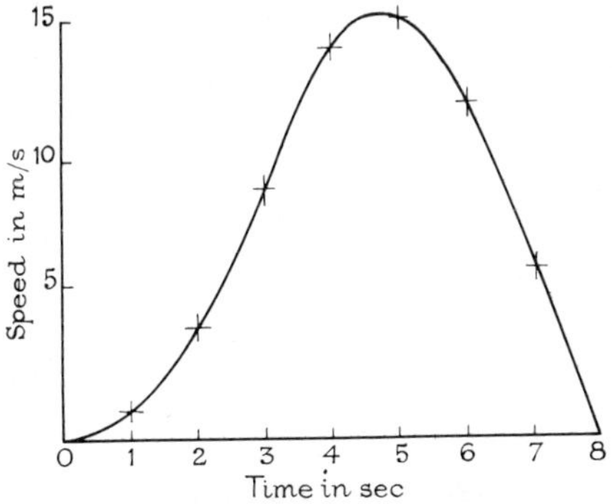

A graph is drawn from the data.

(i) By Simpson's Rule, area from $t = 0$ to $t = 4$ is given as

$$\tfrac{1}{3}[14 + 2 \times 4 + 4 \times 10] = 20\tfrac{2}{3} \text{ units}$$

$\therefore$ Distance in first 4 sec $\simeq$ 20·7 m

(ii) Whole area $= \frac{1}{3}[0 + 2 \times 30 + 4 \times 31] = 61\frac{1}{3}$

$\therefore$ Distance in whole time $\simeq 61{\cdot}3$ m

(iii) By inspection of the gradient of the graph

greatest possible acceleration $\simeq 5$ m/s² (at time 3 sec).

(iv) Greatest negative acceleration $\simeq -6$ m/s² (from 6 sec to 8 sec).

Example 2.—A point moves in a straight line, and its distance s m from its starting point at time t sec is given by $s = t^3 - 23t^2 + 120t$. Find (i) when it is again at its starting point, (ii) with what speed it starts, and when it is momentarily at rest, (iii) for what time its acceleration is negative.

(i) $s = t(t - 8)(t - 15)$. The required times are 8 sec and 15 sec.

(ii) $\dot{s} = 3t^2 - 46t + 120$. When $t = 0$, speed $= 120$ m/s.

$\dot{s} = (3t - 10)(t - 12)$ Point is at rest at $3\frac{1}{3}$ sec and at 12 sec.

(iii) $\ddot{s} = 6t - 46$. Acceleration is negative from $t = 0$ to $t = 7\frac{2}{3}$.

It is instructive to draw distance-time, speed-time, and acceleration-time graphs for this example.

Example 3.—A stone is projected vertically upwards, and its acceleration during the motion is 9·8 m/s² downwards. (i) Prove that the speed at any height is the same on the upward and downward journeys. (ii) Find the speed of projection for a height of 10 m to be just reached, and the time required to reach it.

(i) From the standard formula $v^2 = u^2 + 2fs$, we have in this case $v^2 = u^2 - 19{\cdot}6\,s$, and u is fixed.
For any value of s, this gives two equal and opposite values of v.

(ii) If a height of 10 m is just reached, the speed at this height must be zero.

Hence $0 = u^2 - 19{\cdot}6 \times 10$, $\therefore u = \pm 14$.

The speed of projection is 14 m/s upwards. (The negative value gives the speed at which the stone arrives again at the level of projection.)
From the formula $s = \frac{1}{2}(u + v)t$ we have $10 = 7t$, so that the required time is $1\frac{3}{7}$ sec.
The formula $v = u + ft$ might alternatively be used.

Example 4.—A stone is let fall from rest, and 1 sec later another stone is projected vertically downwards at 10·5 m/s. Taking the acceleration of both at 9·8 m/s² downwards, find when and where the second overtakes the first.

Since the two stones have no relative acceleration, the speed of one relative to the other is constant.

When the second stone is projected, the first is 4·9 m below it, and moving downwards at 9·8 m/s.

$\therefore$ The speed of the second relative to the first remains at 0·7 m/s downwards, and it overtakes the first in 7 sec.

At this instant, the first stone has been falling for 8 sec from rest, and the formula $s = ut + \frac{1}{2}ft^2$ gives the distance fallen as 313·6 m.

EXERCISE 3A

1. The following table gives the speed of a car during 14 sec:

Time in sec	0	2	4	6	8	10	12	14
Speed in m/s	0	5·8	10·8	14·1	15·3	13·2	7·6	0

Find (i) the total distance covered, (ii) the accelerations at times 0, 4, 10, 14 sec.

2. The following table gives the speed of a train during 30 sec:

Time in sec	0	2	4	6	8	10	25	27	29	30
Speed in km/h	0	3·1	10·6	19·4	26·9	30	30	25·9	11·7	0

The speed is constant from 10 sec to 25 sec. Find the distance travelled during (i) the first 10 sec, (ii) the last 5 sec, (iii) the whole time. Find also the greatest positive and negative accelerations, in m/s².

3. A point moves so that its speed at time t sec is ($10 \cos t°$) m/s. Draw the speed-time graph for the first 90 sec of the motion, and find the distance covered (i) in the first 45 sec, (ii) in the whole time. (iii) Find the accelerations at times 0 sec, 30 sec, 60 sec, 90 sec, and sketch the acceleration-time graph.

4. A point moves in a straight line according to the law $v = 12 + 32t$, the units being metres and seconds. Sketch the speed-time and distance-time graphs for the first 5 sec of the motion. Find the distance travelled (i) in the first 4 sec, (ii) during the fourth second.

5. A point moves in a straight line according to the law $v = 12 - 6t$, the units being metres and seconds. Find a formula giving the distance s m from the starting point after t sec. Sketch the speed-time and distance-time graphs for the first 6 sec of the motion. (i) Find the position of the point after 2 sec, 4 sec, 6 sec. (ii) What is the total distance travelled by the point at each of these times?

6. A point moves in a straight line, starting from rest, with acceleration $(20 - 4t)$ m/s². What is (i) its speed, (ii) its distance from the starting point, after 5 sec?

7. A point moves in a straight line with acceleration $(2t - 4)$ m/s². Its speed at starting, when $t = 0$, is 3 m/s. Show that it returns to its starting point, and find when it reaches that point. Which way does it move when it leaves the starting point for the second time?

8. The speed of a car during a run of 7 min is read each minute, and the readings, in order, are 0, 8, 15, 18, 20, 19, 11, 0 km/h. Find the length of the run in km.

9. A point moves in a straight line so that $v = 8t - 2t^2$. (i) When is its speed a maximum? (ii) When does it change its direction of motion? Sketch the graphs of distance, speed, and acceleration against time.

10. A machine is gradually slowing down, so that at any instant its speed of rotation is $\frac{9}{10}$ of what it was one minute earlier. Draw a graph showing the speed of rotation for 5 min, starting with a speed of 100 rev/min, and find the total number of revolutions made during the time.

EXERCISE 3B

1. A car is moving in a straight line. In the following table v m/s is its speed t sec after the start. Using 1 cm for the unit of time and 2 mm for the unit of speed, draw the speed-time graph.

t	0	1	2	3	4	5	6	7
v	0	2·6	5·6	8·8	12·4	16·2	20·4	24·8

Draw on the same diagram the acceleration-time graph, using 2 cm as the unit of acceleration. Find the distance travelled after 7 sec.

2. An electric train moves in a straight line. The table below gives the speed v km/h at a time t sec after starting from rest.

t	5	10	15	20	25	30	35
v	1	3	7	15	26	32	34

Draw the speed-time graph and from it find (i) the total distance in km covered in 35 sec, (ii) the average speed in km/h during the period, (iii) the acceleration in m/s² at the end of 15 sec, (iv) the time after the start at which the acceleration is a maximum.

3. A train, starting from rest, accelerates uniformly, thereby reaching a speed of 45 km/h in 40 sec. This speed is maintained for 3 min, after which steam is shut off and the brakes applied, bringing the train to rest with uniform retardation. The total distance run is 3 km. By consideration of the velocity-time graph, or otherwise, find (i) the acceleration in m/s², (ii) the distance run whilst accelerating, (iii) the distance run whilst retarding, (iv) the time taken for the whole journey.

4. A train starts from rest at one station and comes to rest at the next station 2 km away. It travels with uniform acceleration until its speed is 30 km/h; then with uniform speed until the brakes are applied. The retardation is twice the acceleration, and the total time of the journey is $5\frac{1}{2}$ min. By means of a velocity-time graph, or otherwise, calculate (i) the time in minutes during which the train travelled with uniform speed, (ii) the distance in km travelled by the train whilst slowing down, (iii) the acceleration of the train in m per sec per min.

5. A train runs for 6 min between stops. Its greatest speed is 60 km/h which is attained after 3 min. With scales 1 cm to 1 min, and 1 cm to 60 km/h, the speed-time graph is part of a circle. (i) Draw the graph, and find from it in the initial acceleration, and the acceleration after 2 min, in km/h per min. (ii) Find the whole distance between stops, in km.

6. A train travels from rest at one station to rest at another, at a distance s from the first, in time t. The first part of the journey is made under constant acceleration f, and the last part under constant retardation f', the middle part being run at constant speed v. Each of these symbols represents a positive number. By using a speed-time diagram, or otherwise, show that v is given by the equation $\left(\frac{1}{f}+\frac{1}{f'}\right)v^2 - 2tv + 2s = 0$. Show that the larger root of this equation is inconsistent with the physical conditions of the problem, explaining its algebraic existence with the help of a diagram.

7. A ball moves in an inclined groove, with a constant acceleration of 4 m/s^2 downward. It starts from the bottom of the groove with a speed of 10 m/s upward. Find its distance up the plane, and its speed (i) after 2 sec, (ii) after 3 sec. (iii) When and where is the ball instantaneously at rest? (iv) When, and with what speed, does the ball again reach the bottom of the groove?

8. When $t > 1$, the speed of a body moving in a straight line is $16/t^2$ m/s. When $t = 2$ the body is passing through a point A on the line. Find expressions for its acceleration and its distance from A (i) when $t = 10$, (ii) when $t = 100$. (iii) To what limits do the acceleration, speed. and distance from A tend as t continually increases?

9. A body moves in a vertical line so that its speed after t sec is ($10 \sin t°$) m/s. (i) Find its speed at ½-min intervals for 6 min and draw the speed-time graph. (ii) The body begins to move upward. For how long does it continue to do so? (iii) Between what times is the acceleration downward? (iv) At the end of 6 min, is the body above, below, or at its starting point?

10. From the speed-time graph drawn in question 9, (i) find the acceleration of the body at ½-min intervals, and draw the acceleration-time graph. (ii) Find the distances travelled in successive minutes. Hence find the distance of the body from its starting point after 1, 2, 3, 4, 5, 6 min, and draw the distance-time graph.

EXERCISE 4A

In questions 1–10, u and v are initial and final speeds, f is the acceleration, t the time, and s the distance, from starting point to finishing point.

Find the missing quantities in the following table:

	u	v	f	t	s
1.	25 m/s			20 sec	1500 m
2.			-16 m/s^2	4 sec	-32 m
3.		130 m/s		16 sec	800 m
4.	50 cm/s	-350 cm/s		20 sec	
5.	30 km/h .	75 km/h			$\frac{7}{16}$ km
6.	54 km/h .		-1 m/s^2	½ min	
7.	1500 km/h .	150 km/h	$-6{\cdot}25$ m/s^2		
8.		28 m/s	-6 m/s^2		768 m
9.	150 cm/s		250 cm/s^2		432 m
10.		60 m/min	$-0{\cdot}01$ cm/s^2	1 h	

11. A particle is projected vertically upwards at 24·5 m/s, and its downward acceleration is 9·8 m/s^2. (i) When and where will it cease to move upwards? (ii) When will it pass its starting point, and with what speed? (iii) When will it be 29·4 m below its starting point?

12. A train which usually runs at 100 km/h is required to cover 1 km of track under repair at 20 km/h. It takes 1 km to slow down to this speed, and 1 km to regain the speed of 100 km/h. What time is lost, as compared with normal running? Assume that retardation and acceleration are uniform.

13. A car accelerates uniformly from rest at 0·8 m/s^2, and immediately begins to decelerate to a stop at 1·2 m/s^2. The total distance covered is 1350 m. Find the total time, and the greatest speed attained.

14. A mine cage is lowered, first with uniform downward acceleration, then with constant speed, and finally with uniform upward acceleration. The distances covered during the accelerating periods are equal, and each is $\frac{1}{10}$ of the distance covered at constant speed. The whole distance is 1800 m, and the whole time for the descent is 70 sec. Find the rates of acceleration and uniform speed.

15. With the notation of questions 1-10, find a formula giving s in terms of v, f, t.

16. A train running with uniform acceleration covers successive distances of 750m in 30 sec and 20 sec. Find the acceleration in m/s^2, and the initial and final speeds in km/h.

17. Prove that the distances covered in any number of successive seconds by a point moving with uniform acceleration form an arithmetic progression. If the common difference of this A.P. is 18 cm, find the acceleration.

18. With the notation of questions 1-10, find, in terms of u and v, the speeds at time $\frac{1}{2}t$ and at distance $\frac{1}{2}s$, and prove that the latter is always greater than the former.

EXERCISE 4B

Questions 1–10 are to be worked as in Exercise 4A.

	u	v	f	t	s
1.		0	-32 cm/s²	$2\frac{1}{2}$ sec	
2.	20 m/s		60 cm/s²	1 min	
3.	36 km/h			5 sec	75 m
4.	1000 m/s	0			50 cm
5.	25 cm/s	-25 cm/s		10 sec	
6.			-6 m/s²	11 sec	605 m
7.	180 cm/s		-980 cm/s²		$-38·7$ m
8.		80 m/min		2 h	12,000 m
9.		2000 m/s	80,000 m/s²		$2\frac{1}{2}$ m
10.	50 cm/s	80 cm/s	1 mm/s²		

11. A motorist accelerates from 36 km/h to 72 km/h in 5 sec to pass another car, and at once slows down to 36 km/h during the next 5 sec. How much farther along the road is he than if he had continued at 36 km/h?

12. A ball dropped on to a floor is found to bounce to $\frac{9}{16}$ of the height from which it was dropped. How long does it take to drop to the floor from a height of 4·9 m? (Take $g = 9·8$ m/s^2 in this and other questions, unless another value is specified.) How long after this does it hit the floor for the second time? And for the third time? Show that bouncing must cease within 7 sec from the time the ball was first dropped.

13. A paper strip on a moving platform passes under an apparatus which marks the paper every $\frac{1}{5}$ sec. The distances between marks are 1 cm, 1·1 cm, 1·2 cm, and so on, regularly increasing by 0·1 cm. Show that the acceleration of the platform is constant, and find it in cm/s^2. Find also the speed at the instant when the first mark was made.

14. A particle drops from the ceiling of a lift, 4 m above the floor, when the lift has a downward acceleration of 1·8 m/s^2. How long will it take to reach the floor?

15. A stone is let fall from the top of a tower 39·2 m high, and at the same instant another stone is projected vertically upward from the bottom of the tower at 19·6 m/s. When and where will the stones pass one another?

16. A lift ascends from rest at ground level with acceleration 1·8 m/s^2. After 7 sec it dislodges a stone which falls from rest. How long will the stone take to reach the ground?

3

Circular Motion. Simple Harmonic Motion

It is important to realize that acceleration is not always uniform, and the formulæ for uniform acceleration given in the last chapter must not be thoughtlessly used in cases to which they do not apply. We therefore consider now two kinds of motion whose importance will be seen later: *circular motion*, which may be uniform or non-uniform, and *simple harmonic motion*. It will be seen that in all these cases the acceleration varies in magnitude or direction, or both, and in none of them must the formulæ of the last chapter be used.

1. Uniform circular motion

When a point moves in a circle its velocity cannot be constant, even though the speed may be uniform, for the direction of motion is changing. The rate of change of the direction of a moving line is called its *angular velocity*, and is usually measured in revolutions per minute (rev/min) or in radians per second (rad/s).

Let P be moving in a circle of radius r, with centre O, and let OP be turning at the rate of ω rad/s.

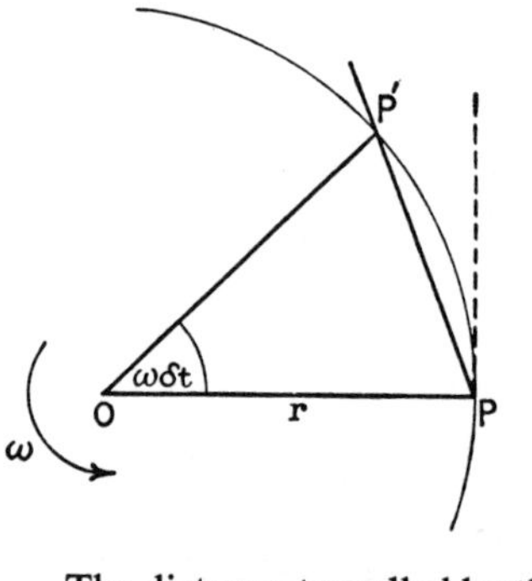

In a time interval δt P moves to P′, and angle POP′ $= \omega\,\delta t$.

As $\delta t \to 0$, $P' \to P$, and the direction of $\overline{PP'}$ approaches that of the tangent to the circle at P. This is then the direction of the velocity at P.

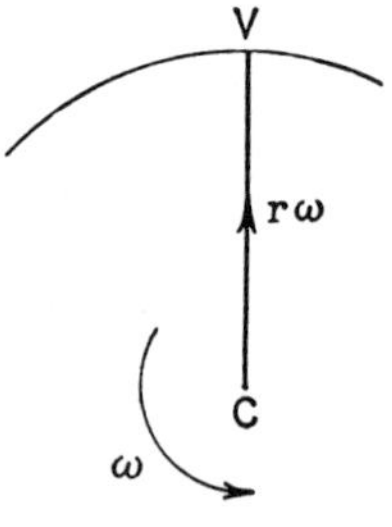

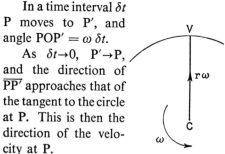

The distance travelled by the moving point in the interval is the length of arc PP′, that is $r\omega\,\delta t$, and thus the magnitude of the velocity is $r\omega$.

Now let $\overline{CV}$ represent the velocity of P, and suppose that ω is constant. As P moves round the circle with centre O, V will move with the same angular velocity round the circle with centre C and radius $r\omega$, CV being always 90° ahead of OP.

rate of change of the velocity vector $\overline{CV}$ gives the acceleration , which is therefore of magnitude $r\omega^2$ and in the direction perpendicular to $\overline{CV}$, that is, parallel to $\overline{PO}$.

Thus, if v is the speed of a point moving in a circle of radius r with angular velocity ω about the centre, $v = r\omega$; and if ω is constant the acceleration of the point is $r\omega^2$, which may also be expressed as v^2/r, along the inward radius.

When these formulæ are used, the angular velocity must be expressed in radians per unit of time. This is necessary because the radian formula for length of arc has been used in obtaining the expression for the velocity.

2. Non-uniform circular motion

When the angular velocity is not constant, the work of the previous section still applies as far as the velocity of the moving point is concerned, but the diagram of velocity vectors is no longer a circle, so that another method must be used to find the acceleration. We shall find separately its components along the tangent at P and along the radius PO.

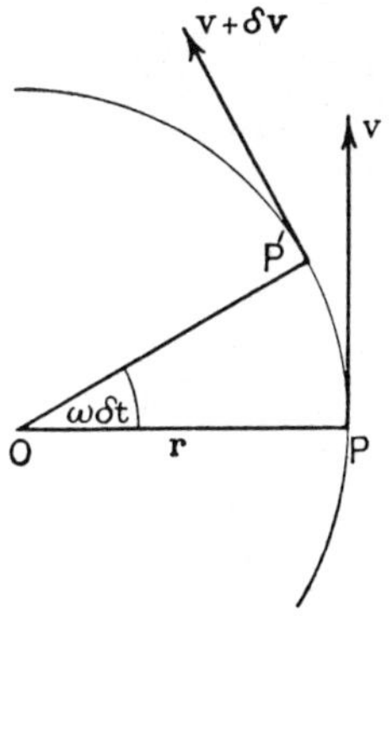

Let v be the speed at P, and $v+\delta v$ the speed at P′, the angle POP′ being $\omega\,\delta t$ as before.

Change in velocity in direction of tangent at P

$$= (v+\delta v)\cos\omega\,\delta t - v = \delta v$$

if we neglect small quantities of the second and higher orders, such as $(\delta t)^2$, etc.

∴ Acceleration along tangent at P

$$= \underset{\delta t\to 0}{\mathrm{Lt}}\ \frac{\delta v}{\delta t} = \frac{dv}{dt} = r\frac{d\omega}{dt}$$

Increase in velocity in direction PO

$$= (v+\delta v)\sin\omega\,\delta t = v\,\omega\,\delta t$$

neglecting second-order small quantities $\delta v\,.\,\delta t$, etc.

∴ Acceleration along PO $= v\omega = v^2/r = r\omega^2$, just as in the case of uniform circular motion, where, however, there is no tangential acceleration.

Example 1.—A point moves uniformly round a circle at 120 rev/min. If the acceleration to the centre is 32 m/s², find the radius of the circle, and the speed of the point in m/s.

Angular velocity $\omega = 2$ rev/ s

$= 4\pi$ rad/s

Let radius $= r$ m

Acceleration to centre $= r \, . \, 16\pi^2 = 32$

$\therefore \; \underline{r = 2/\pi^2}$

Speed $= \dfrac{2}{\pi^2} \, . \, 4\pi = \underline{8/\pi \text{ m/s}}$

Example 2.—A point moves in a circle of radius 50 cm. At time t sec its angular velocity is $\pi(10 - 2t)$ rad/s. Find its speed and acceleration after 4 sec.

We work in metres, so that $r = \frac{1}{2}$.

When $t = 4$, $\omega = 2\pi$.

$\therefore$ Speed $= r\omega = \underline{\pi \text{ m/s}}$

Acceleration to centre $= r\omega^2 = 2\pi^2$ m/s²

Acceleration along tangent $= r \dfrac{d\omega}{dt} = r \dfrac{d}{dt}(10\pi - 2\pi t)$

$= r(-2\pi) = -\pi$ m/s²

(The sign indicates that the acceleration is opposite to the motion.)

Combining these accelerations we find a total acceleration of $\underline{\pi\sqrt{(4\pi^2 + 1)} \text{ m/s}^2}$, at an angle $\underline{\tan^{-1} 2\pi \text{ to the backward tangent}}$.

EXERCISE 5A

Do not substitute a numerical value for π except in question 7.

1. Find the speed and acceleration of a point moving in a circle of radius 2 m with angular velocity (i) 4 rad/s, (ii) 300 rev/min.
2. A point is moving round a circle at 20 m/s. Find its angular velocity, in rad/s and rev/min, and its acceleration if the radius is (i) 10 m, (ii) 2 m.
3. A point moves in a circle of radius 1 m. What must be its speed to give an acceleration of 4 m/s² towards the centre? Find the angular velocity in rev/min.
4. A stone at the end of a string moves in a vertical circle, and the greatest speed, which is at the lowest point, is twice the least speed, which is at the highest point. Compare the accelerations at these two points.
5. A point travels uniformly round a circle in 3 sec. (i) If the radius is 5 m, find the acceleration. (ii) If the acceleration is 10 m/s², find the radius of the circle.
6. A motor cycle runs at a steady 36 km/h round an S bend made up of parts of two circles. The radius of the first is 200 m, and that of the second 400 m. Find the change in acceleration on passing from one circle to the other.

7. An artificial satellite goes round the Earth in a circular orbit of radius r km in t hr. Prove that its acceleration towards the centre is approximately $0{\cdot}003\ r/t^2$ m/s². If this acceleration must be equal to $\left(\frac{6350}{r}\right)^2 \times 9{\cdot}8$ m/s², (i) find t when $r = 8000$ and when $r = 7000$, (ii) find r when $t = 1{\cdot}5$. Give answers correct to 2 figures.

8. A point moves in a vertical circle of radius 10 m. When it is level with the centre, and descending, its speed is 14 m/s, and this speed is increasing at 9·8 m/s². Find the magnitude of the total acceleration, and the point on the vertical diameter towards which it is directed.

9. A point moves in a vertical circle. When it is level with the centre its acceleration is towards the lowest point of the circle. The angular velocity at that point is 5 rad/sec. Find the angular acceleration.

10. A car is crossing a hump-backed bridge which is in the form of an arc of a circle of radius 16 m. Its acceleration at the highest point is 9 m/s² vertically downwards. Find the speed of the car in km/h.

EXERCISE 5B

Do not substitute for π except in question 6.

1. Find the speed and acceleration of a point moving in a circle of 20 cm radius, (i) at 5 rad/s, (ii) at 120° per min.

2. A point moves in a circle at a steady speed of 30 cm/s. Find the angular velocity in rev/min if the radius is (i) 3 cm, (ii) 30 cm.

3. A point moves in a circle at a steady speed of 90 cm/s, and its acceleration towards the centre is 900 cm/s². Find the angular velocity in rad/s, and the radius of the circle.

4. The coordinates of a moving point at time t are $(a \cos \omega t,\ a \sin \omega t)$. Prove that its locus is a circle and, by differentiation, that its velocity is $a\omega$ along the tangent, and its acceleration $a\omega^2$ towards the centre.

5. A particle fastened by a string 2 m long to a fixed point on a smooth horizontal table is performing uniform circular motion when the string catches on a pin 1·5 m from the fixed point. What is the effect on the acceleration of the particle?

6. Find the acceleration of a point on the Equator (radius 6350 km) due to the daily rotation of the Earth. Answer in m/s², to 2 figures.

7. An aircraft is turning in a circular arc at 252 km/h. If the acceleration towards the centre is not to exceed $2g$, find the least radius of the circle, taking g as 9·8 m/s².

8. A point moves in a circle of radius 1 m with angular velocity $(10 - 4t)$ rad/s. Find the total acceleration, in magnitude and direction, when $t = 2$.

9. A point moves in a circle of 2 m radius, with angular velocity $(4t - 2t^2)$ rad/s. Find its acceleration when $t = 1$ and when $t = 2$.

10. O is the centre, and OA a radius, of a circle of 10 cm radius. A point P moves on the circumference so that at time t sec, the angle AOP is $(\sin \pi t)$ radians. Find the velocity and acceleration of P when $t = \frac{1}{2}, 1, 1\frac{1}{2}, 2$.

3. Simple harmonic motion

A common type of motion is that called *simple harmonic*. It can be pictured as the projection of uniform circular motion on to a straight line.

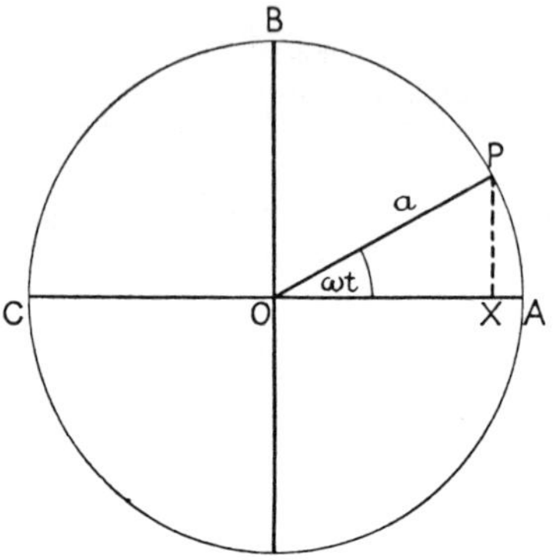

Consider a point P moving in a circle of radius a with constant angular velocity ω.

Let X be its projection on COA. The motion of X is that called simple harmonic.

Let P start anticlockwise from A at time $t = 0$. Then

$$\text{angle AOP} = \omega t$$

$$\text{OX} = a \cos \omega t = x \text{ say}$$

$$\text{Then} \quad \dot{x} = -a\omega \sin \omega t \quad (1)$$

so that X moves backwards and forwards along COA. Its speed at any point is ω . PX, and has the maximum value $a\omega$ at the point O, which may be called the centre of the motion.

Differentiating (1) again, we have

$$\ddot{x} = -a\omega^2 \cos \omega t \quad . \quad . \quad . \quad . \quad . \quad . \quad (2)$$

$$\therefore \ddot{x} = -\omega^2 x \quad . \quad . \quad . \quad . \quad . \quad . \quad . \quad . \quad . \quad (3)$$

so that the acceleration is towards O and proportional to the distance OX.

A result to be noted is
$$\begin{aligned} \dot{x}^2 &= a^2\omega^2 \sin^2 \omega t \\ &= a^2\omega^2(1-\cos^2 \omega t) \\ &= \omega^2(a^2-x^2) \quad . \quad . \quad . \quad . \quad . \quad . \quad . \quad . \quad (4) \end{aligned}$$

giving the speed at any point.

If the motion starts from any point other than A, the angle ωt in (1) and (2) must be replaced by $(\omega t+\alpha)$, where α is some fixed angle. If the motion of P is clockwise, the sign of ω must be changed. In either case, equations (3) and (4) will be unaltered.

In particular, if the point X starts from O and begins its motion towards A, it is convenient to suppose that P begins at B and ω is reversed. In this case, equations (1) and (2) become $x = a \sin \omega t$, and $\dot{x} = a\omega \cos \omega t$.

It is to be observed that the position and velocity at time t are repeated at times $t+2\pi/\omega$, $t+4\pi/\omega$, etc. This is clear from equations (1) and (2), and also from the fact that $2\pi/\omega$ is the time for one revolution of P in its uniform circular motion.

$2\pi/\omega$ is called the *period* of the simple harmonic motion, and the distance a between the centre and an extreme point is called the *amplitude.*

Example 3.—A point in simple harmonic motion is moving outwards from the centre. At distance 1 m, its speed is 5 m/s; at distance 1·25 m, 4 m/s. Find the period, the amplitude, the maximum speed, and the time taken between the two points mentioned.

Using standard notation, we have, by a standard formula,

$$\omega^2(a^2 - 1) = 25$$
$$\omega^2(a^2 - \tfrac{25}{16}) = 16$$
$$\therefore \tfrac{9}{16}\omega^2 = 9$$
$$\therefore \omega = 4$$
$$\therefore \text{Period} = 2\pi/\omega = \underline{\tfrac{1}{2}\pi \text{ sec}}$$

Again, from the first equation, $a^2 - 1 = \frac{25}{16}$

$$\therefore a = \tfrac{1}{4}\sqrt{41} \simeq \underline{1{\cdot}60 \text{ m}}$$

Maximum speed $= a\omega \simeq \underline{6{\cdot}4 \text{ m/s}}$.

The appropriate equation connecting distance and time for motion starting at the centre is $x = a \sin \omega t$.

Hence
$$1 = \tfrac{1}{4}\sqrt{41} \sin \omega t_1$$
$$1\tfrac{1}{4} = \tfrac{1}{4}\sqrt{41} \sin \omega t_2$$

t_1 and t_2 being the times from the centre to the points mentioned.

The first equation gives $\omega t_1 = 38° \, 40' = 0{\cdot}6748$ rad.

But $\omega = 4$ rad/ s $\quad \therefore t_1 = 0{\cdot}1687$ sec.

From the second equation, $\omega t_2 = 51° \, 21' = 0{\cdot}8962$ rad.

$$\therefore t_2 = 0{\cdot}2240 \text{ sec.}$$
$$\therefore \text{Interval required} \simeq \underline{0{\cdot}055 \text{ sec}}$$

As a check, note that 0·25 m at 5 m/s takes 0·05 sec, and 0·25 m at 4 m/s takes 0·0625 sec.

EXERCISE 6A

Do not substitute approximate values for π or for surds.

1. The maximum speed of a point moving with simple harmonic motion (S.H.M.) is 4 m/s, and its period is $\frac{1}{4}\pi$ sec. Find the amplitude of the motion.
2. The amplitude of an S.H.M. is 50 cm, and 6 complete oscillations per sec are performed. Find the speeds (i) at the centre of the motion, (ii) at a point midway between the centre and an extreme position.

3. The maximum speed in an S.H.M. is 5 m/s, and the speed at 25 cm from the centre is 4 m/s. Find the period and amplitude of the motion.

4. A point moving with S.H.M. passes from one extreme position to the mean position, a distance of 2 m, in 3 sec. Find its speeds after 1 sec, 2 sec, 3 sec.

5. A particle moving with S.H.M. passes a particular point at 3 m/s. After 1 sec it passes the same point in the opposite direction, and after a further 2 sec passes again in the original direction. Find the period and amplitude of the motion.

6. The amplitude of an S.H.M. is 25 cm, and the period 2 sec. Find the maximum acceleration.

7. An S.H.M. is completely performed 4 times per sec. Find the greatest allowable amplitude if the maximum acceleration is not to exceed 32 m/s^2.

8. A point moving with S.H.M. passes its mean position every $\frac{1}{4}$ sec at a speed of 5 cm/s. Find the acceleration in the extreme positions.

9. The piston of an engine, whose stroke is 1·5 m, may be assumed to perform S.H.M. Find its greatest speed, and greatest acceleration, when the speed of the engine is 150 rev/min.

10. In an S.H.M. the distance between extreme and mean positions is 6 cm, and is covered in 9 sec. What time is taken to cover (i) the first centimetre, (ii) the sixth centimetre, starting from the extreme position?

11. The tide tables for a port show that on a certain day high water will be at noon, and low water at 6 a.m. and 6 p.m., and that the total rise of the tide will be 8 m. If there is 1 m of water at the harbour entrance at low tide, during what time before and after noon will there be at least 6 m of water at the harbour entrance? The change in depth of water is to be taken as accurately S.H.M.

12. If v denotes the maximum speed in S.H.M., find in terms of v (i) the speed at a point midway between extreme and mean positions, (ii) the speed when half the time from extreme to mean position has elapsed.

EXERCISE 6B

1. A point moves in S.H.M. with amplitude 1 mm, and makes 5 vibrations per second. Find its greatest acceleration in cm/s^2.

2. A point moves in S.H.M. with amplitude 10 cm. A and B are points on its path, which it passes at intervals in the order AABBAA . . . . Find the distances of A and B from the centre of the motion if these time-intervals are all equal.

3. A point moves in S.H.M. with amplitude 10 cm and period 2 sec. Find its speed at a point 8 cm from the centre of the motion.

4. A point moving in S.H.M. has acceleration $1{\cdot}44\pi^2$ m/s^2 when it is 1 cm from the centre of the motion. Find its period.

5. A point moving in S.H.M. has speed 4 cm/s when it is 2 cm from the centre, and 2 cm/s when it is 4 cm from the centre. Find the amplitude and period.

6. A point moving in S.H.M. is at one end of its travel at time 0 sec, 5 cm from that end at time 1 sec, and 15 cm from it at time 2 sec. Find the amplitude and period.

7. A point moving in S.H.M. passes a fixed point at intervals of 1 sec and 4 sec alternately, at a speed of 4π cm/s. Find the amplitude of the motion.

8. A point moving in S.H.M. with a frequency of 10 vibrations per minute has velocity 10 cm/s one second after leaving an extreme point of the motion. Find the amplitude.

9. A point moving in S.H.M. travels 30 cm from its mean position in 1 sec and 48 in 2 sec. How far from its mean position does it move in $\frac{1}{2}$ sec?

10. A particle is oscillating with S.H.M. of frequency 8 per minute. The velocity at a point P, 1 m away from the centre of motion is 2 m/s. Find the amplitude of the oscillation. If the particle is passing through P and moving away from the centre of oscillation, find the time that elapses before the particle next passes through P.
(A numerical answer is required: take $\log \pi = 0{\cdot}4971$.)

11. A moving platform does S.H.M. over a table, with amplitude 4 cm. A particle on the platform does S.H.M. relative to the platform, along the same line and with the same period, but with amplitude 3 cm, and is in its mean position when the platform is at one end of its travel. Prove that the motion of the particle relative to the table is S.H.M. and find its amplitude.

12. A moving platform carrying a sheet of paper does S.H.M., and a pen travels over it in a fixed line perpendicular to its motion, doing S.H.M. of the same period and amplitude. What is the trace of the pen on the paper (i) if the pen and the platform are at the ends of their paths simultaneously, (ii) if the pen is at its mean position when the platform is at one end of its path?

4

Forces Acting at a Point

1. Force

A force acting on a particle is a pull or push or attraction or repulsion of a certain size and in a certain direction. A particle, of course, we shall regard as a point, i.e. we shall regard it as having dimensions which are negligible in comparison with other dimensions in a particular problem.

A force thus may be represented by a straight line AB drawn in the direction of the force and of length proportional to the size of the force according to some chosen scale. In other words a force is a *vector*. If the force acts on a particle this vector passes through the point representing the particle and it is then called a *localized vector*.

2. Units

When a force acts on a body which is free to move it produces an acceleration, and the size of the force depends on this acceleration and on the mass, or quantity of matter, in the body. Note that mass is a scalar quantity; it has magnitude, but no direction. The unit mass in the SI system is the *kilogramme*, which is the mass of a certain piece of platinum kept at Sèvres, near Paris. Consequently, the unit force in the SI system is that which gives to a mass of 1 kg an acceleration of 1 m/s^2, and this force is called a *newton* (N).

When a body is influenced only by its weight, which is the attraction of the Earth upon it, it has an acceleration which is given the symbol g, and is approximately 9·8 m/s^2. Thus the downward force of the Earth's attraction on a body of mass m kg is mg N, or approximately 9·8 m N.

3. Comparison of forces. Hooke's Law

It can be shown experimentally that if an elastic string or a spring is stretched, the extension of the string is proportional to the force stretching it. This is usually stated in the form:

Tension is proportional to extension. This is Hooke's Law.

If l is the natural length of a string and l' is the stretched length due to a tension T, then

$$T = \frac{\lambda}{l}(l' - l)$$

where λ is a constant depending on the thickness and material of the string, and called the *modulus of elasticity*. By putting $l' = 2l$ in the formula above it is clear that λ is the force required to stretch the string by an amount equal to its natural length.

It must be pointed out that the above law is only true for a string which is not strained, i.e. when the force ceases to act the string must return to the same unstretched length.

Hence to compare two forces we compare the extensions they produce in a spring and these are proportional to the forces.

4. Composition of forces

A force is a vector, like a displacement, a velocity, or an acceleration, so that the size and direction of the resultant of two forces is found from the vector law of addition, called in this instance the *parallelogram of forces*. This states that if two forces acting at a point are represented in magnitude and direction by the sides of a parallelogram drawn from that point, then their resultant is represented by the diagonal of the parallelogram drawn from that point.

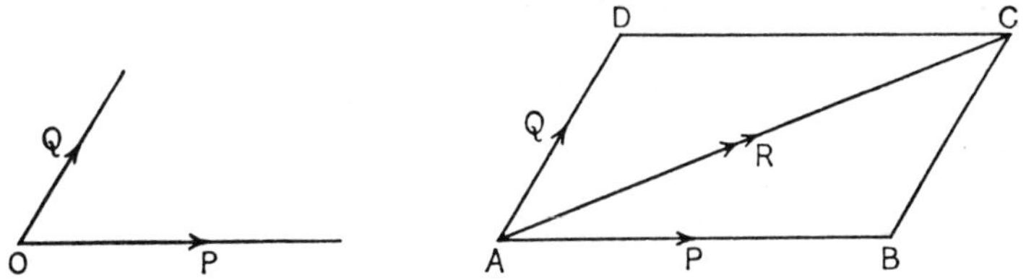

Thus, if P and Q are two forces acting at a point O and we draw AB parallel and proportional to P and AD parallel and proportional to Q and we complete the parallelogram ABCD, then AC is parallel and proportional to R, the resultant of P and Q, and of course R acts through O.

It should be noted that in the figure above it is not necessary to draw the whole parallelogram. The triangle ABC will suffice. AB is drawn to represent P and from B the line BC is drawn to represent Q; then AC represents the resultant R. The directions indicated by the arrows are most important. The arrows for P and Q follow each other, that for R meets them. $\overline{AB}+\overline{BC} = \overline{AC}$. Notice also that single arrows are used for P and Q and a double arrow for their resultant R.

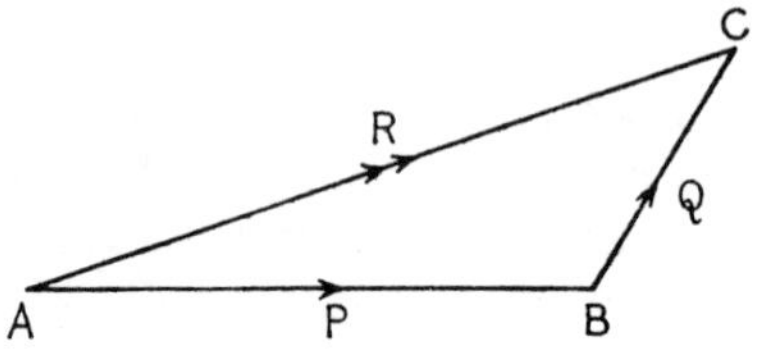

5. To calculate the magnitude and direction of the resultant R we solve the triangle ABC

If the angle between the forces P and Q is α then

$$\angle ABC = 180°-\alpha \text{ and so } \cos ABC = -\cos\alpha$$

Hence from the Cosine Formula

$$R^2 = P^2+Q^2-2PQ\cos(180°-\alpha)$$
$$= \underline{P^2+Q^2+2PQ\cos\alpha}$$

Then from the Sine Formula

$$\frac{R}{\sin(180°-\alpha)} = \frac{Q}{\sin A} = \frac{P}{\sin C} \quad \ldots\ldots \quad \text{(i)}$$

Alternatively if we drop the perpendicular from C to AB

$$\tan A = \frac{Q\sin\alpha}{P+Q\cos\alpha} \quad \ldots\ldots \quad \text{(ii)}$$

Either (i) or (ii) will give the direction of R.

6. The Triangle of Forces

In the figure on p. 29, let $\overline{AB}$, $\overline{BC}$, $\overline{CA}$ represent in magnitude and direction three forces P, Q, R, acting at the point A. Note the order of the letters; if arrows were drawn on the sides of the triangle to show the directions of the vectors, they would follow one another round the figure. The sides of the triangle are said to be " taken in order."

We have shown in para. 4 that $\overline{AC}$ represents the resultant of P and Q. Thus the third force R, which is represented by $\overline{CA}$ is equal and opposite to the resultant of P and Q and so P, Q, and R are in equilibrium. Hence: *when three forces acting at a point can be represented by the sides of a triangle taken in order the forces are in equilibrium.*

The converse theorem is also true; namely that if three forces acting at a point are in equilibrium, and a triangle is drawn with its sides parallel to the directions of the forces, then the lengths of the sides of the triangle will be proportional to the sizes of the corresponding forces.

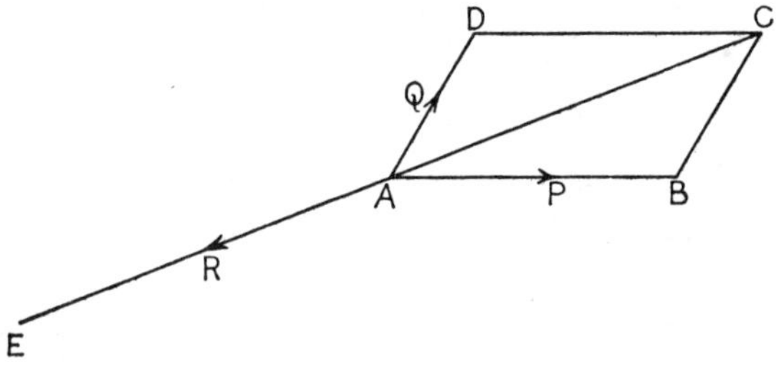

Using the same figure, since P, Q, and R are in equilibrium R must balance the resultant $\overline{AC}$ of P and Q. Therefore $\overline{CA}$ represents R while $\overline{AB}$ and $\overline{BC}$ represent P and Q. Hence the triangle ABC is a triangle with its sides parallel to the forces P, Q, R and also proportional to their magnitudes.

7. Lami's Theorem

From the last paragraph it follows that if P, Q, R are three forces acting at a point and in equilibrium then

$$\frac{P}{AB} = \frac{Q}{BC} = \frac{R}{CA}$$

and using the Sine Formula

$$\frac{P}{\sin ACB} = \frac{Q}{\sin CAB} = \frac{R}{\sin ABC}$$

but since $\angle ACB = \angle DAC = 180° - \angle EAD$,

$$\angle CAB = 180° - \angle EAB, \quad \angle ABC = 180° - \angle DAB$$

$$\therefore \frac{P}{\sin EAD} = \frac{Q}{\sin EAB} = \frac{R}{\sin DAB}$$

If three forces acting at a point are in equilibrium then each force is proportional to the sine of the angle between the other two forces.

8. Resolution of forces

We can replace a given force R by two forces P and Q in any given directions.

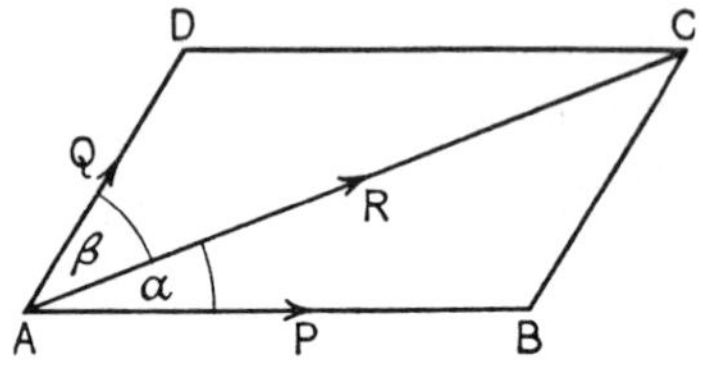

Given the force R suppose the two directions in which we want the components make α and β with R.

Draw a parallelogram with diagonal R and sides making α and β with R.

The sides AB and AD of this parallelogram are then proportional to the required components P and Q. Their sizes can be found from the Sine Formula.

$$\frac{P}{\sin\beta} = \frac{Q}{\sin\alpha} = \frac{R}{\sin(\alpha+\beta)}$$

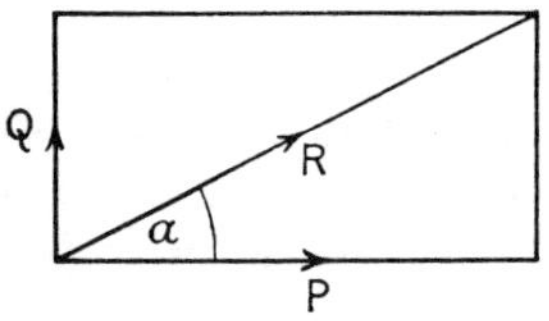

The above is most often used when P and Q are at right angles, in which case R may be replaced by the forces P and Q where $\underline{P = R\cos\alpha}$ and $\underline{Q = R\sin\alpha}$.

9. The Polygon of Forces

Consider several forces $P, Q, S, T, \ldots$, acting on a particle O.

Draw AB parallel and proportional to P, draw BC parallel and proportional to Q, CD parallel and proportional to S, and so on.

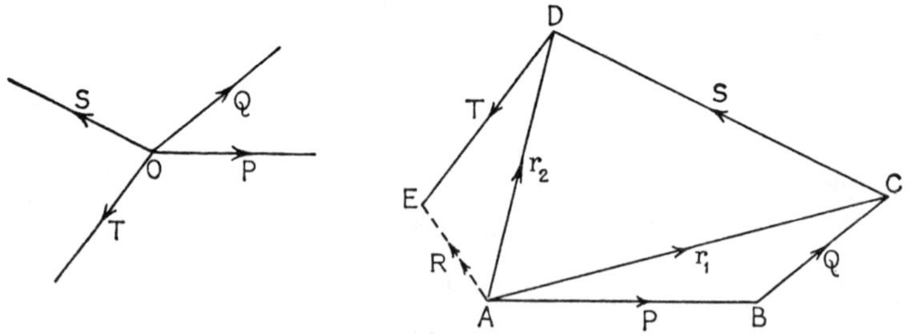

The resultant of P and Q is given in magnitude and direction by r_1 and acts through O. The resultant of r_1 and S is given by r_2 and acts through O. The resultant of r_2 and T is parallel and proportional to R and acts through O. More concisely

$$P+Q = \overline{AB}+\overline{BC} = \overline{AC}$$
$$P+Q+S = \overline{AC}+\overline{CD} = \overline{AD}$$
$$P+Q+S+T = \overline{AD}+\overline{DE} = R$$

and the method could be used for any number of forces acting at a point. The right-hand figure is called a *force polygon.*

10. Equilibrium of a set of forces acting at a point

If the set of forces acting at O were in equilibrium, their resultant R would be zero, i.e. the points E and A in the figure would coincide, i.e. *the force polygon would close.*

11. Resultant of any number of concurrent forces

Let AB, BC, . . . , represent a set of forces.

The force polygon is ABCDE and the closing line AE represents the resultant of the forces.

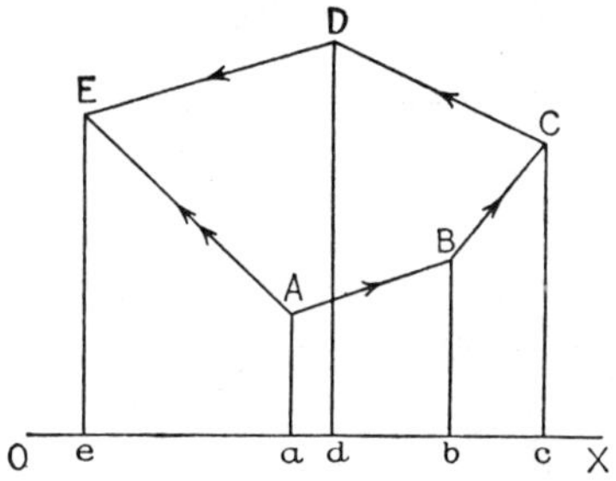

Let $a, b, c, \dots$, be the orthogonal projections of A, B, C, . . . , on to any line OX.

Then with due regard to signs

$$ab+bc+cd+de = ae$$

But these are the resolved parts of the forces $\overline{AB}$, $\overline{BC}$, . . . , in the direction OX. Hence the sum of the resolved parts of the forces $\overline{AB}$, $\overline{BC}$, . . . , is equal to the resolved part of their resultant $\overline{AE}$, provided both are taken in the same direction. For convenience of drawing the figure shows four forces only, but the method is perfectly general.

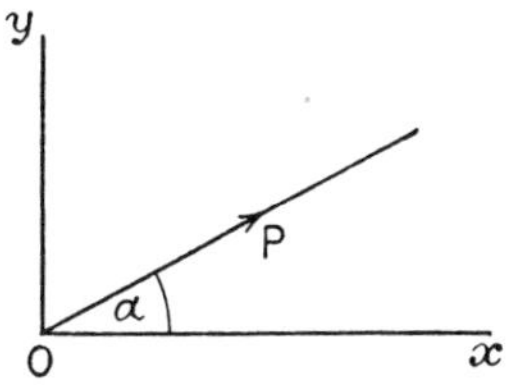

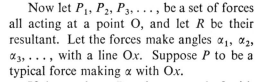

Now let P_1, P_2, P_3, . . . , be a set of forces all acting at a point O, and let R be their resultant. Let the forces make angles α_1, α_2, α_3, . . . , with a line Ox. Suppose P to be a typical force making α with Ox.

If the resultant R makes an angle θ with Ox then

$$P_1 \cos \alpha_1+P_2 \cos \alpha_2+ \dots = \Sigma P \cos \alpha = R \cos \theta$$

and $$P_1 \sin \alpha_1+P_2 \sin \alpha_2+ \dots = \Sigma P \sin \alpha = R \sin \theta$$

Suppose $\Sigma P \cos \alpha = X$ (a single force along Ox or xO)

and $\Sigma P \sin \alpha = Y$ (a single force along Oy or yO)

Hence the original system of forces is reduced to the two forces X and Y shown and we have $R \cos \theta = X$ and $R \sin \theta = Y$

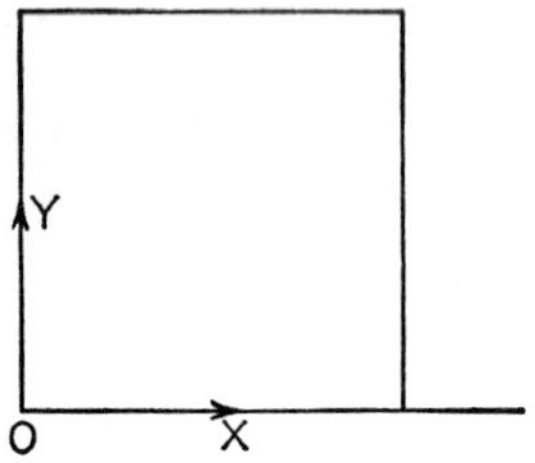

squaring and adding $R^2 = X^2 + Y^2$

i.e. $R = \sqrt{(X^2 + Y^2)}$

and dividing $\tan \theta = \dfrac{Y}{X}$

12. Conditions for equilibrium for a set of coplanar forces which concur

If the original system of forces was in equilibrium $R = 0$ and hence since $R^2 = X^2 + Y^2$ we must have both $X = 0$ and $Y = 0$.

Hence if a set of concurrent coplanar forces is in equilibrium, the sums of their resolved parts in two perpendicular directions must each be zero.

13. Worked examples

Example 1.—Find the magnitude and direction of the resultant of two forces of 7 N and 8 N which act at an angle of 40°.

By drawing.

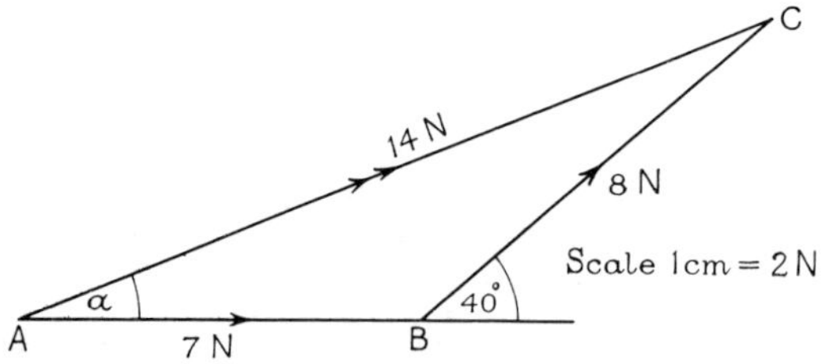

By calculation. $AC^2 = 7^2 + 8^2 - 2 \times 7 \times 8 \cos 140°$

$= 49 + 64 + 112 \cos 40° = 198{\cdot}79$

$AC = 14{\cdot}1$

$$\frac{8}{\sin \alpha} = \frac{14{\cdot}1}{\sin 140°} \qquad \therefore \sin \alpha = \frac{8 \sin 40°}{14{\cdot}1} = 0{\cdot}3648$$

$$\alpha = 21°\ 24'$$

By drawing the resultant is 14 N making $21\frac{1}{2}°$ with the 7 N.
By calculation the resultant is 14·1 N making 21° 24′ with the 7 N.

Example 2.—A mass of 12 kg is hung by two strings of length 3 m and 4 m, the other ends of the strings being attached to points in a horizontal line 5 m apart. Calculate the tension in each string.

Notice that the $\triangle$ABC is right-angled.

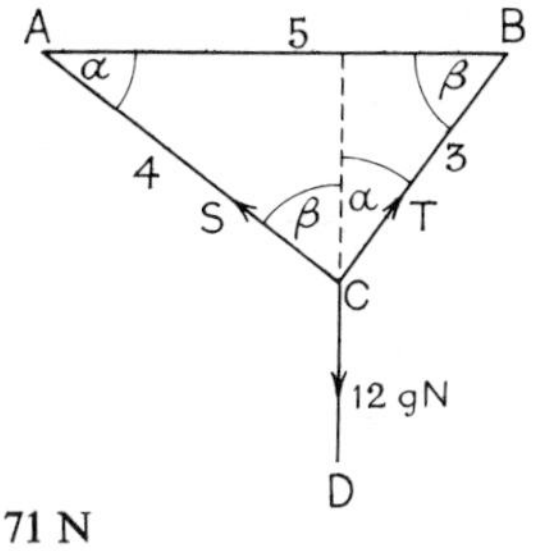

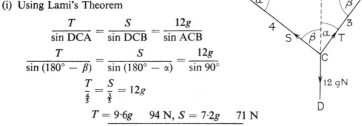

(i) Using Lami's Theorem

$$\frac{T}{\sin DCA} = \frac{S}{\sin DCB} = \frac{12g}{\sin ACB}$$

i.e. $$\frac{T}{\sin (180^\circ - \beta)} = \frac{S}{\sin (180^\circ - \alpha)} = \frac{12g}{\sin 90^\circ}$$

$$\frac{T}{\frac{4}{5}} = \frac{S}{\frac{3}{5}} = 12g$$

$$\underline{T = 9{\cdot}6g \quad 94\text{ N}, S = 7{\cdot}2g \quad 71\text{ N}}$$

(ii) Resolving the forces vertically and horizontally.

Vertically, $T\cos\alpha + S\cos\beta = 12g$, i.e. $\frac{4}{5}T + \frac{3}{5}S = 12g$.

Horizontally, $T\sin\alpha = S\sin\beta$, i.e. $\frac{3}{5}T = \frac{4}{5}S$.

Solving these equations, $\underline{T = 9{\cdot}6g\text{ N}, S = 7{\cdot}2g\text{ N}.}$

Example 3.—A bead whose weight is 1 N slides on a smooth circular wire whose plane is vertical. The bead is attached to the highest point of the wire by an inextensible string. In the equilibrium position the string is inclined to the vertical at 30°. Calculate the tension in the string and the reaction of the wire on the bead.

Since the wire is smooth the reaction of the wire on the bead is normal to the wire, i.e. R passes through the centre of the wire O.

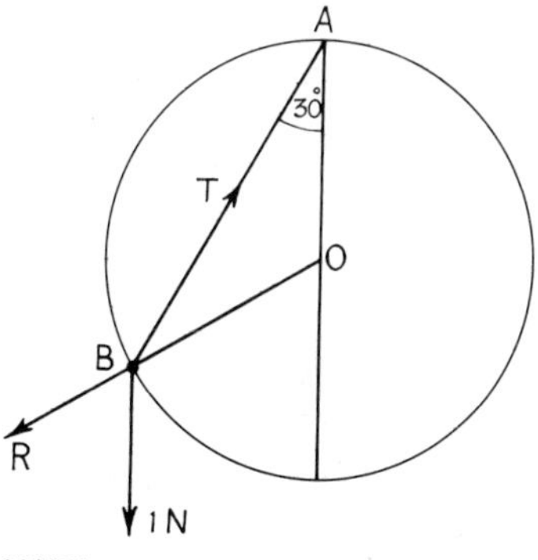

(i) From Lami's Theorem

$$\frac{T}{\sin 60^\circ} = \frac{R}{\sin 150^\circ} = \frac{1}{\sin 150^\circ}$$

from which $\underline{T = \sqrt{3} = 1{\cdot}73\text{ N and } R = 1\text{ N}}$

(ii) Resolve vertically, $T\cos 30^\circ = R\cos 60^\circ + 1$, $\therefore\ T\sqrt{3} = R + 2$

Resolve horizontally, $R\cos 30^\circ = T\cos 60^\circ$, $\therefore\ \sqrt{3}R = T$

from which $\underline{R = 1\text{ N and } T = \sqrt{3}\text{ N}}$

(iii) The triangle BOA has its sides parallel to the three forces T, R, 1. Hence

$$\frac{T}{AB} = \frac{R}{OB} = \frac{1}{OA}$$

i.e.

$$\frac{T}{2a \cos 30°} = \frac{R}{a} = \frac{1}{a}$$

where a = radius of wire.

$$\therefore \frac{T}{\sqrt{3}} = \frac{R}{1} = 1$$

i.e. $\underline{T = \sqrt{3} = 1{\cdot}73 \text{ N and } R = 1 \text{ N}}$

Example 4.—A small ring of weight W, free to slide on a smooth vertical circle, is tied to the highest point of the circle by an elastic string of natural length equal to the radius of the circle and of modulus λW. Show that there is an equilibrium position in which the string makes an angle $\cos^{-1} \dfrac{\lambda}{2\lambda - 2}$ with the vertical, provided $\lambda > 2$.

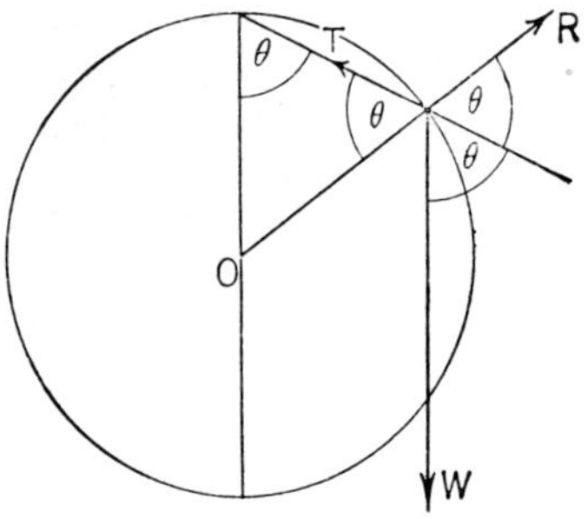

In the position of equilibrium let the string make θ with the vertical. Three forces act on the ring, T, R, and W.

By Lami's Theorem

$$\frac{T}{\sin 2\theta} = \frac{W}{\sin (\pi - \theta)}$$

$$\therefore T = \frac{2W \sin \theta \cos \theta}{\sin \theta} = 2W \cos \theta$$

From Hooke's Law

$$T = 2W \cos \theta = \frac{\lambda W}{a} (2a \cos \theta - a)$$

$$\therefore 2 \cos \theta = \lambda(2 \cos \theta - 1)$$

$$\cos \theta = \frac{\lambda}{2\lambda - 2}$$

Since $\cos \theta < 1$

$$\frac{\lambda}{2\lambda - 2} < 1, \quad \therefore \lambda < 2\lambda - 2, \quad \therefore \boldsymbol{\lambda > 2}$$

Example 5.—Find the magnitude and direction of the resultant of the forces (in newtons) shown in the diagram.

By drawing.

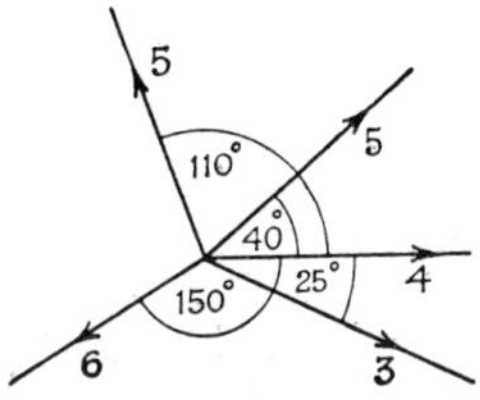

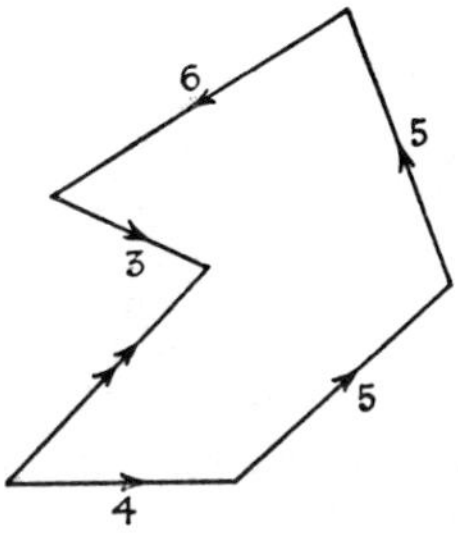

The resultant is 5·15 N at an angle of 45° with the 4 N.

By calculation.

$$\begin{aligned} R\cos\theta &= 4 + 5\cos 40^\circ + 5\cos 110^\circ + 6\cos 210^\circ + 3\cos 335^\circ \\ &= 4 + 5\cos 40^\circ - 5\cos 70^\circ - 6\cos 30^\circ + 3\cos 25^\circ \\ &= 3{\cdot}643 \\ R\sin\theta &= 5\sin 40^\circ + 5\sin 110^\circ + 6\sin 210^\circ + 3\sin 335^\circ \\ &= 5\sin 40^\circ + 5\sin 70^\circ - 6\sin 30^\circ - 3\sin 25^\circ \\ &= 3{\cdot}645 \end{aligned}$$

Squaring and adding

$$R^2 = 13{\cdot}27 + 13{\cdot}29, \quad \therefore\ R = 5{\cdot}154$$

Dividing $\tan\theta = \dfrac{3{\cdot}645}{3{\cdot}643} \simeq 1, \quad \therefore\ \theta = 45^\circ.$

Hence the resultant is 5·154 N at 45° to the 4 N.

EXERCISE 7A

1. Find by calculation the resultant of two forces of 7 N and 5 N acting at an angle of 90°. Calculate the angle the resultant makes with the larger force. Check by drawing.
2. Calculate the magnitude and direction of the resultant of two forces of 5 N and 3 N which act at an angle of 100°.
3. Calculate the resultant of two forces each 10 N inclined at an angle of 120°.
4. Forces of 6 N and 4 N act at an angle of 60°. Find graphically and by calculation their resultant in magnitude and direction.
5. Two forces inclined to one another at 50° have a resultant of 8 N. If one force is 5 N, calculate the other force.
6. A force of 12 N is compounded into two forces; one is 5 N making 30° with the 12 N. Calculate the magnitude and direction of the other force.
7. If the resultant of two forces of 10 N and 6 N is 14 N, calculate the angle between the two forces.

8. ABCD is a square. Calculate the resultant of 5 N along AB, $3\sqrt{2}$ N along AC, and 3 N along AD. Find also the angle the resultant makes with AB.

9. ABCDEF is a regular hexagon. Forces of 1, 2, 3, 4, 5 N act along AB, AC, AD, AE, and AF respectively. Find graphically and by calculation the size of the resultant and the angle it makes with AB.

10. ABCDEF is a regular hexagon whose centre is O. Forces of 1, 2, 3, 4, 5, and 6 N act along OA, OB, OC, OD, OE, and OF respectively. Calculate the size of the resultant and the angle it makes with AB.

11. Using a scale of 1 cm to 5 N, find graphically the magnitude and direction of the resultant of 20 N due E, 25 N in direction 040°, 20 N in direction 330°, and 40 N in direction 250°.

12. Four horizontal wires are attached to the top of a post and exert the following tensions on it: 10 N due N, 15 N due E, 20 N SW and 25 N SE. Calculate the resultant pull on the post and the direction in which it acts.

13. ABC is an equilateral triangle. Three forces act at a point O: 2 N parallel to AB, 3 N parallel to AC, and 4 N parallel to BC. Calculate the size of the resultant and the angle its direction makes with AB.

14. The medians of an equilateral triangle ABC intersect at G. Forces of 2, 2, and 4 N act along GB, GC, and GA respectively. Find the magnitude and direction of their resultant.

15. A mass of 10 kg is hung from a point by an inextensible string. It is pulled by a horizontal force P until the string makes 30° with the vertical. Calculate the tension in the string and the size of P.

16. Repeat question 15 when P is perpendicular to the string.

17. AB and BC are two strings of length 12 m and 5 m joined at B and fastened to two points A and C in the same horizontal line so that angle ABC is 90°. If a mass of 26 kg hangs from B, find the tensions in AB and BC.

18. Two strings AB and BC support a mass B of 10 kg. AB and BC are inclined to the vertical at 30° and 60° respectively. Calculate the tensions in the strings.

19. A string of length 2 m is attached to two points A and B at the same level and 1 m apart. A ring of weight 12 N, slung on the string, is acted upon by a horizontal force P which holds it in equilibrium vertically below B. Calculate the tension in the string and the size of P.

20. A mass of weight 13 N is suspended by two strings of length 5 m and 12 m from two points in the same horizontal line and 13 m apart. Calculate the tensions in the strings.

21. A particle of weight W rests on a smooth plane inclined at 30° to the horizontal. It is held there by a force P parallel to a line of greatest slope of the plane. Calculate P and the reaction of the plane on the particle.

22. A string ABCD has its ends attached to fixed points. The string carries a weight of 15 N at B and a weight W at C. AB is inclined at 60° and CD is inclined at 30° to BC, which is horizontal. Calculate W.

23. A bead of weight W can slide on a smooth circular wire whose plane is vertical. The bead is tied to the highest point of the wire by a light thread. If the thread is inclined at $\theta°$ to the vertical in the position of equilibrium, calculate in this position the tension in the thread and the reaction of the wire on the bead.

24. A and C are two pulleys in the same horizontal line 2 m apart. A light string passes over the pulleys and carries masses of 3 kg at each end and 2 kg at its midpoint B. Calculate the depth of B below AC.

25. ABCD is a light string tied to vertical walls at A and D with A above D. At B and C masses of weights W and w are attached. If AB is inclined at θ and BC at ϕ to the vertical and CD is horizontal, prove $T\cos\theta = W + w$ and $w\tan\phi = (W + w)\tan\theta$, where T is the tension in AB.

26. A small bead of weight W slides on a smooth vertical circular wire of radius a. It is attached by a string of length $2l(l < a/\sqrt{2})$ to a point on one extremity of the horizontal diameter. Show that the tension in the string is $\dfrac{W(a^2 - 2l^2)}{a\sqrt{(a^2 - l^2)}}$

27. An elastic string of natural length 4·8 m stretches 60 cm when pulled by a force of 13 N. The ends of the string are fastened to two points A and B at the same level and 4·8 m apart and a mass of weight W is attached to the middle point of the string. It is found that the mass rests in equilibrium at a depth of 1 m below the level of AB. Calculate the tension in the string and the magnitude of W.

28. A mass of weight 30 N is hung by two similar and equal elastic strings from two points at the same horizontal level and 4 m apart. The mass is at rest 1·5 m below the level of AB. Calculate the tension in each string. If each string is such that a force of 100 N would be required to stretch it by an amount equal to its natural length, calculate the natural length of each string.

29. Two smooth rings of weight W and W_1 connected by a light string slide on two smooth fixed wires; W slides on a vertical wire and W_1 on a wire inclined downwards at α to the horizontal. A particle P of weight w is tied to the string. If θ and ϕ are the angles which the portions of the string make with the vertical and θ refers to the weight W and ϕ to W_1, prove

$$\frac{W}{\cot\theta} = \frac{w + W}{\cot\phi} = \frac{w + W + W_1}{\cot\alpha}$$

EXERCISE 7B

1. Find the magnitude and direction of the resultant of two forces of 7 N and 8 N whose directions are inclined to each other at 60°.

2. Find the magnitude and direction of the resultant of two forces, each 8 N, whose directions are inclined at 120°.

3. A particle of weight 2·5 N is hung up by two strings. If the tensions in the strings are 1·5 N and 2 N, find the inclination of the strings to the vertical.

4. The resultant of two forces of 5 N and 7 N is 9 N. Find the angle between the forces of 5 N and 7 N.

5. The resultant of two forces P and Q of magnitude 1 and 2 N respectively is perpendicular to P. Find the angle between P and Q.

6. A force of 10 N is the resultant of two forces. One is 5 N at 30° to the 10 N. Find the other force.

7. A mass of $8\frac{1}{2}$ kg is suspended by two strings of length 8 m and 15 m attached to the weight and to points in a horizontal line 17 m apart. Calculate the tensions in the strings.

8. A mass of 12 kg is on a smooth inclined plane inclined at 30° to the horizontal. Find the force parallel to the plane to hold the weight in equilibrium. Find also the force at an angle of 45° to the plane to do this.

9. Two particles each weight W repel each other with a force of μ/x^2, where x is their distance apart. They are suspended from the same point by strings of length l and are in equilibrium $2d$ apart, prove that $4d^3W = \mu\sqrt{(l^2 - d^2)}$.

10. Three strings AB, BC, CD are knotted together at B and C. A and D are fastened to points in a horizontal line and masses of 5 kg and 7 kg hang from B and C. If AB is inclined to the horizontal at 60° and BC is horizontal, find the inclination of CD to the horizontal.

11. A string of length a is attached to two points A and B in a horizontal line b apart. A ring of weight W slides on the string. The ring is pulled by a horizontal force F and is in equilibrium vertically below B. Prove $F = Wb/a$ and find the tension in the string.

12. A particle can be supported on a smooth plane whose inclination is α by either a force P parallel to the plane or a force Q horizontally. Prove $\cot \alpha = \dfrac{P}{\sqrt{(Q^2-P^2)}}$

13. AB and BC are two smooth fixed rails each inclined to the horizontal at 30°. Smooth rings each of mass 1 kg slide on the rails. To these rings is attached a light string on which is another ring of mass $1\frac{1}{2}$ kg. In the position of equilibrium calculate the inclination of the parts of the string to the vertical.

14. Two beads of weight W and w are attached by a light string of length $2l$. They slide on a smooth vertical wire circle of radius a. If $W > w$ and $a > l$, and if in the position of equilibrium the string is inclined to the horizontal at θ, prove

$$\tan \theta = \frac{W - w}{W + w} \cdot \frac{l}{\sqrt{(a^2 - l^2)}}$$

15. ABCD is a square. Forces of 1, $3\sqrt{2}$, and 2 N act along BA, DB, and BC respectively. Find the magnitude of the resultant and the angle it makes with AB.

16. ABCDEF is a regular hexagon. Forces of 1, $2\sqrt{3}$, 6, $\sqrt{3}$ and 2 N act at a point in the directions AB, AC, DA, AE, and AF respectively, the senses of the forces being given by the order of the letters. Show that the forces are in equilibrium.

17. ABC is an equilateral triangle; D, E, F are the midpoints of BC, CA and AB. Forces of 1, 2, 3, 1, 1, 1 N act at A parallel to AB, BC, CA, DA, EB, and FC respectively. Calculate the magnitude and direction of their resultant.

18. ABCDEF is a regular hexagon. Forces of 4, $2\sqrt{3}$, 2, $2\sqrt{3}$, 4 N act along AB, AC, AD, AE, and AF respectively. Find the magnitude and direction of their resultant.

19. A light elastic string has a natural length of $1\frac{1}{2}$ m. It is suspended from one end A and a mass of 4 kg is attached to its other end B. When this mass hangs at rest the length of AB is 2 m. B is now pulled aside by a horizontal force until the string makes 60° with the vertical. Calculate in this position the tension in AB, the length of AB, and the horizontal force.

20. Two springs, alike in every respect, each have a natural length of 20 cm. A force of g N would stretch each spring 2 cm. One end of each spring is attached to a heavy particle O; the other ends of the springs are attached, one to each of two points in a horizontal line 30 cm apart. The particle hangs in equilibrium with the length of each spring 25 cm. Calculate the tension in each spring and the mass of the particle.

21. A light string ABCD is fixed at the ends A and D. A is above D; AB and BC are inclined to the downward vertical at 30° and 60° respectively; and CD is horizontal. The string has weights W_1 and W_2 attached to it at B and C respectively. Prove $W_1 = 2W_2$.

14. Friction

Consider a heavy particle resting on a horizontal plane, the particle being pulled by an increasing horizontal force P. The forces acting on the particle are as shown: (i) the pull P, (ii) the weight W of the particle, (iii) the reaction R of the plane on the particle.

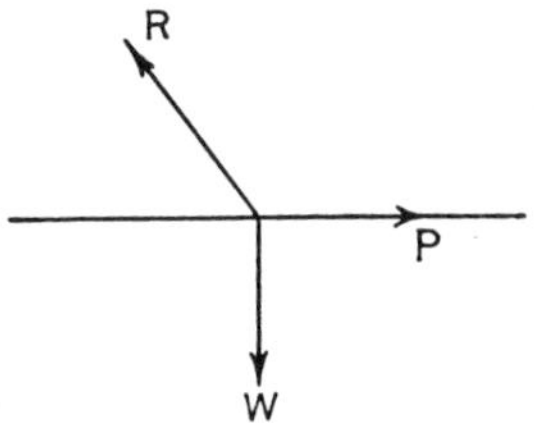

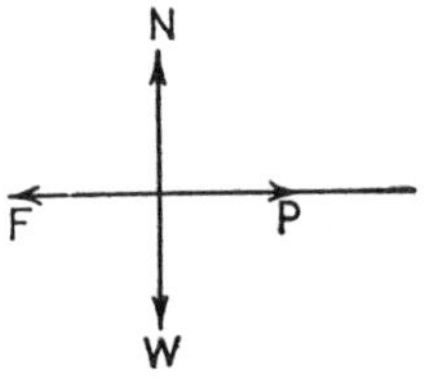

We can replace R by two forces, N normal to the plane and F along the plane. N is called the normal reaction of the plane on the particle and F the frictional force.

If P is not large enough to move the particle we shall have $P = F$ and $N = W$ since the particle is in equilibrium.

Suppose now that P is increased until the particle is just about to move, i.e. P is just equal to the maximum value F can have (the limiting value of F); the particle is then said to be in limiting equilibrium.

The following points concerning a frictional force can be shown experimentally.

(i) Friction is a passive force, i.e. up to a certain point $F = P$.

(ii) Its direction is always opposite to the direction in which the particle tends to move.

(iii) Only a certain amount of friction can be called into play. The maximum amount is called the *limiting friction.*

(iv) The limiting friction (F) divided by the normal reaction (N), i.e. $\frac{F}{N}$ is a constant usually denoted by μ. This ratio μ depends on the nature of the surfaces and is called the *coefficient of limiting friction.*

Hence when we are told that the friction is limiting, and only then, we may assume $F = \mu N$. If in doubt do not assume that friction is limiting friction.

15. Angle and cone of friction

We have said that the resultant reaction R can be replaced by a normal reaction N and a tangential force F called the friction.

From the figure $\tan\theta = \dfrac{F}{N}$

Hence as F increases, $\tan\theta$, and therefore θ, increases until when F reaches its limiting value θ is maximum.

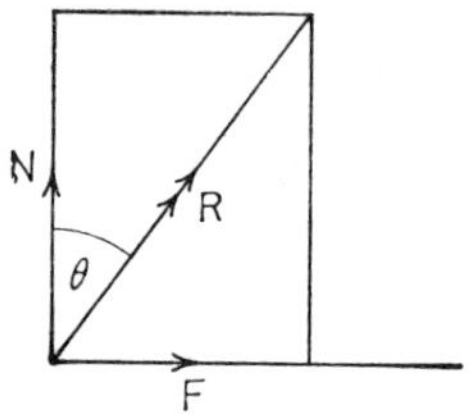

In this case the angle θ is usually denoted by λ and is called the *angle of friction*, i.e. we have

$$\tan\lambda = \mu = \frac{F}{N}$$

where F is the limiting friction.

Notice that θ may take any numerical value from 0 to λ, i.e. R may take any position on each side of the normal so long as its inclination to the normal does not exceed λ.

In other words the resultant reactions may act anywhere inside a cone whose axis is the normal and whose semi-vertical angle is λ the angle of friction.

Example 6.—A particle of weight W is on a rough plane inclined to the horizontal at $\alpha°$. If the angle of friction is λ calculate the least force P required to pull the particle up the plane and the direction in which P should act.

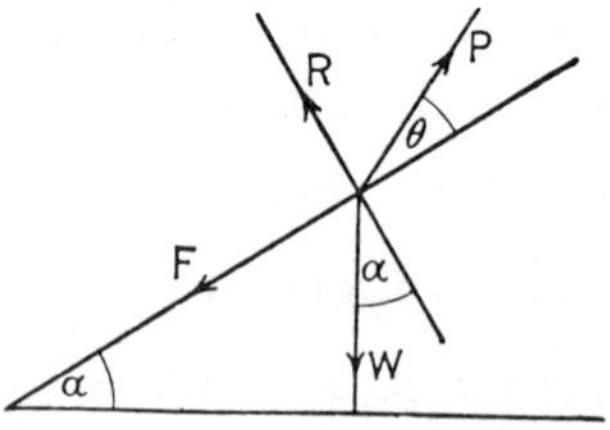

The diagram shows the four forces acting on the particle. Suppose P acts at θ to the plane. For the weight to be on the point of sliding up the plane

$$P\cos\theta \geqslant W\sin\alpha + F$$

where F is the limiting friction. Resolving perpendicular to the plane

$$R = W\cos\alpha - P\sin\theta$$

Hence
$$P\cos\theta \geqslant W\sin\alpha + \mu R$$
$$\geqslant W\sin\alpha + W\cos\alpha\tan\lambda - P\sin\theta\tan\lambda$$
$$P\left(\cos\theta + \frac{\sin\theta\sin\lambda}{\cos\lambda}\right) \geqslant W\left(\sin\alpha + \cos\alpha\,\frac{\sin\lambda}{\cos\lambda}\right)$$
$$P(\cos\theta\cos\lambda + \sin\theta\sin\lambda) \geqslant W(\sin\alpha\cos\lambda + \cos\alpha\sin\lambda)$$
$$P \geqslant \frac{W\sin(\alpha+\lambda)}{\cos(\theta-\lambda)}$$

We may vary θ, and so P is least when $\cos(\theta-\lambda) = 1$, i.e. when $\theta = \lambda$.

Hence the least value of P is $W\sin(\alpha+\lambda)$ and P then acts at an angle of λ with the plane.

EXERCISE 8A

1. A particle of weight 15 N is pulled along a rough horizontal plane. If the coefficient of limiting friction is $\frac{3}{5}$ calculate (i) the least horizontal force and (ii) the least force inclined to the plane at 53° 8′ required.

2. A particle of weight W is pulled along a horizontal rough plane by a force P inclined to the plane at $\theta°$. If the angle of friction is λ prove that the least value of P is $\dfrac{W\sin\lambda}{\cos(\theta-\lambda)}$.

3. A body of weight 100 N rests on a rough plane inclined to the horizontal at 40°. A force P N acts on the weight parallel to a line of greatest slope. If the coefficient of limiting friction is $\frac{1}{4}$, calculate (i) the least value of P to prevent the weight sliding down the plane and (ii) the least value of P to pull the weight up the plane.

4. A body of weight 20 N rests in equilibrium on a rough plane inclined at 20° to the horizontal. Calculate the frictional force when (i) a force of 5 N acts on the body up the plane, (ii) a force of 3 N acts on the body at 20° to the plane and up the plane, and (iii) a horizontal force of 3 N acts on the body towards the plane.

5. A block whose weight is 120 N is just prevented from slipping down a rough inclined plane of inclination 45° by a force P N acting up the plane. If the coefficient of limiting friction is $\frac{1}{3}$, find P.

6. An inclined plane is of length 5 m and height 3 m. A body of weight 10 N is on the plane and a force of 3 N acting along the plane just prevents it from sliding down the plane. Find the coefficient of limiting friction between the weight and the plane, and calculate the least force parallel to the plane required to pull the weight up the plane.

7. A block of wood weight W just begins to slide down a plane when the angle of the plane is α. Find the coefficient of limiting friction in terms of α. Find in terms of W and α the least force parallel to the plane required to pull the block up the plane.

8. Two planes sloping at angles of 53° 8′ and 36° 52′ intersect in a horizontal line. A weight of $2W$ rests on the former and a weight of W on the latter. The weights are connected by a light inextensible string which passes over a small smooth pulley at the common edge with the string lying along lines of greatest slope. If the heavier particle is on the point of sliding down the plane, find the coefficient of limiting friction if the planes are equally rough.

9. Two equal weights attached by a light string rest on the surface of a rough sphere of radius r, one of the weights being at the highest point of the sphere. Find the greatest length of the string if the angle of limiting friction is λ.

10. Two planes sloping at angles of 60° and 30° to the horizontal intersect in a horizontal line. Particles of weight $2W$ and W are placed one on each plane, $2W$ being on the plane whose slope is 60°. The weights are connected by a light inextensible string which passes over a small light pulley at the common edge of the planes with the string along lines of greatest slope. If the heavier particle is on the point of slipping, find the coefficient of limiting friction assuming it the same for both planes.

11. A rough rod is fixed at 60° to the vertical. A ring of weight W is on the rod and a force P acts on the ring, P being inclined to the rod at 30°. If the coefficient of friction is μ and the ring is on the point of sliding up the rod, find the least value of P.

12. A particle of weight W is on a rough horizontal plane. Find the least force P which will pull the particle along the plane and state the direction in which P then acts if the angle of friction is λ.

13. A body of weight W rests on a rough plane of inclination α. Show that the least force that will move it up the plane is $W \sin(\alpha + \lambda)$, where λ is the angle of friction.

EXERCISE 8B

1. A body resting on a rough horizontal table is pulled by two strings at right angles to each other. The tensions in the strings are 8 N and 6 N. If the body does not move find the frictional force.

2. A body is on a rough horizontal plane. It is pulled by a force which makes angle θ with the plane. If the coefficient of limiting friction is μ, find the value of θ for the magnitude of the force to be least when the body is on the point of moving.

3. The angle of a rough plane is 40°. Calculate the least force parallel to the plane required to pull a mass of weight 4 N up the plane if the coefficient of limiting friction is $\frac{1}{4}$.

4. A body of weight 10 N is just prevented from sliding down a rough inclined plane of inclination 30° to the horizontal by a force of 4 N parallel to the plane. Find the coefficient of friction. Find also the least force parallel to the plane required to pull the body up the plane.

5. The least force which will move a body up an inclined plane is P. Show that the least force parallel to the plane is $P \sec \lambda$, where λ is the angle of friction.

6. Two rough planes of inclination to the horizontal 53° 8′ and 36° 52′ meet in a common horizontal line. Masses of weight 2 N, W N respectively rest on the planes connected by a string parallel to lines of greatest slope which passes over a smooth pulley at the junction of the planes. If both coefficients of friction are $\frac{1}{2}$ and the first mass is on the point of sliding up its plane, calculate W.

7. Find the least horizontal force which will move a body of weight W up a rough inclined plane of inclination α to the horizontal if the angle of friction is λ.

8. The least force which will move a body up an inclined plane of inclination α is twice the least force which will prevent it from sliding down the plane. Find the coefficient of friction in terms of α.

9. Particles of mass 2 kg and 6 kg rest on a double inclined plane. They are connected by a light string which passes along lines of greatest slope and over a smooth pulley at the junction of the two planes. If the inclinations of the planes are 30° and 45° respectively and the coefficient of friction is the same for both planes, find it if the weights are on the point of sliding.

10. The angle of a rough inclined plane is 30°. A particle of mass 4 kg is on the plane and is connected by a light string which passes along a line of greatest slope over a pulley to a mass M kg hanging freely. If the particle is on the point of moving up the plane and if the coefficient of friction is $\sqrt{3}/6$, calculate M.

11. Two masses, of weights W_1 and W_2 rest on a rough inclined plane of inclination α. They are connected by a light string which is along a line of greatest slope. If μ_1 and μ_2 are the respective coefficients of limiting friction, prove that if the masses are on the point of slipping

$$\tan \alpha = \frac{\mu_1 W_1 + \mu_2 W_2}{W_1 + W_2}$$

provided $\mu_1 > \tan \alpha > \mu_2$.

12. Two rough rings of equal weight W are on a horizontal rod. They are connected by a light string of length $2l$ on which slides another ring of weight $2W$. Prove that if μ is the coefficient of friction between the rail and the rings, and the string is smooth, the greatest possible distance between the rings on the rod is

$$\frac{4\mu l}{\sqrt{(1 + 4\mu^2)}}$$

5

Parallel Forces and Centre of Gravity

1. A rigid body

So far we have considered only forces acting at a point. We now go on to deal with forces which are not concurrent. Such forces must act on a body of some kind. We shall assume that the bodies—rods, laminæ, ladders, wheels, etc.—on which our forces act are *rigid*, i.e. that the shape of the body is unaltered by the forces acting on it.

2. The moment of a force about a point

The moment of a force about a point is defined as the product of the force and the length of the perpendicular from the point on to its line of action; i.e. if P is the force and d the length of the perpendicular from the point O then the moment of P about O is Pd.

If the force P is represented by the line AB then $Pd = \text{AB} \cdot d =$ twice the area of triangle OAB. In other words the moment of a force about a point may be represented by the area of a triangle.

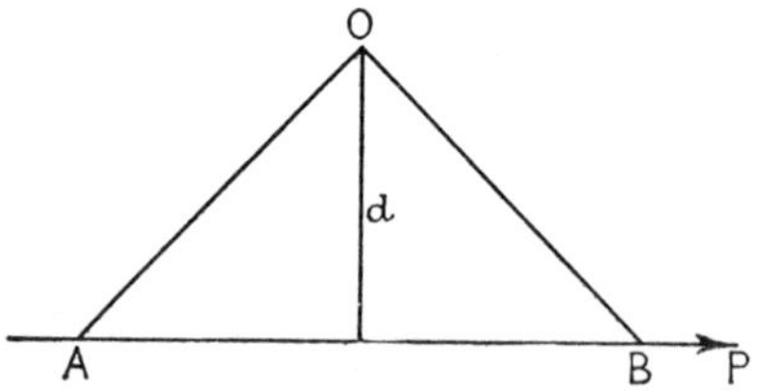

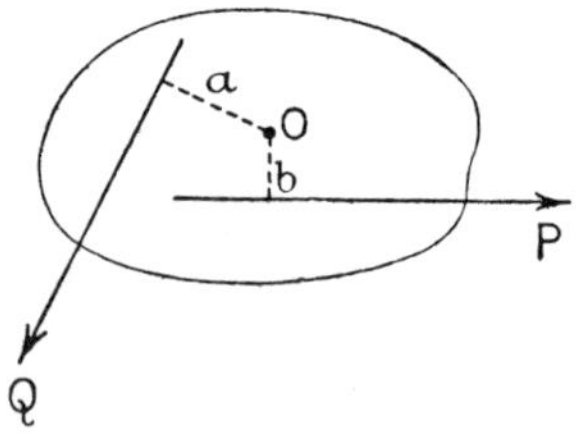

The moment of a force is the turning effect of the force about the point.

In the diagram on the right, suppose the two forces P and Q are applied to a thin flat body—a lamina—which is free to turn about a pivot O, then P and Q both turn the lamina in the same sense, i.e.

anticlockwise about O. Their turning effects are measured by Qa and Pb, and so the total turning effect is $Qa+Pb$.

If the direction of P were reversed the total effect would be $Qa-Pb$.

3. Parallel forces

Parallel forces are said to be *like* when they act in the same sense and *unlike* when their senses are opposite.

4. The resultant of two like parallel forces

Suppose P and Q are the forces acting at points A and B of a rigid body. Let P and Q be represented by AZ and BN respectively. At A and B introduce two equal and opposite forces S acting in the line AB. These equal and opposite forces will have no effect on the body

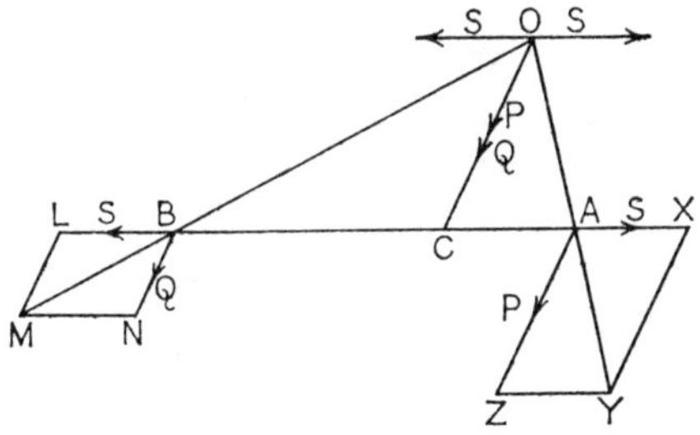

which is rigid. Complete the parallelograms AXYZ and BLMN, then AY and BM represent the resultants of P and S and Q and S respectively. Produce MB and YA to meet at O. Draw OC parallel to AZ to meet AB at C. The forces $\overline{\text{BM}}$ and $\overline{\text{AY}}$ may be considered to act at O and may be resolved into a force S parallel to BL and a force Q along OC for $\overline{\text{BM}}$, and a force S parallel to AX and a force P along OC for $\overline{\text{AY}}$. The two forces S at O are equal and opposite, and so we are left with a force $P+Q$ along OC, i.e. a force $P+Q$ parallel to each of P and Q.

To find the position of C.

The triangles OCA and AZY are similar,

$$\therefore \frac{\text{OC}}{\text{CA}}=\frac{\text{AZ}}{\text{ZY}}=\frac{P}{S} \qquad \ldots\ldots \quad \text{(i)}$$

In the same way the triangles OCB and BNM are similar,

$$\therefore \frac{OC}{CB} = \frac{BN}{NM} = \frac{Q}{S} \quad \ldots\ldots \text{(ii)}$$

Hence dividing (i) by (ii), $\frac{CB}{CA} = \frac{P}{Q}$

i.e. $P\,.\,CA = Q\,.\,CB$

i.e. C is the point dividing AB internally in the inverse ratio of the forces.

5. The resultant of two unlike parallel forces

The construction is similar to that for two like parallel forces.

The resultant is a force $P-Q$ $(P>Q)$ parallel to each of P and Q and cutting BA produced at the point C so that $Q\,.\,BC = P\,.\,AC$.

Again C is the point dividing AB in the inverse ratio of the forces but this time externally.

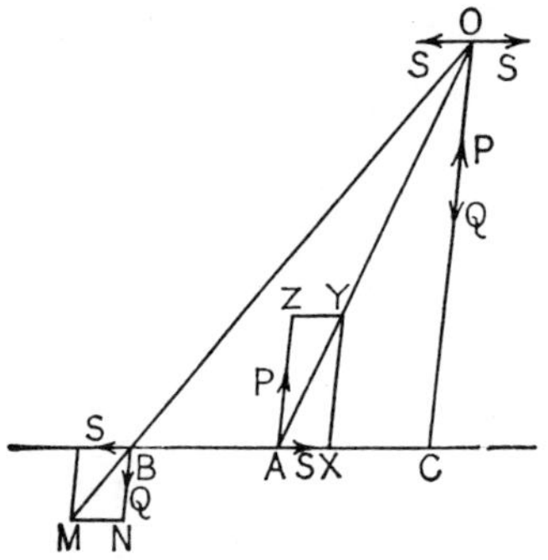

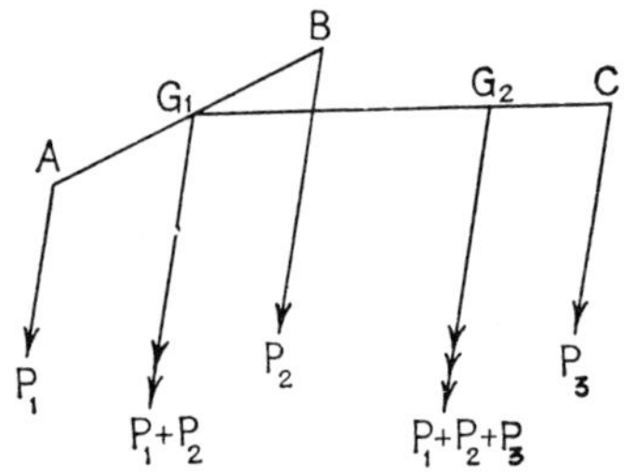

6. Centre of a system of parallel forces

Suppose $P_1, P_2, P_3, \ldots, P_n$ are a set of parallel forces, not necessarily coplanar.

The resultant of P_1 and P_2 is a force P_1+P_2 acting at G_1 so that $P_1 \times AG_1 = P_2 \times BG_1$, i.e. G_1 is a fixed point.

The resultant of P_1+P_2 at G_1 and P_3 is a force $P_1+P_2+P_3$ at G_2 so that $(P_1+P_2)G_1G_2 = P_3 \times G_2C.$, i.e. G_2 is a fixed point.

Continuing in this way we see that the resultant of all the forces will be $(P_1+P_2+P_3+\ldots+P_n)$ acting at a fixed point.

Clearly this point is independent of the direction of the given forces and depends only on their sizes.

7. A couple

When the two unlike forces are parallel and *equal* the lines MB and AY in the left-hand figure are parallel so there is no point O and so the construction breaks down.

Hence there is no single force equivalent to two equal unlike parallel forces.

Such a pair of forces constitutes *a couple.*

Clearly a couple tends to rotate a body on which it acts but does not translate the body. We will postpone further consideration of couples until later in the chapter.

8. The Principle of Moments

Consider first a set of forces P_1, P_2, P_3, . . . , of which a typical force is P, all acting at a point O. Suppose their resultant is R, and X the point about which we take moments. Let the forces R, P_1, P_2, P_3, . . . , be at distances r, p_1, p_2, p_3, . . . from X, and let them make

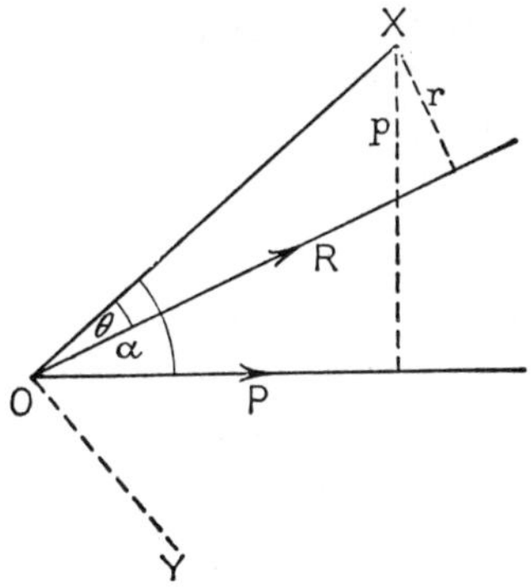

angles θ, α_1, α_2, α_3, . . . , with OX. Draw OY perpendicular to OX. The algebraic sum of the moments about X of P_1, P_2, P_3, . . . , is ΣPp.

$$\begin{aligned}\Sigma Pp &= \Sigma P\,.\,\text{OX} \sin \alpha \\ &= \text{OX}\,.\,\Sigma P \sin \alpha \\ &= \text{OX} \times (\text{the sum of the resolved parts of the forces along OY}) \\ &= \text{OX} \times (\text{the resolved part of } R \text{ along OY}) \\ &= \text{OX} \times R \sin \theta = R\,.\,\text{OX} \sin \theta = Rr \\ &= \text{the moment of } R \text{ about X.}\end{aligned}$$

Next consider two parallel forces P and Q.

Let O be any point in the plane of the parallel forces, and let R be their resultant. From O draw OAZB perpendicular to the forces to meet their lines of action in A, B, and Z.

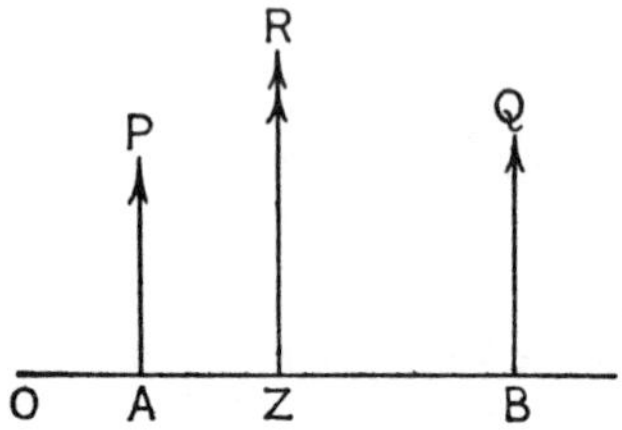

The algebraic sum of the moments of P and Q about O

$$\begin{aligned} &= P\,.\,\mathrm{OA}+Q\,.\,\mathrm{OB} \\ &= P(\mathrm{OZ}-\mathrm{AZ})+Q(\mathrm{OZ}+\mathrm{ZB}) \\ &= (P+Q)\mathrm{OZ} \\ &= \text{the moment of } R \text{ about O.} \end{aligned}$$

(since $P\,.\,\mathrm{AZ} = Q\,.\,\mathrm{BZ}$)

A similar proof holds for two unlike parallel forces and similarly by addition we can show that the sum of the moments of any number of coplanar parallel forces about any point in their plane is equal to the moment about this point of their resultant.

Hence we have established the *Principle of Moments*, namely that the sum of the moments about any point in their plane of any system of coplanar forces is equal to the moment of their resultant about the same point.

It follows that if a system of coplanar forces is in equilibrium, i.e. they have a resultant zero, then the sum of their moments about any point in their plane is zero.

9. Couples

As we have said, a couple consists of two equal and opposite parallel forces. The algebraic sum of the resolved parts of two such forces in any direction is zero. Hence there is no tendency for a couple to produce a translation of a body on which it acts. It does however tend to produce rotation of the body.

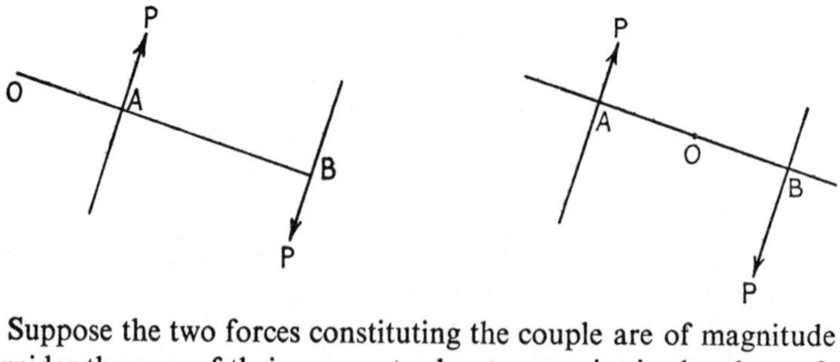

Suppose the two forces constituting the couple are of magnitude P. Consider the sum of their moments about any point in the plane of the forces.

From O draw OAB perpendicular to the forces to meet their lines of action in A and B. Then the algebraic sum of the moments of the forces about O is

$P \,.\, \text{OB} - P \,.\, \text{OA} = P \,.\, \text{AB}$ clockwise in the figure on the left.

$P \,.\, \text{OB} + P \,.\, \text{OA} = P \,.\, \text{AB}$ clockwise in the figure on the right.

Hence the moment of a couple about any point in its plane is constant, and it is equal to the force P multiplied by the perpendicular distance between the two forces d, i.e. Pd.

It follows that two coplanar couples are equal provided their moments are equal in magnitude and sign; for, if we reverse one of the couples, we have a system of forces in equilibrium, since the sum of the moments about any point in their plane is zero.

It follows also that any number of couples in the same plane may be replaced by a single couple whose moment is the algebraic sum of the moments of the given couples. None of the couples produces any tendency to translatory displacement, and each of the couples may be replaced by equal and opposite forces at a distance d apart, in each case, provided we adjust the forces of each couple so that its moment is equal to the fixed distance multiplied by the force, i.e. so that

$$F_1 = \frac{G_1}{d},\ F_2 = \frac{G_2}{d}$$

and so on, where G_1, G_2, G_3, etc., are the moments of the given couples.

In this way we have a set of forces as shown below

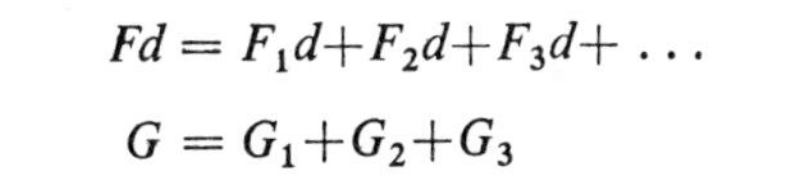

If the algebraic sum of $F_1+F_2+F_3+\ldots$ is F then

$$Fd = F_1d+F_2d+F_3d+\ldots$$

i.e. $$G = G_1+G_2+G_3$$

Hence the moments of couples may be added provided they are coplanar, and of course regard is paid to the sense of the couples.

10. Centre of gravity

The *centre of gravity* of a body is the point in the body through which its weight acts. We shall postpone the finding of centres of gravity until later in the chapter and at this stage will assume that the centre of gravity of a uniform rod is at its midpoint.

11. Worked examples

Example 1.—Two unlike parallel forces of 6 N and 4 N act at A and B. If AB is 1 m, find the size of the resultant and the point at which it cuts AB.

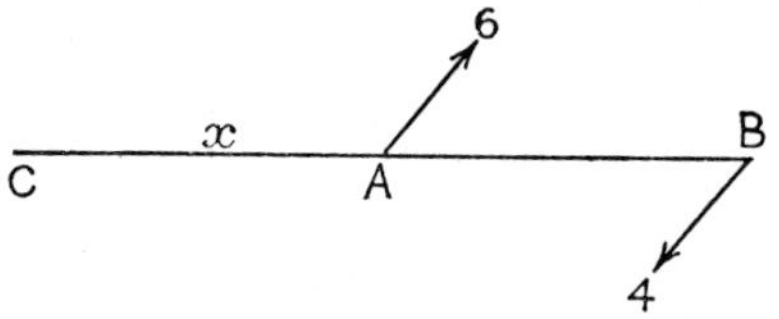

The resultant is a force of 2 N parallel to and in the same sense as the 6 N force. Let it cut BA produced at C. Then, if

$$AC = x, \quad 6x = 4(x + 1) \quad \therefore \ x = 2$$

∴ The resultant cuts BA produced 2 m from A.

Example 2.—A uniform beam AB is 6 m long and weighs 5 N. The beam is pivoted at a point 1 m from A. A mass of weight 30 N hangs at A and the beam is held inclined to the horizontal at 60° by a force P perpendicular to the beam at B. Find P.

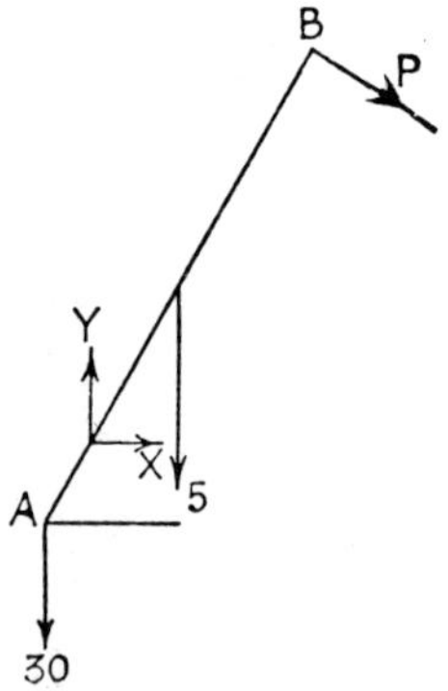

The forces acting on the beam are as shown.
Taking moments about the pivot

$$30 \times 1 \cos 60^\circ = 5 \times 2 \cos 60^\circ + 5P$$
$$15 = 5 + 5P$$
$$\therefore \ 2 = P$$

Notice we have not found the reaction on the beam at the pivot and have avoided it by taking moments about that point.

Example 3.—ABCD is a square of side a. Forces of P, Q, P, Q, act along the sides BA, BC, DC, DA, of the square. Find the moment of the couple required to keep the square in equilibrium.

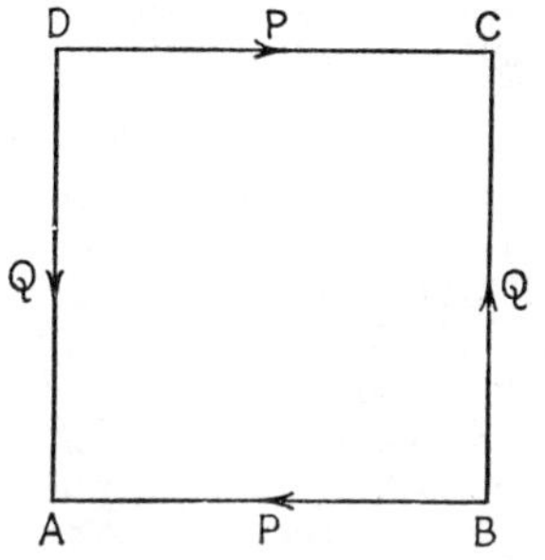

The square is acted on by two couples of moment Pa clockwise and Qa anti-clockwise, i.e. $(Q - P)a$ anticlockwise.

Hence to keep the square in equilibrium we must apply a couple of moment $(Q - P)a$ clockwise or $(P - Q)a$ anticlockwise.

EXERCISE 9A

1. Two like parallel forces of 5 N and 3 N act at points A and B which are 16 cm apart. Find the magnitude of their resultant and the distance from A at which it cuts AB. Repeat the question when the forces are unlike.
2. P and Q are two like parallel forces acting at A and B. Their resultant is 8 N and cuts AB at a point 1 m from A and 3 m from B. Find P and Q. Repeat this question when P and Q are unlike.
3. Masses of 10 kg and 20 kg are hung at the ends of a light rod of length 3 m. Find the point about which the rod will balance. If the rod were heavy, of mass 6 kg, and were uniform, find the distance from its centre of the point about which it would balance.
4. Four equal parallel like forces act at the corners of a square. Show that their resultant passes through the centre of the square. Would the same result be true for six such forces acting at the corners of a regular hexagon?
5. ABC is a triangular lamina. Three equal like parallel forces act through the vertices of the triangle. Prove that their resultant acts through the point of intersection of the medians of the lamina. Prove the same result when the three forces act through the midpoints of the sides of the triangle.
6. ABCD is a straight line. AB is 2 m, BC is 2 m, and CD is 3 m. Forces of 4 N and 3 N act vertically upwards at A and C, and forces of 2 N and 1 N act vertically downwards at B and D. Find the distance from A of the point where the resultant cuts AD.
7. A balance has arms of unequal length. When a body is weighed in one scale-pan it appears to have mass M_1 and when in the other it appears to have mass M_2 kg. Find the true mass of the body.
8. A uniform plank is 6 m long and has mass 60 kg. It rests on two supports one of which is 50 cm from one end and the other of which is 1 m from the other end. Calculate the mass of the heaviest man who can walk along the plank without tilting the plank.

9. The lever of a safety valve ABC is pivoted at A. The thrust of the steam acts vertically upwards at B and AB is 2 cm. If BC is 14 cm, find the force to be applied to C for the valve to be about to open when the thrust at B is 400 N.

10. A heavy beam ABCD rests on two supports at B and C. If AB = BC = CD, find the mass of the beam if it will just tilt when a mass of x kg is hung from A or when a mass of y kg is hung from D. Prove also that the centre of gravity of the beam divides it in the ratio $2x + y:x + 2y$.

11. A horizontal bar ABC 12 m long and of mass 60 kg has its centre of gravity at D where AD = 5 m. The bar is hung from spring balances at A, B, and C, where AB = 8 m. (i) If the balance at B reads 25 kg, what will be the readings of the balances at A and C? (ii) How far must the balance at B be moved in order that all three readings may be the same?

12. A uniform beam ACEDB 10 m long and weighing 40 N rests on two supports C and D which are 6 m apart, C being 1·5 m from A. A load is carried at E 3 m from A. (i) Calculate the supporting forces at C and D if the load at E is 20 N. (ii) Find the load at E if the supporting force at C is twice the supporting force at D.

13. A and B are two points 6 m apart and P is a force of 10 N in a plane through AB. The moments of P about A and B are 30 N m and 20 N m, both anti-clockwise. Calculate (i) the distance from B of the point where the line of action of P cuts AB, and (ii) the angle P makes with AB.

14. A uniform rod AB of length $2a$ and weight W is freely hinged at A. A mass of weight $\frac{1}{2}W$ is attached to B and the rod is held in equilibrium by a horizontal string attached to B and to a point C vertically above A. If AC = $2b$ calculate the tension in the string.

15. Two uniform rods AB and BC of lengths a and b are rigidly connected at B so that angle $B = 90°$. The system is freely suspended from A. If the weights per unit length of AB and BC are the same and AB makes an angle θ with the vertical prove $\tan\theta = \dfrac{b^2}{a(a + 2b)}$

16. AB is a light rod which hangs by two equal strings from a point C so that angle ACB = 90°. Weights W_1 and W_2 are attached to A and B. If $W_1 > W_2$ show that the inclination of AB to the horizontal is given by $\tan\theta = \dfrac{W_1 - W_2}{W_1 + W_2}$

17. A circular tray has a single coaxial cylindrical leg of radius a. The total weight of the tray and leg is W. Find the greatest distance from the centre of the tray at which a weight w may be placed for the tray not to topple over.

18. A table has a circular top of radius 1 m. It rests on 4 vertical legs attached to the rim of the top at the points A, B, C, and D, and ABCD is a square. If the weight of the table is 30 N acting through the centre of the top, calculate the greatest weight which can be placed on the rim of the table at the midpoint of the minor arc AB so that the table shall not overturn. What would then be the thrust on the floor by the leg at A in this case?

19. A heavy uniform rod of weight W is hung from a point by two equal strings, one attached to each end of the rod. A mass of weight w is hung halfway between the centre and the end of the rod. Calculate the ratio of the tensions in the strings.

20. A uniform rod AB of length 4 m and weight 6 N is freely pivoted at A. It has a mass of weight 12 N attached to it at a point 1 m from A. It is held in a horizontal position by a light rod CB, C being 3 m vertically below A. Find the thrust in CB.

21. A uniform horizontal beam is pivoted about a horizontal axis through its middle point, so that the beam can turn in a vertical plane, but in order to move the beam a couple of 6 N m is necessary to overcome friction at the pivot. A weight of 20 N is suspended on the right-hand half of the beam $1\frac{1}{2}$ m from the pivot. Find the greatest and least values of a load W N which will keep the beam in equilibrium in a horizontal position when the load is suspended 2 m from the pivot on its left-hand side. If $W = 15$ N, find how many cm it may be moved along the beam without disturbing equilibrium.

22. Three forces are represented completely by the sides of a triangle taken in order. Show that they are equivalent to a couple whose moment is represented by twice the area of the triangle.

23. Two elastic strings AB and DC, each of natural length l and modulus λ, are attached to the ends of two equal straight rods AD and BC which lie on a smooth horizontal table with the strings parallel to each other. Two equal and opposite forces P act outwards in directions parallel to the strings at points H and K in AD and BC respectively. H and K divide AD and BC internally so that AH/HD = BK/KC = 2/3. If in the position of equilibrium HK $= l + x$, prove $P = \dfrac{25\lambda x}{13l}$

EXERCISE 9B

1. Two like parallel forces of 7 N and 3 N act at points A and B which are 20 cm apart. Find the size of their resultant and the distance from A at which it cuts AB. Repeat the question when the forces are unlike.

2. P and Q are two like parallel forces acting at A and B and AB is 24 cm. If their resultant is 6 N and acts 8 cm from A, find P and Q. Repeat this question when P and Q are unlike.

3. A uniform rod AB 10 m long rests on two supports at C and D, where AC = 2 m and BD = 3 m. If the rod weighs 20 N, find the pressure on each support. What weight must be placed at B so that the reaction at C is zero?

4. A uniform plank AB of mass 28 kg and length 9 m lies on a horizontal roof in a direction at right angles to the edge of the roof. The end B projects 2 m over the edge. A man of mass 70 kg walks out along the plank. Find how far along the plank he can walk without causing the plank to tip up. Find also the mass which must be placed at A so that the man can reach B without upsetting the plank.

5. A uniform beam AB of length 6 m and weight 60 N rests horizontally on props C and D where AC = 1 m and BD = 2 m. Find the reactions at C and D. If a weight of 10 N is now hung from B, what weight must be hung from A so that the reaction at C is unaltered?

6. P and Q are two like parallel forces. If Q is moved a distance x farther from P, prove that the resultant of P and Q moves a distance $\dfrac{Qx}{P+Q}$ from its original position.

7. A uniform rod AB of length 12 m passes under a smooth peg at C and over a smooth peg at D. If the rod weighs 30 N and has a weight of 10 N attached at B, and if AC = CD = 2 m, find the reactions between the rod and the pegs.

8. Four forces act along the sides of a square. If they are in equilibrium prove that they are equal in size.

9. A uniform rod AB 2 m long is hinged to a wall at A and is held in equilibrium in a horizontal position perpendicular to the wall by a string attached to B and to a point $1\frac{1}{2}$ m vertically above A. If the weight of the rod is 5 N and a weight of 2 N is attached to B, find the tension in the string.

10. AC and BC are two light rods. The ends A and B are hinged to a vertical wall and the rods are joined at C. If a weight of 10 N is hung from C, calculate the thrust or tension in each rod when AC = 2 m, AB = 1 m, and angle ABC is 90°.

11. A uniform rod AB can turn freely about a hinge at A. To B is attached a string which passes over a small smooth pulley at C vertically above A and carries a weight W. If the weight of the rod is w prove that if the system is in equilibrium

$$\frac{BC}{AC} = \frac{2W}{w}$$

12. A uniform rod AB length $2a$ and weight W is hung by a string attached to A. If the rod has a couple of moment G applied to it, find its inclination to the vertical when it is in equilibrium.

13. A and B are two points 4 m apart. A force of 10 N coplanar with A and B meets BA produced at C. The moments of the force about A and B are 10 N m and 30 N m, both clockwise. Calculate AC and the angle at which the force cuts AB.

14. A uniform ladder 10 m long and weight 120 N rests against a smooth vertical wall with its lower end on rough horizontal ground and 4 m from the wall. Calculate the reaction of the wall on the ladder.

15. Two uniform rods AB and BC of length 3 m and 4 m respectively are rigidly connected at B so that angle $B = 90°$. The system is freely suspended from A. If the weight per unit length of the two rods is the same, find the inclination of AB to the vertical.

16. Two uniform rods AB and BC are rigidly connected at B so that angle ABC = 120°. The weight per unit length of each rod is w. The system is freely pivoted at B and is in equilibrium with BC horizontal. When a weight W is attached to C the system is in equilibrium with AB horizontal. Prove $W = \frac{3}{2}w \, . \, BC$.

17. A circular table has 3 legs attached symmetrically to the rim of the top. If the radius of the top is a and the weight of the table W acts through the centre of the top, find the greatest weight which can be placed on the rim of the table diametrically opposite a leg if the table is not to overturn.

18. A uniform rod AB of weight W and length $2a$ is freely hinged at A and hangs at 30° to the horizontal. It is held in this position by a string BC inclined at 45° to the horizontal. If $\sin 75° = \dfrac{\sqrt{3}+1}{2\sqrt{2}}$ prove that the tension in the string is $\dfrac{W\sqrt{6}}{2(\sqrt{3}+1)}$

19. A uniform rectangular lamina ABCD of weight W hangs freely from A with a weight w attached to it at B. If AB = a and BC = b and θ is the inclination of AC to the vertical, prove $\tan\theta = \dfrac{2wab}{W(a^2+b^2)+2wa^2}$

20. Two vertical strings AB and CD each of natural length a and with moduli 2λ and λ are tied to fixed points A and C in a horizontal line and support horizontally and in equilibrium a non-uniform bar BD with the strings vertical and distant d apart. Prove that the weight of the bar acts at a distance $\frac{1}{3}d$ from B.

12. Centre of gravity

The weights of the constituent particles of any body are a set of parallel forces—they are all directed towards the centre of the Earth. Since they are all like parallel forces, we have shown that they have a resultant, equal to their sum, called the *weight* of the body and that this resultant acts through a definite point in the body, independent of how the body is turned with respect to the vertical. This point is called the *centre of gravity* of the body.

13. Symmetry

The centre of gravity of a uniform body which has an axis of symmetry must lie on this axis; for, corresponding to a particle P_1 of weight W there will be another similarly placed equal particle P_2, and the resultant of these two equal forces W will pass through AB; and similarly for all other pairs of particles and so the centre of gravity must be on AB.

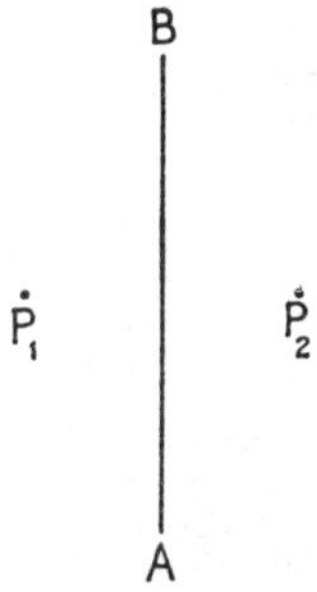

If a lamina has two axes of symmetry the centre of gravity must lie on both and so will be at their point of intersection.

Hence the centre of gravity of a uniform rod is at its midpoint; the centre of gravity of a circle is at its centre; the centre of gravity of an equilateral triangle is at the point of intersection of its medians, and so on.

14. Centre of gravity of a triangle

Imagine the triangle divided into an infinitely large number of infinitely thin strips by lines parallel to one side. The figure shows one such strip parallel to BC. This strip may be regarded as a very thin rectangle, and its centre of gravity will be at its midpoint. This point lies on the median of the triangle (AD). Similarly the centres of gravity of all strips parallel to BC lie on the median AD. Hence the centre of gravity of the triangle lies on the median AD. In the same way the centre of gravity can be shown to lie on each of the other two medians.

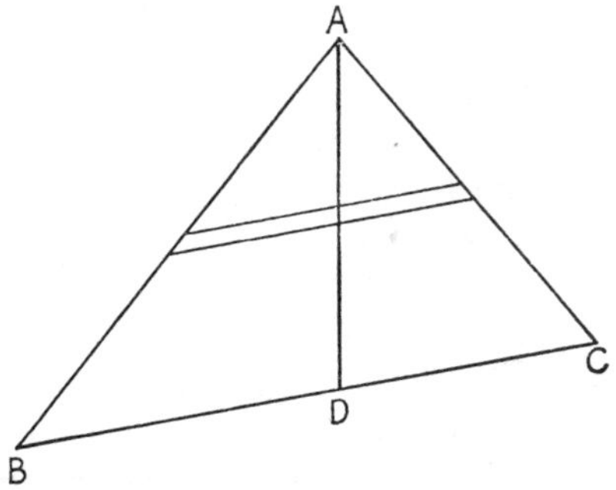

Hence the centre of gravity of a triangle is at the point of intersection of its medians. This point can be shown to be $\frac{1}{3}$ of the length of the median along the median from the point where the median cuts the side of the triangle, i.e. G is on AD and GD $= \frac{1}{3}$AD.

A uniform triangular lamina is statically equivalent to three particles each one-third of the weight of the triangle and placed one at each vertex of the triangle.

Weights $\frac{1}{3}W$ at B and $\frac{1}{3}W$ at C may be replaced by $\frac{2}{3}W$ at D, where D is the midpoint of BC. $\frac{2}{3}W$ at D and $\frac{1}{3}W$ at A may be replaced by

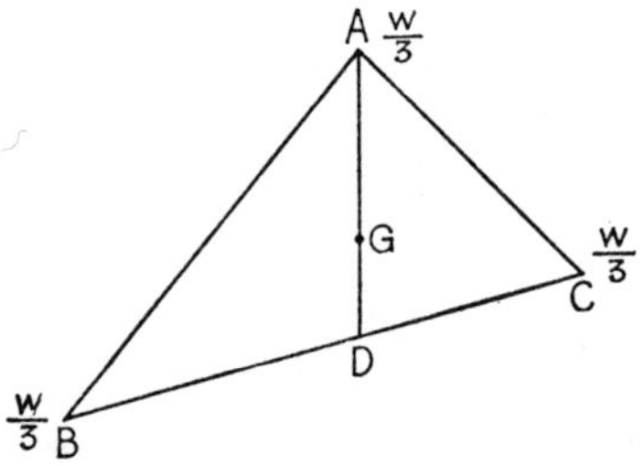

W at G, where G is the point on AD such that AG $=$ 2GD, and this point is the centre of gravity of the triangular lamina ABC.

This result will be found useful for finding the centre of gravity of other laminæ which can be divided into triangles.

15. Centre of gravity of the curved surface area of a right circular cone

Imagine the circumference of the base of the cone divided into a very large number of small parts, and the ends of each part joined to the vertex of the cone. In this way the curved surface is divided into a large number of ' triangles '.

The centres of gravity of all these triangles lie on a plane parallel to the base of the cone and $\frac{1}{3}$ the height of the cone from the base, and from symmetry the centre of gravity is on the axis of the cone. Hence it is on the axis and $\frac{1}{3}h$ from the base.

16. Centre of gravity of a number of particles

Let the particles A_1, A_2, A_3, . . . , have weights w_1, w_2, w_3, . . . , and let them be situated at (x_1, y_1, z_1), (x_2, y_2, z_2), . . . , referred to axes Ox, Oy, Oz. Let the coordinates of their centre of gravity be $(\bar{x}, \bar{y}, \bar{z})$ referred to these axes.

Since the direction of the weights of the particles is independent of the direction of the axes we may suppose them parallel to Oz.

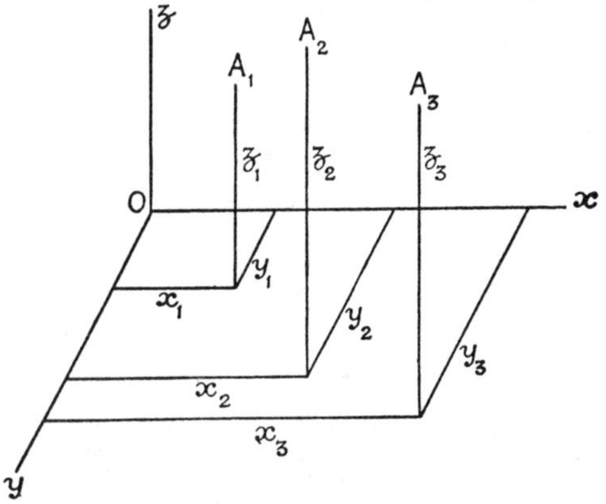

Hence taking moments about Oy

$$(w_1+w_2+w_3+\ldots)\bar{x} = w_1x_1+w_2x_2+\ldots$$

$$\therefore \bar{x} = \frac{\Sigma w_1x_1}{\Sigma w_1}$$

similarly

$$\bar{y} = \frac{\Sigma w_1y_1}{\Sigma w_1} \text{ and } \bar{z} = \frac{\Sigma w_1z_1}{\Sigma w_1}$$

Usually the finding of Σw_1x_1 involves an integration as in the following examples.

17. Centre of gravity of a uniform right solid cone

Imagine the cone placed as shown with the axis of the cone as x-axis. The centre of gravity of the cone lies on this axis from sym-

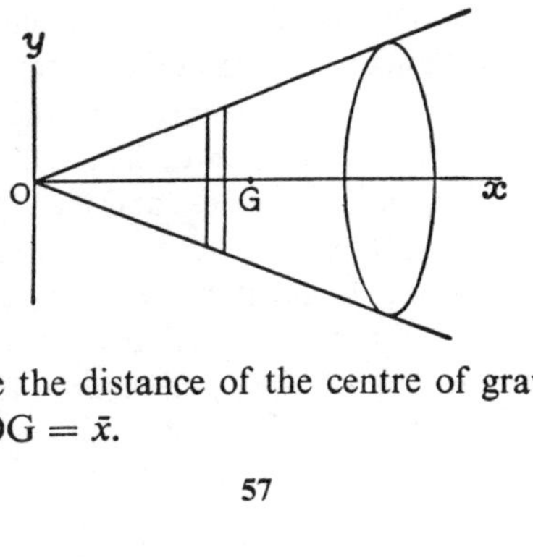

metry. We require the distance of the centre of gravity from O. Let it be at G where OG $= \bar{x}$.

Divide the cone into a series of thin discs by planes parallel to its base. A typical one is shown. Let each disc be very thin and of thickness δx. Let the typical disc have radius y and be x from O.

The weight of the typical disc is $\pi y^2 w \delta x$, where w is the weight per unit volume, and its centre of gravity is at the centre of the disc, i.e. it is x from O. Let r and h be the radius and height of the cone. Taking moments about a line through O perpendicular to Ox we have

$$\tfrac{1}{3}\pi r^2 h w \bar{x} = \Sigma \pi y^2 x w \, \delta x$$

the summation extending for values of x from 0 to h.

We can evaluate this sum by integrating $\pi y^2 x w \, dx$ from 0 to h. Hence

$$\tfrac{1}{3}\pi r^2 h w \bar{x} = \int_0^h \pi y^2 x w \, dx$$

From similar triangles

$$\frac{y}{x} = \frac{r}{h}, \quad \therefore y = \frac{xr}{h}$$

$$\text{i.e. } \frac{\pi}{3} r^2 h w \bar{x} = \int_0^h \pi \frac{x^2 r^2}{h^2} x \, w \, dx = \frac{\pi r^2}{h^2} w \left[\tfrac{1}{4} x^4\right]_0^h$$

$$\frac{\pi}{3} r^2 h w \bar{x} = \frac{\pi r^2 w}{h^2} \tfrac{1}{4} h^4$$

$$\bar{x} = \tfrac{3}{4} h$$

Hence the centre of gravity of a right circular uniform cone is on its axis and $\frac{1}{4}h$ from its base.

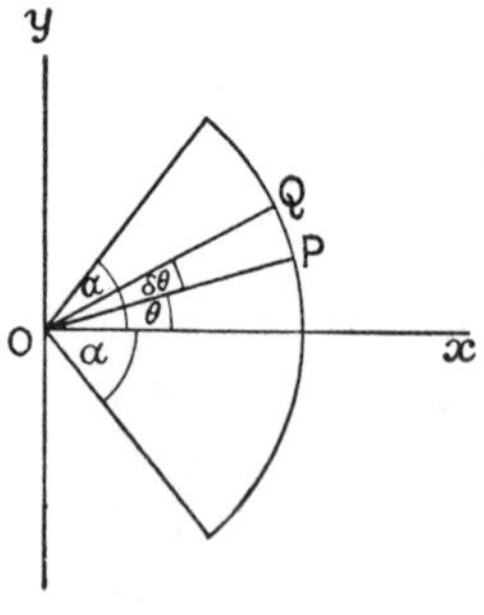

18. Centre of gravity of an arc of a circle

Consider an arc of a circle which subtends an angle of 2α at the centre O. From symmetry the centre of gravity of the arc lies on the line Ox which bisects the angle 2α. Draw Oy perpendicular to Ox and let PQ be an infinitesimal element of the arc so that angle POx is θ and angle POQ $= \delta\theta$. The centre of gravity of PQ is $r \cos\theta$ from Oy and its weight is $wr\delta\theta$, where r is the radius of the circle and w the weight per unit length of the arc.

Taking moments about Oy

$$2r\alpha w\bar{x} = \int_{-\alpha}^{\alpha} r\,\delta\theta\, w \,.\, r\cos\theta = \int_{-\alpha}^{\alpha} r^2 w\cos\theta\, d\theta$$

$$= 2r^2 w\sin\alpha$$

$$\therefore\ \bar{x} = r\frac{\sin\alpha}{\alpha}$$

If $\alpha = 90°$ the arc is a semicircle and $\bar{x} = \dfrac{2r}{\pi}$

19. Centre of gravity of a sector of a circle

Using the previous figure but this time considering the sector subtending 2α at the centre and taking w to be the weight per unit area, by a similar argument since the centre of gravity of the element POQ is $\frac{2}{3}r$ from O,

$$\tfrac{1}{2}r^2 \,.\, 2\alpha \,.\, w \,.\, \bar{x} = \int_{-\alpha}^{\alpha} \tfrac{1}{2}r^2\,\delta\theta \,.\, w \,.\, \tfrac{2}{3}r\cos\theta$$

$$= \tfrac{1}{3}r^3 w \int_{-\alpha}^{\alpha} \cos\theta\, d\theta = \tfrac{2}{3}r^3 w\sin\alpha$$

$$\therefore\ \bar{x} = \tfrac{2}{3}r\frac{\sin\alpha}{\alpha}$$

Again when $\alpha = 90°$ the sector becomes a semicircular lamina and

$$\bar{x} = \frac{4r}{3\pi}$$

20. Centre of gravity of a solid hemisphere

Suppose YAY_1 is a section of the hemisphere perpendicular to its plane base (diagram on p. 60). YOY_1 is a diameter of the base where O is the centre and OA the radius perpendicular to the base. Let r be the radius and w the weight per unit volume.

Divide the hemisphere into very thin discs by planes parallel to its base. A typical disc is shown in the figure.

From the symmetry the centre of gravity of the hemisphere lies on OA and the centre of gravity of each slice lies on OA. Let the thickness of the slices be dx and the distance of the typical slice from YY_1 be x. Let the radius of the typical slice be y.

Then $$\tfrac{2}{3}\pi r^3 w\bar{x} = \int_0^r \pi y^2\, dx\, .\, wx$$

but $$x^2+y^2 = r^2,$$

$$\therefore \tfrac{2}{3}\pi r^3 w\bar{x} = \int_0^r \pi w x(r^2-x^2)\, dx$$
$$= \pi w\left[\tfrac{1}{2}r^2x^2-\tfrac{1}{4}x^4\right]_0^r$$
$$= \tfrac{1}{4}\pi w r^4$$
$$\therefore \bar{x} = \tfrac{3}{8}r$$

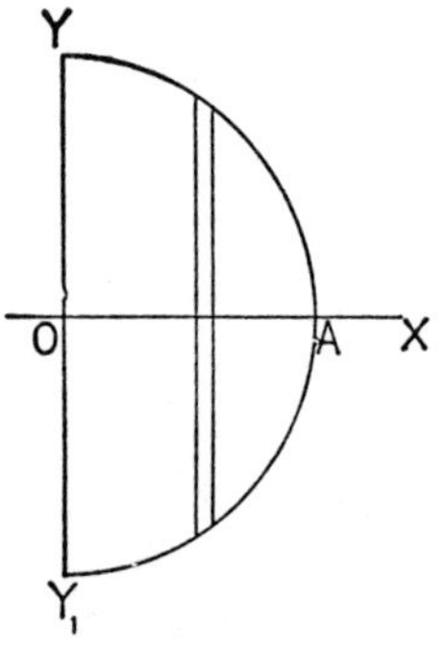

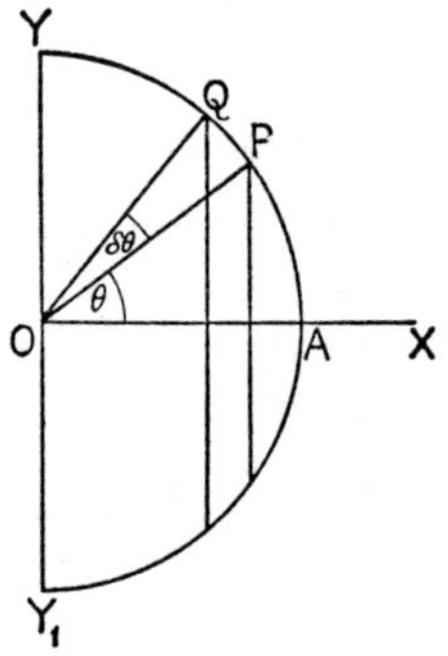

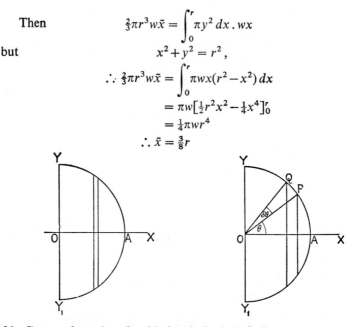

21. Centre of gravity of a thin hemispherical shell

Let YAY_1 be a section of the hemisphere perpendicular to its base. Let YOY_1 be a diameter of the base, O the centre of the sphere, and r its radius. Divide the surface into very narrow bands by planes parallel to the base. A typical band is shown in the figure. Let angle POA be θ and POQ be $\delta\theta$. The typical band is approximately the curved surface of a cylinder of radius $r\sin\theta$ and height $r\delta\theta$ and its centre of gravity is on OA and $r\cos\theta$ from O.

Hence, if w is the weight per unit area,

$$2\pi r^2 w\bar{x} = \int_0^{\pi/2} 2\pi\, .\, r\sin\theta\, .\, r\, d\theta\, .\, w\, .\, r\cos\theta$$
$$\bar{x} = \int_0^{\pi/2} r\sin\theta\cos\theta\, d\theta = \int_0^{\pi/2} \tfrac{1}{2}r\sin 2\theta\, d\theta$$
$$= \tfrac{1}{2}r\left[-\tfrac{1}{2}\cos 2\theta\right]_0^{\pi/2}$$
$$\therefore \bar{x} = \tfrac{1}{2}r$$

22. The centre of gravity of two bodies combined

Suppose we have two bodies whose weights are W_1 and W_2, and whose centres of gravity are at the points (x_1, y_1, z_1) and (x_2, y_2, z_2), then the centre of gravity of the two bodies combined lies at a point in the line joining (x_1, y_1, z_1) to (x_2, y_2, z_2) divided in the ratio $W_2 : W_1$.

If $\bar{x}$ is the x-coordinate of the centre of gravity of the combined body then

$$\Sigma w \,.\, \bar{x} = \Sigma wx$$

$$\therefore\ (W_1+W_2)\bar{x} = W_1x_1+W_2x_2 \quad . \quad . \quad . \quad . \quad . \quad \text{(i)}$$

$$\therefore\ \bar{x} = \frac{W_1x_1+W_2x_2}{W_1+W_2}$$

In the same way, if from a body whose centre of gravity is at (x_1, y_1, z_1) and whose weight is W_1 we take away a part of weight W_2 whose centre of gravity is at (x_2, y_2, z_2), then the centre of gravity of the remaining body is at $(\bar{x}, \bar{y}, \bar{z})$ where

$$\bar{x} = \frac{W_1x_1-W_2x_2}{W_1-W_2} \quad . \quad . \quad . \quad . \quad . \quad . \quad . \quad . \quad \text{(ii)}$$

This follows from line (i) above if we imagine the part taken away replaced; we then have

$$W_1x_1 = W_2x_2+(W_1-W_2)\bar{x}$$

which reduces to (ii) above.

Similar results follow in the same way for $\bar{y}$ and $\bar{z}$.

23. Centre of mass

It has been pointed out that the centre of gravity of a body is independent of the position of the body with respect to the vertical, and in section 16 the system was at first supposed to be placed so that the axes of x and y were horizontal. (To obtain the value of $\bar{z}$ a different supposition must be made.)

If in that section we write $m_1, m_2, \ldots$ for the masses of the particles, then $w_1 = m_1 g$, etc., and we get

$$\bar{x} = \frac{\Sigma m_1\, g\, x_1}{\Sigma m_1\, g} = \frac{\Sigma m_1\, x_1}{\Sigma m_1}$$

with similar results for $\bar{y}$, $\bar{z}$.

For this reason the centre of gravity of a body is also described as its *centre of mass*. The idea of mass-centre is in fact more fundamental than that of centre of gravity. If a system of particles were entirely removed from the influence of gravity, it would still have the same centre of mass, defined by

$$\bar{x} = \frac{\Sigma m_1\, x_1}{\Sigma m_1}$$

and two similar equations.

24. Worked examples

Example 4.—ABCD is a square. Masses of 1, 2, 3, 4 kg respectively are placed at the corners A, B, C, and D. If AB = 1 m, calculate the distances of the centre of mass from AB and AD.

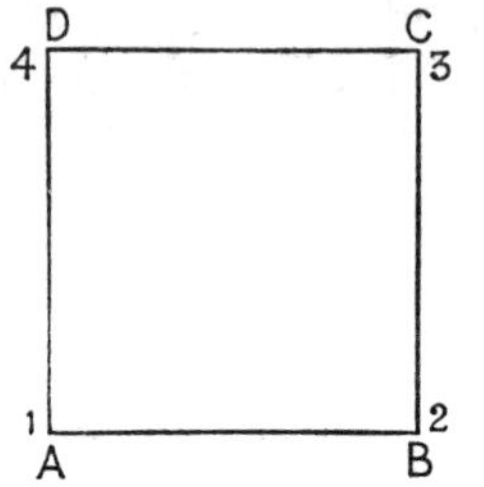

Let the coordinates of the centre of mass referred to AB and AD as axes be (x, y). Take distances from AD.

$$3 \times 1 + 2 \times 1 = 10x, \qquad \therefore x = 0{\cdot}5.$$

Take distances from AB.

$$4 \times 1 + 3 \times 1 = 10y, \qquad \therefore y = 0{\cdot}7.$$

Example 5.—A uniform cylinder of height h and radius r is surmounted by a uniform hemisphere of radius r whose place base coincides with the top of the cylinder. If the weights per unit volume of cylinder and hemisphere are the same and if the centre of gravity of the composite body is in the plane face of the hemisphere, find r/h.

Let w = weight per unit volume. Consider distances from the plane face of the hemisphere.

$$\tfrac{2}{3}\pi r^3 w \times \tfrac{3}{8}r = \pi r^2 h w \times \tfrac{1}{2}h$$

$$r^2 = 2h^2$$

$$\therefore \frac{r}{h} = \frac{\sqrt{2}}{1}$$

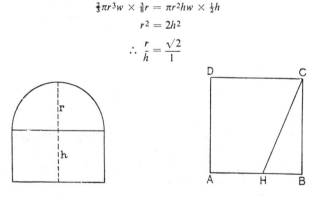

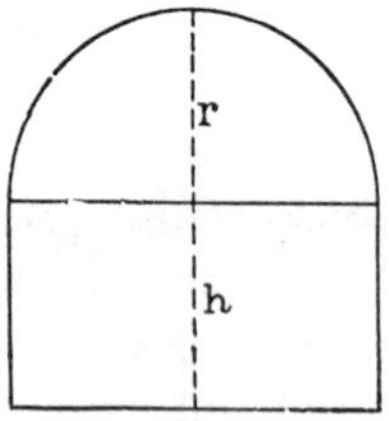

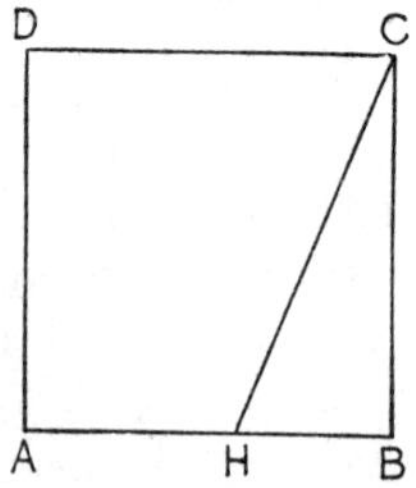

Example 6.—ABCD is a uniform square lamina. H is a point in AB. The portion CHB is removed. If the side of the square is a and AH = x, find the least value of x which allows the lamina AHCD to stand on a smooth horizontal table with AH in contact with the table.

Let w = weight per unit area of ABCD.

Figure	Weight	Distance of C. of G. from AD
ABCD	a^2w	$\frac{1}{2}a$
$\triangle$CHB	$\frac{1}{2}a(a-x)w$	$x+\frac{2}{3}(a-x)=\frac{1}{3}(2a+x)$
AHCD	$\frac{1}{2}a(a+x)w$	$\bar{x}$

Taking moments about AD

$$\tfrac{1}{2}a^3 - \tfrac{1}{2}a(a-x)\,.\,\tfrac{1}{3}(2a+x) = \tfrac{1}{2}a(a+x)\bar{x}$$

$$\frac{a^2+ax+x^2}{3(a+x)} = \bar{x}$$

For equilibrium to be possible $\bar{x}$ must be $\leqslant x$. There will be only two forces acting, the reaction of the plane which acts between A and H, and the weight, and these must be collinear.

$$\therefore\ x \geqslant \frac{a^2+ax+x^2}{3(a+x)}$$

$$2x^2+2ax \geqslant a^2$$

$$(x+\tfrac{1}{2}a)^2 \geqslant \tfrac{3}{4}a^2$$

$$x \geqslant \tfrac{1}{2}a(\sqrt{3}-1)$$

$\therefore$ the least value of x is $\frac{1}{2}a(\sqrt{3}-1)$

EXERCISE 10A

1. Masses of 1, 2, 3, 4 kg are placed at the corners A, B, C, D respectively of a square of side 6 cm. Find the distances of their centre of mass from AB and AD.
2. Masses 1, 5, 3, 4, 2, and 6 kg are placed at the angular points A, B, C, D, E, and F respectively of a regular hexagon. Prove that their centre of mass is at the centre of the hexagon.
3. ABCD is a rectangle in which AB is 10 cm and BC is 5 cm. Masses 1, 2, 3, 4 kg are placed at A, B, C, D respectively. Find the distances of their centre of mass from AB and AD.
4. ABC is an equilateral triangle of side $2a$. Masses of 3, 4, 5 kg are placed at A, B, and C respectively. Calculate the distance of their centre of mass from A.
5. Find the centre of gravity of 5 equal uniform rods forming 5 sides of a regular hexagon of side $2a$. Give its distance from the centre in terms of a.
6. OA and OB are two thin rods of lengths 10 cm and 15 cm respectively rigidly connected at O so that AOB is a right angle. The weight per unit length of the two rods is the same. Calculate the distances of the centre of gravity of the whole rod AOB from each of OA and OB. If AOB is freely suspended from A, calculate the inclination of OA to the vertical.
7. ABCD is a trapezium made of thin sheet metal. AB is parallel to DC, AB = 6 cm, BC = 4 cm, CD = 3 cm, and angle ABC = 90°. Calculate the distances of the centre of gravity of the lamina from AB and BC. If the lamina is freely suspended from B, calculate the inclination of AB to the vertical.
8. A circular metal disc of radius 9 cm has a circular hole of radius 3 cm bored in it. The centre of the hole is 4 cm from the centre of the disc. Find the position of the centre of gravity of the remainder.
9. A thin uniform wire of length 30 cm is bent into the shape of a triangle ABC with AB = 5 cm, BC = 12 cm, and angle ABC = 90°. Calculate the distances from AB and BC of the centre of gravity of the wire triangle. Hence show that if the wire hangs from A in equilibrium the inclination of AB to the vertical is about 55°.
10. A lamina ABCE consists of a rectangle ABCD and a triangle DCE right-angled at D. AB is 9 cm, BC is 4 cm, and DE is 12 cm. The weight per unit area of ABCD is 3 times the weight per unit area of DCE. Prove that the centre of gravity of the lamina lies in DC and calculate its distance from AE. If the lamina is freely suspended from E, calculate the inclination of AE to the vertical.
11. ABCD is a uniform square plate of side 2·32 m. Points Q and R are taken in BC and CD so that BQ = CR = 58 cm. The part CQR of the plate is cut off. Find the distances from AB and BC of the centre of gravity of the remainder.
12. A sheet of metal is in the form of an isosceles triangle ABC in which AB = AC = 20 cm and BC = 32 cm. The corner A is folded down until it comes to the middle of BC, and the metal is flattened down so that the fold is parallel to BC. Find the centre of gravity of the folded sheet.
13. A uniform plate ABCD consists of two congruent triangles on opposite sides of AC, each having angle C obtuse. If C is 3 cm from the line BD, find AC if the centre of gravity of the lamina coincides with C.

14. ABC is a uniform triangular lamina in which angle $C = 90°$, $AC = b$, and $BC = a$. The middle points of AC and BC are E and D respectively. Triangles are cut away by cuts EK and DH perpendicular to AB. Prove that the centre of gravity of the remainder CEKHD is $\frac{7}{18} \cdot \frac{ab}{\sqrt{(a^2 + b^2)}}$ from KH.

15. The outer radius of a hollow hemisphere is 6 cm, and the inner radius is 4 cm. Find the distance of the centre of gravity from the centre of the hemisphere.

16. A uniform wire ABC is bent at right angles at B, and when suspended from B the parts AB and BC make angles of 30° and 60° respectively with the vertical. Find the ratio AB : BC.

17. A lamina in the form of an obtuse-angled triangle ABC in which angle B is obtuse rests in a vertical plane with AB on a horizontal plane. Show that the triangle will not topple over if $3c^2 > b^2 - a^2$.

18. A solid cone of height h and radius r is placed on a plane which is gradually tilted. If the plane is sufficiently rough to prevent sliding, find the greatest possible inclination of the plane to the horizontal.

19. A rigid framework is made from 4 equal uniform rods in the form of 4 consecutive sides of a regular hexagon ABCDE. If the framework is suspended from A, show that the tangent of the angle AB makes with the vertical is $\frac{4\sqrt{3}}{7}$.

20. A uniform solid consists of a solid hemisphere and a right circular cone with a common plane base of radius r. Find the greatest height of the cone if the solid can rest on a horizontal plane in stable equilibrium with the cone uppermost.

21. A piece of wire is bent into a circular quadrant of radius a and its two bounding radii. Find the distance of the centre of gravity from the centre of the circle.

22. Find the centre of gravity of a rod whose density at any point varies as the square of the distance from one end.

23. ABCD is a square uniform lamina of side a. H is a point in AB. The triangular portion CBH is cut away and the remaining lamina DCHA stands in a vertical plane on a horizontal table with AH on the table. Prove that the lamina will topple over if $AH < \frac{1}{2}a(\sqrt{3} - 1)$.

EXERCISE 10B

1. ABCD is a square of side 36 cm. Masses of 2, 3, 4, 5 kg are placed at A, B, C, and D respectively and 3 kg and 1 kg are placed at the midpoints of AD and CD respectively. Find the distances of the centre of mass from AB and AD.
2. At the corners A, B, C, D of a rectangle masses 2, 3, 4, 5 kg are placed. The sides AB and AD are of lengths 70 cm and 42 cm. Calculate the distances of the centre of mass from AB and AD.
3. ABCDEF is a regular hexagon. Masses of 2, 4, 4, 4, 4, 2 kg are placed at A, B, C, D, E, F respectively. Calculate the distances of their centre of mass from AC and AF when the side of the hexagon is a.
4. From a circular lamina of radius a a circular hole is cut out of radius $\frac{1}{2}a$. If the centre of the circle cut out bisects the distance between the centre of the lamina and its circumference, find the distance of the centre of gravity of the remainder from the centre of the lamina.

5. A solid cylinder and a hemisphere have the same radius r. They are made of the same uniform material, and the base of the hemisphere is fixed to the top of the cylinder so that the two fit. If the height of the cylinder is $2r$ find the height of the centre of gravity of the solid above the bottom of the cylinder.

6. ABC is a triangle whose medians AD, BE, and CF cut at G. The part AFGE is cut away. Show that the centre of gravity of the remainder is on GD and is $\frac{7}{12}$ GD from D.

7. The radii of the circular ends of a frustum of a cone are a and b, $(a > b)$ and its height is h. Calculate the distance of the centre of gravity of its curved surface above the end of radius a.

8. The radii of the ends of a frustum of a cone are $2a$ and $3a$. Prove that the centre of gravity divides the axis in the ratio 43 : 33.

9. ABCD is a lamina in which AD is parallel to BC; AB = $6\frac{1}{2}$ cm, BC = $4\frac{1}{2}$ cm, CD = 10 cm, and DA = 15 cm. Find the distance of its centre of gravity from AD.

10. A rectangular lamina of sides 24 cm and 32 cm is folded so that one short side lies entirely along one long side. Find the distance of the centre of gravity of the folded lamina from the other two sides.

11. The centre of gravity of a quadrilateral coincides with that of 4 equal particles placed one at each vertex. Prove that the quadrilateral is a parallelogram.

12. ABCD is a square lamina of side a. A part HKD is cut away by a cut HK through K the midpoint of CD and through a point H in AD so that HD = x. Find the greatest value of x so that the lamina ABCKH may stand on AH with its plane vertical without toppling over.

13. ABC is a triangular lamina in which angle $A = 90°$. D is the midpoint of AC. The triangle BAD is cut away. Show that if the remainder BDC rests in a vertical plane with DC on a horizontal table it will be on the point of toppling over.

14. A cylindrical vessel has a base but no top. It is made of thin material which weighs $\frac{1}{2}$ g/cm^2. The radius of the cylinder is 5 cm, and its height is 10 cm. Find the distance of the centre of gravity from the base. The vessel rests with its base on a horizontal table and water is slowly poured in. Prove that the centre of gravity G of the water and vessel together is lowest when the height of the water is such that G lies in the water surface.

15. Find the centre of gravity of a cap of height h of a sphere of radius a.

16. AB, BC, CD are 3 equal uniform rods rigidly connected so that they are 3 consecutive sides of a regular hexagon. The system is suspended from B. Show that the rod AB is horizontal in the equilibrium position.

17. A uniform rectangular lamina ABCD of weight W with a weight W attached to it at B is freely suspended from A. If θ is the inclination of the diagonal AC to the vertical, prove $\tan\theta = \dfrac{2ab}{3a^2 + b^2}$, where a = AB, b = BC.

6

Coplanar Forces acting on a Rigid Body

1. Two forces

When two forces keep a body in equilibrium it is clear that they must be equal in magnitude, opposite in direction, and must act in the same line.

2. Three forces

When three forces keep a body in equilibrium they must either all three meet in a point or all three must be parallel. This is a necessary condition; it is not a sufficient one.

Suppose the three forces are P_1, P_2, P_3 and suppose first that two of them, say P_1 and P_2 are not parallel. Since they are coplanar they must therefore meet in a point, say A. The resultant of P_1 and P_2, found from the parallelogram of forces, must therefore also pass through A. For equilibrium the third force P_3 must balance the resultant of P_1 and P_2. It must therefore have the same line of action as this resultant and so must also pass through A, i.e. all three forces pass through A.

Next suppose two of the forces, say P_1 and P_2 are parallel. Their resultant will be a force parallel to them, and for equilibrium it must be balanced by P_3. Hence P_3 must be parallel to P_1 and P_2, for its line of action is the same as that of the resultant of P_1 and P_2, i.e. all three forces must be parallel if two of them are parallel.

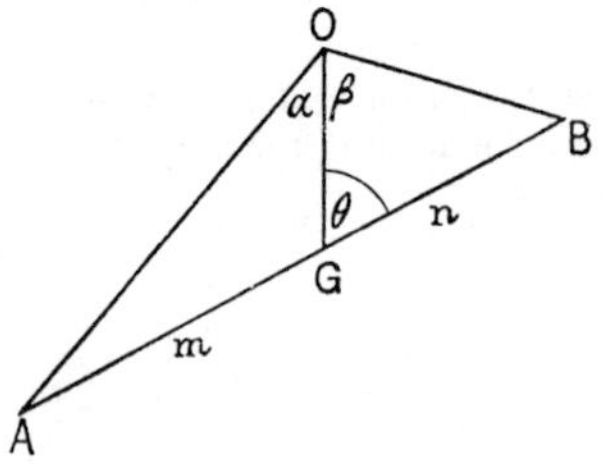

When the three forces keeping a body in equilibrium meet in a point their magnitudes and directions must be consistent with the triangle of forces or, what is the same idea, Lami's Theorem. The following is a useful geometrical theorem in these cases.

If OAB is a triangle and G is a point dividing AB in the ratio $m : n$, then if angle AOG is α and angle BOG is β and angle OGB $= \theta$

$$(m+n) \cot \theta = m \cot \alpha - n \cot \beta$$

and
$$(m+n) \cot \theta = n \cot A - m \cot B$$

From the sine formula

$$\frac{OG}{AG} = \frac{\sin(\theta-\alpha)}{\sin\alpha}, \quad \frac{OG}{BG} = \frac{\sin(\theta+\beta)}{\sin\beta}$$

Dividing
$$\frac{BG}{AG} = \frac{n}{m} = \frac{\sin(\theta-\alpha)\sin\beta}{\sin\alpha\sin(\theta+\beta)}$$

$$\therefore\ n \sin\alpha(\sin\theta\cos\beta+\cos\theta\sin\beta)$$
$$= m \sin\beta(\sin\theta\cos\alpha-\cos\theta\sin\alpha)$$

$$(m+n)\sin\alpha\sin\beta\cos\theta = m\sin\beta\sin\theta\cos\alpha - n\sin\theta\sin\alpha\cos\beta$$

$$\therefore\ (m+n)\cot\theta = m\cot\alpha - n\cot\beta$$

A similar method will show the other result true. In fact these results are most often used in the special cases when $n = m$, i.e.

$$2\cot\theta = \cot\alpha - \cot\beta = \cot A - \cot B$$

3. Worked examples on three forces

Example 1.—A heavy bead is threaded on a smooth circular wire whose plane is vertical. The bead is tied to the highest point of the wire by a string which in the position of equilibrium is inclined to the vertical at 30°. Calculate the tension in the string and the reaction of the wire on the bead.

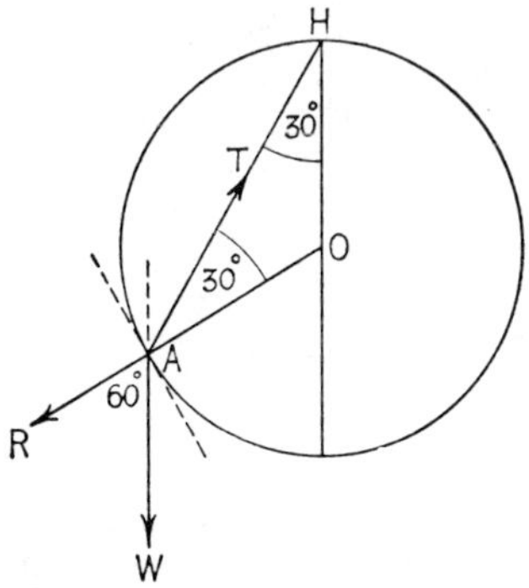

There are 3 forces acting on the bead: the weight of the bead, the tension in the string, and the reaction of the wire on the bead, which, because the wire is smooth, must pass through O, the centre of the circle. The three forces all go through A, the bead.

(i) OAH is a triangle of forces

$$\therefore \frac{T}{AH} = \frac{R}{OA} = \frac{W}{OH}$$

i.e.
$$\frac{T}{a\sqrt{3}} = \frac{R}{a} = \frac{W}{a}$$

$$\therefore T = W\sqrt{3} \text{ and } R = W$$

(ii) From Lami's Theorem

$$\frac{T}{\sin 60^\circ} = \frac{R}{\sin 150^\circ} = \frac{W}{\sin 150^\circ}$$

i.e.
$$\frac{T}{\frac{1}{2}\sqrt{3}} = \frac{R}{\frac{1}{2}} = \frac{W}{\frac{1}{2}}$$

$$\therefore T = W\sqrt{3} \text{ and } R = W$$

(iii) Resolving the forces along OA and the perpendicular to AO,

$$R + W\cos 60^\circ = T\cos 30^\circ \quad \text{and} \quad W\sin 60^\circ = T\sin 30^\circ,$$

i.e.
$$T = W\sqrt{3} \quad \text{and} \quad R = W$$

Example 2.—A uniform sphere of radius a and weight W rests on a smooth inclined plane. It is held by a string of length l attached to the sphere and to a point of the plane. If the inclination of the plane to the horizontal is α, prove that T the tension in the string is

$$\frac{W(a + l)\sin\alpha}{\sqrt{(l^2 + 2al)}}$$

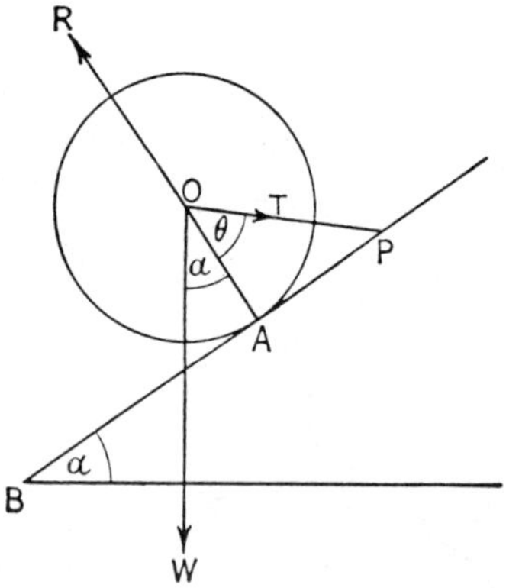

There are three forces acting on the sphere. Its weight W which acts through O, the reaction of the plane which acts through O—the plane is smooth—and the third force T which must therefore also go through O.

(i) From Lami's Theorem

$$\frac{R}{\sin(\theta + \alpha)} = \frac{W}{\sin\theta} = \frac{T}{\sin\alpha} \quad \ldots\ldots\ldots\ldots \text{(i)}$$

From $\triangle$OAP, $AP^2 = (a + l)^2 - a^2 = l^2 + 2al$

$$\therefore \sin\theta = \frac{\sqrt{(l^2 + 2al)}}{a + l}$$

Hence from (i) $$\frac{T}{\sin\alpha} = \frac{W(a + l)}{\sqrt{(l^2 + 2al)}}$$

$$\therefore T = \frac{W(a + l)\sin\alpha}{\sqrt{(l^2 + 2al)}}$$

(ii) Resolving perpendicular to OA, $T\sin\theta = W\sin\alpha$ and so on as before.

Example 3.—Two smooth planes are inclined to the horizontal at angles α and β. A non-uniform rod whose centre of gravity is a from one end and b from the other rests on the planes inclined at θ to the horizontal. Find an equation for $\tan\theta$ in terms of a, b, α, β, and find the reactions of the planes on the rods.

There are 3 forces acting on the rod: R, S, and W. These must meet in a point P.

$$\angle PGB = 90^\circ - \theta,\ \angle APG = \alpha,\ \angle BPG = \beta$$

$$\therefore (a + b)\cot(90^\circ - \theta) = a\cot\alpha - b\cot\beta$$

$$\text{i.e. } (a + b)\tan\theta = a\cot\alpha - b\cot\beta$$

From Lami's Theorem

$$\frac{S}{\sin\alpha} = \frac{R}{\sin\beta} = \frac{W}{\sin(\alpha + \beta)}$$

$$\therefore S = \frac{W\sin\alpha}{\sin(\alpha + \beta)},\quad R = \frac{W\sin\beta}{\sin(\alpha + \beta)}$$

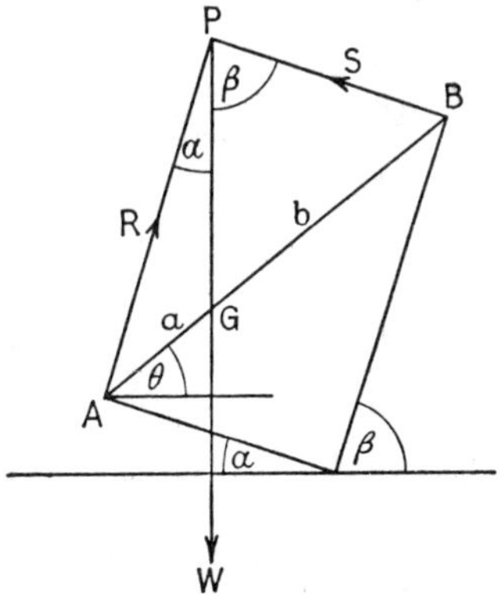

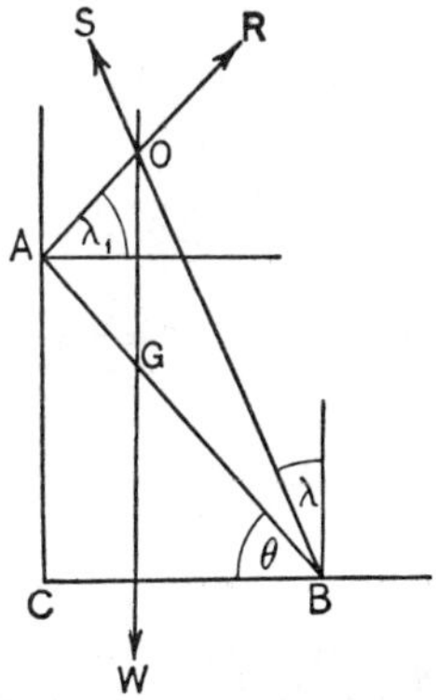

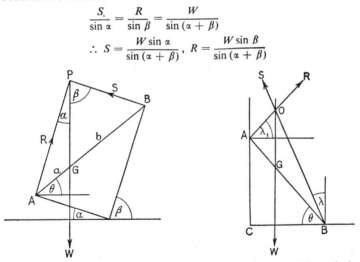

Example 4.—A uniform ladder of weight W rests against a rough vertical wall and stands on rough horizontal ground. It is inclined to the horizontal at θ°. The angles of limiting friction at the ground and the wall are λ and λ_1. Find an equation for θ when the ladder is on the point of slipping.

There are three forces acting on the ladder: its weight W and the total reactions S and R at the ground and the wall. When the ladder is in limiting equilibrium these must meet in a point and the angles between the lines of action of S and R with the normals will be λ and λ_1.

$$\angle AOG = 90^\circ - \lambda_1,\ \angle BOG = \lambda,\ \angle AGO = 90^\circ - \theta$$

Hence
$$2 \cot (90° - \theta) = \cot \lambda - \cot (90° - \lambda_1)$$
$$2 \tan \theta = \cot \lambda - \tan \lambda_1$$

Or, if $\tan \lambda = \mu$ and $\tan \lambda_1 = \mu_1$,
$$2 \tan \theta = \frac{1}{\mu} - \mu_1 = \frac{1 - \mu\mu_1}{\mu}$$
$$\therefore \tan \theta = \frac{1 - \mu\mu_1}{2\mu}$$

Example 5.—Two forces equal to λ . OA and μ . OB act along lines OA and OB. Prove that their resultant is a force $(\lambda + \mu)$OG where G is a point in AB so that λ . GA $= \mu$. GB.

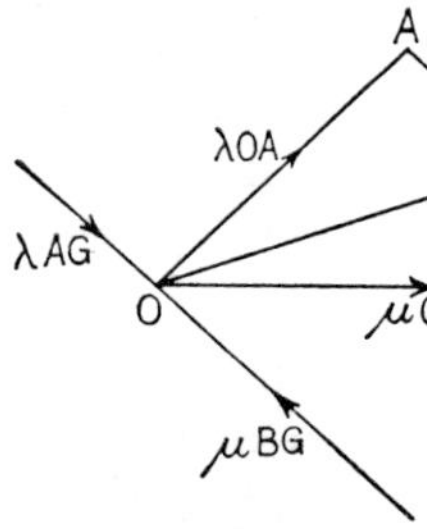

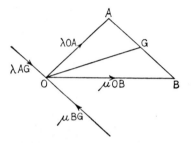

At O introduce equal and opposite forces parallel to AB equal to λ . AG or μ . BG.

From triangle OAG
$$\lambda . \overline{OA} + \lambda . \overline{AG} = \lambda . \overline{OG}$$

From triangle OGB
$$\mu . \overline{OB} + \mu . \overline{BG} = \mu . \overline{OG}$$

Adding and remembering that λ . AG $= \mu$. BG, so that $\lambda . \overline{AG} + \mu . \overline{BG} = 0$,
$$\lambda . \overline{OA} + \mu . \overline{OB} = (\lambda + \mu)\overline{OG}$$

EXERCISE 11A

1. A uniform rod AB of weight W is freely hinged at A. The rod is in equilibrium at an angle θ to the vertical when a horizontal force $\frac{1}{2}W$ acts at B. Calculate θ and the reaction of the pivot on the rod in magnitude and direction.

2. A uniform rod AB of weight W is freely hinged at A. The rod is held at an angle θ to the vertical by a force $\frac{1}{4}W$ at B acting perpendicular to the rod. Calculate θ and the reaction of the pivot on the rod.

3. A uniform rod AB of weight W is pivoted freely at A and held in a horizontal position by a string attached to B and to a point vertically above A. If the string is inclined to the horizontal at 30°, calculate the tension in the string and the reaction on the rod at A.

4. A uniform ladder AB of weight 300 N rests at an angle of 30° to a smooth vertical wall with the other end B on rough horizontal ground. Find the reaction of the wall on the ladder and the least value of the coefficient of friction at B.

5. A smooth uniform sphere of radius 9 cm and weight 12 N rests against a smooth vertical wall supported by a string 6 cm long tied to the wall and the surface of the sphere. Find the tension in the string and the reaction of the wall on the sphere.

6. A uniform ladder AB of weight 120 N and length 12·5 m rests with the end A against a smooth vertical wall and the end B on rough horizontal ground. Calculate the reaction of the wall on the ladder, the total reaction of the ground on the ladder and the least value of the coefficient of friction at B when the foot of the ladder is 3·5 m from the wall.

7. A uniform rod AB of weight 200 N and length 4 m rests with the end A on a rough horizontal plane. The rod is inclined at 60° to the plane and is held in equilibrium by a force P N acting at B perpendicular to AB and in the same vertical plane as the rod. Calculate P, the normal reaction at A, the frictional force at A, and show that the least value of the coefficient of friction is $\sqrt{3}/7$.

8. A ladder whose centre of gravity is at G rests on level ground at A and against a vertical wall at B, the coefficients of friction at A and B being $\frac{1}{2}$ and $\frac{7}{24}$ respectively. If AG = 5 m and GB = 6 m, find the inclination to the ground at which the ladder will be on the point of slipping.

9. A non-uniform rod AB has its centre of gravity at G where AG = a m and BG = b m. The rod rests inside a smooth vertical hoop and subtends an angle of 2α at the centre of the hoop. If θ is the inclination of the rod to the horizontal, prove $\tan\theta = \dfrac{b \sim a}{b + a}\tan\alpha$.

10. A uniform rod 20 cm long and weight 20 N is suspended by a light string attached to the ends of the rod and hung over a smooth peg. If the string is 30 cm long find the tension in it.

11. A triangular lamina ABC is suspended by two strings from a point O attached to the lamina at A and B. If the triangle hangs with the side BC vertical and with the strings OA and OB making α and β respectively with the vertical prove $2\cot\alpha - \cot\beta = 3\cot B$.

12. A uniform heavy rod of length $2a$ rests over a rough peg with one end against a rough vertical wall. If the distance of the peg from the wall is c and the rod is in limiting equilibrium inclined to the vertical at θ, with the peg below the point of contact with the wall, show that if the angle of friction at both points of contact is λ then $a\sec^2\lambda\sin^3\theta = c$.

13. A uniform rod rests in limiting equilibrium equally inclined to two planes one of which is horizontal and the other inclined to it at 120°. If the angle of friction between the rod and the inclined plane is 30°, calculate the coefficient of limiting friction between the rod and the horizontal plane.

14. A uniform rod of length l rests on a rough horizontal plane and is held at $\alpha°$ to the vertical by a rope tied to its top end and to a point l vertically above the bottom of the rod. Prove that for equilibrium to be possible the coefficient of friction must exceed $\tan\frac{1}{2}\alpha$.

EXERCISE 11B

1. A uniform rod AB of length $2l$ and weight 60 N is freely pivoted at A and is held inclined at 30° to the vertical by a horizontal force P at B. Calculate P and the reaction on the rod at the pivot.

2. A uniform rod AB of weight 40 N is freely pivoted at A and is held inclined to the downward vertical by a force P perpendicular to the rod. If the inclination to the vertical is 30°, calculate P and the reaction of the pivot on the rod.

3. A sphere of weight 24 N is hung by a string 18 cm long from a point in a smooth vertical wall and attached to the surface of the sphere. If the radius of the sphere is 7 cm, find the tension in the string and the reaction between the wall and the sphere.

4. A uniform rod of length 80 cm and weight 25 N rests in equilibrium with one end A in contact with a rough horizontal floor and with a point of its length resting on a smooth peg B fixed at a height 30 cm above the floor. If AB = 50 cm find the size of the reaction at B and the least value of the coefficient of friction at A consistent with equilibrium.

5. A uniform ladder rests with one end on rough ground and the other end against a rough vertical wall. If the coefficients of limiting friction at the ground and the wall are $\frac{1}{2}$ and $\frac{1}{3}$ respectively, calculate the inclination of the ladder to the vertical when on the point of slipping.

6. A uniform stick of length l rests with one end on a smooth horizontal floor and a point of its length against a rough peg at a height h above the floor. Show that an inclined position of equilibrium is impossible unless the coefficient of friction exceeds $\dfrac{h}{\sqrt{(l^2 - h^2)}}$

7. A uniform ladder rests with one end on rough horizontal ground and the other end against an equally rough vertical wall. If the angle of friction at both contacts is λ find the greatest possible inclination of the ladder to the vertical.

8. A uniform rod AB of weight W and length a rests with the end A against a rough vertical wall. It is held in a horizontal position by a light string tied to B which passes over a smooth peg at C, h m vertically above A, and carries a weight P hanging freely. Prove that $P = \dfrac{W(a^2 + h^2)^{1/2}}{2h}$ and find the least value of the coefficient of friction at A.

9. Two smooth planes whose inclinations to the horizontal are α and β meet in a horizontal line. A smooth sphere of weight W rests in contact with the two planes. Find the reactions of the planes on the sphere.

10. A light rod AB of length a m has a weight W attached to it at B. Its lower end A rests against a smooth vertical wall, and the rod goes over a smooth peg C which is c m from the wall. Prove that in the equilibrium position the inclination, θ, of the rod to the vertical is given by $\sin^3 \theta = c/a$ and find the reaction of the peg on the rod.

11. ABCD is a rectangular lamina in which AB $= a$ and BC $= b$. It rests in a vertical plane with A on rough ground which is horizontal and with B against a smooth vertical wall. If the angle of friction at A is λ and when the equilibrium is limiting AB is inclined to the vertical at θ, prove $2a \tan \lambda = a \tan \theta - b$.

12. A uniform rod of length l rests over the rim of a fixed hemispherical bowl of radius r with one end in contact with the surface of the bowl. If all contacts are smooth and the inclination of the rod to the horizontal is θ, prove

$$4r \cos 2\theta = l \cos \theta$$

4. Any number of forces

First we prove a most important theorem, namely that

a system of coplanar forces acting on a rigid body can be replaced by a single force acting at any chosen point in the plane of the forces together with a couple.

Let $P_1, P_2, P_3, P_4, \ldots, P_n$ be the coplanar forces of which P is a typical force and O the arbitrarily chosen point.

At O introduce two equal and opposite forces, each equal to P. This does not affect the body. The three forces may now be regarded

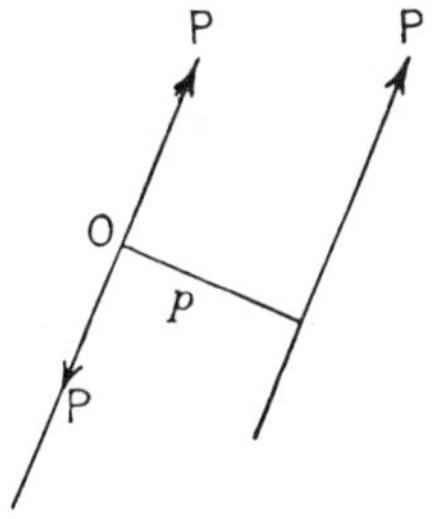

as a force P at O parallel to and in the same sense as the original force P, together with a couple of moment Pp. Notice that the moment of the couple Pp is equal to the moment of the original force P about O. Hence any force P can be transferred to act at a point O in a direction parallel to its original direction, provided we introduce a couple whose moment is the moment of the original force about O.

If now we repeat the above transference for all the forces $P_1, P_2, P_3, \ldots$, we have a system of forces $P_1, P_2, P_3, \ldots$, all acting at O in their original directions and a set of couples $P_1p_1, P_2p_2, P_3p_3, \ldots$, equal to the moments of the original forces about O. The forces, which meet at a point, can be compounded into a single force R acting through O and the couples can be added, with their appropriate signs, to give a single couple of moment G.

R is the resultant of all the given forces moved parallel to themselves to act at O, and so is the same for all positions of O, but G is the algebraic sum of the moments of all the forces about O and therefore does depend on the position of O.

There are now four possibilities:

(i) $R = 0$ and $G = 0$. When both the couple is zero and there is no resultant force we have equilibrium.

(ii) $R \neq 0$ and $G = 0$. The system reduces to a single force R.

(iii) $R = 0$ and $G \neq 0$. The system of forces is equivalent to a couple.

(iv) $R \neq 0$ and $G \neq 0$. The system can be reduced to a single force R. We can replace the couple by two forces each R so that one of them balances R at O provided we make $Rd \doteq G$. Hence we are left with a single force R acting at a distance G/R from O.

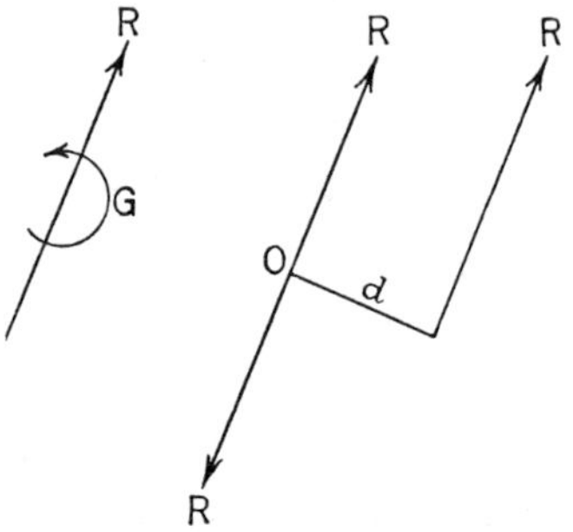

It follows that we may now say that any coplanar system of forces may be reduced to either a single force or a couple.

Hence when dealing with a general system of forces we may start with either a force and a couple or separately a force or a couple.

5. General conditions of equilibrium for coplanar forces

The conditions for the equilibrium of a body under the action of a system of coplanar forces are that there shall be neither resultant force nor couple acting on it.

If the system reduces to a single force R, since R represents the resultant of all the forces moved parallel to themselves, R will be zero if and only if the resolved parts of the forces in two non-parallel directions are separately zero. If the system reduces to a couple G, this will be zero if the algebraic sum of the moments of the forces about any point in their plane is zero. Hence one set of necessary and sufficient conditions for equilibrium is that

the algebraic sum of the resolved parts of the forces in two non-parallel directions in the plane shall be zero and the algebraic sum of the moments of the forces about a point in the plane shall be zero.

Again if the system reduces to a single force R and the algebraic sum of the moments about O is zero, then either $R = 0$ or the force R passes through O. Suppose $R \neq 0$. Then, if the sum of the moments of the forces about a second point O_1 is zero, then either $R = 0$ or R is along OO_1. Again suppose $R \neq 0$ and suppose the sum of the moments about a third point O_2, which is not collinear with O and O_1, is also zero, since we have previously established that if R is not zero its line of action is OO_1 and now the sum of the moments about O_2 is zero, then R must be zero.

If the system reduces to a couple G then G will be zero if the algebraic sum of the moments about one point is zero. Hence in all cases we have a system in equilibrium if *the algebraic sums of the moments of the forces about three non-collinear points in their plane are separately zero.*

A third set of necessary and sufficient conditions for the equilibrium of a set of coplanar forces is that *the algebraic sums of the moments about two points in their plane shall separately be zero and the sum of the resolved parts of the forces in any direction except that perpendicular to the line joining the two points about which moments were taken is also zero.* The establishment of this set of conditions is left to the reader.

We will now repeat the same work analytically.

6. Analytical reduction of a plane system of forces

Let O be any point in the plane of the set of forces $P_1, P_2, P_3, \ldots,$ acting at $A_1, A_2, A_3, \ldots$ Take rectangular axes Ox and Oy through O.

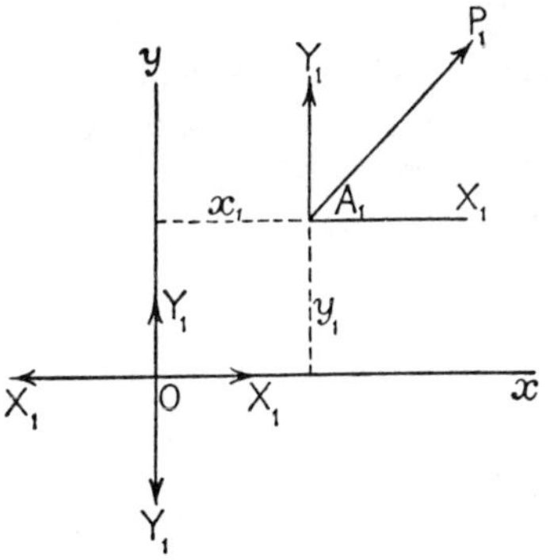

Replace P_1 by its components X_1 and Y_1 parallel to these axes and let the coordinates of $A_1, A_2, \ldots,$ be $(x_1, y_1), (x_2, y_2), \ldots,$ referred to these axes.

If at O we introduce equal and opposite forces X_1 and Y_1 as shown, we see that the forces X_1 and Y_1 at A_1 are equivalent to equal and parallel forces X_1 and Y_1 at O, together with couples of moments $x_1 Y_1$ and $-y_1 X_1$.

Dealing with all the forces $P_1, P_2, P_3, \ldots,$ in this way we reduce the given system to the following:

A force along Ox of ΣX_1. Call this X.

A force along Oy of ΣY_1. Call this Y.

A couple of moment $\Sigma(x_1 Y_1 - y_1 X_1)$. Call this G. And so for equilibrium we must have

$$\Sigma X_1 = 0,\ \Sigma Y_1 = 0,\ \Sigma(x_1 Y_1 - y_1 X_1) = 0$$

i.e.
$$\underline{X = 0,\ Y = 0 \text{ and } G = 0}$$

When the resultant force is not zero we can change the origin O to a new origin O_1 so that the couple vanishes.

If O_1 is a point whose coordinates are (α, β) then the moment of the couple referred to the new origin O_1 is G_1 where

$$\begin{aligned} G_1 &= \Sigma[(x_1-\alpha)Y_1-(y_1-\beta)X_1] \\ &= G-Y\alpha+X\beta \end{aligned}$$

This will be zero provided O_1 lies on the straight line

$$Y\alpha - X\beta = G$$

where α and β are now regarded as current coordinates. In other words *the equation of the line of action of the single resultant force is*

$$Xy - xY = G$$

unless of course $X = 0$ and $Y = 0$, in which case there is no single resultant force but a single couple.

The other sets of conditions for equilibrium

(i) If the sum of the moments about two points, say O and O_1, are each zero and the resolved parts of the forces in one direction, not perpendicular to OO_1, are zero then

$$G = 0,\ G_1 = G - Y\alpha + X\beta = 0 \text{ and } X = 0$$

and solving these we have $X = 0$, $Y = 0$, and $G = 0$ provided $\alpha \neq 0$, i.e. provided X is not perpendicular to OO_1.

(ii) If the sums of the moments about three points are zero, let the points be O (0, 0), O_1 (α_1, β_1), O_2 (α_2, β_2). Then we have

$$\begin{aligned} &G = 0,\ G_1 = G - Y\alpha_1 + X\beta_1 = 0 \\ &G_2 = G - Y\alpha_2 + X\beta_2 = 0 \\ &\therefore\ -Y\alpha_1 + X\beta_1 = 0 \text{ and } -Y\alpha_2 + X\beta_2 = 0 \end{aligned}$$

From these $X = 0$ and $Y = 0$ unless $\alpha_1\beta_2 - \alpha_2\beta_1 = 0$ in which case O, O_1, O_2 are collinear.

$\therefore$ unless O, O_1, O_2 are in a straight line we have

$$X = 0,\ Y = 0, \text{ and } G = 0$$

7. Worked examples

Example 6.—Forces of 1, 2, 3, 4 N act along the sides AB, BC, CD, and DA respectively of a square ABCD of side a. Find the size of the resultant, its direction, and the distance from B at which it cuts AB and BC.

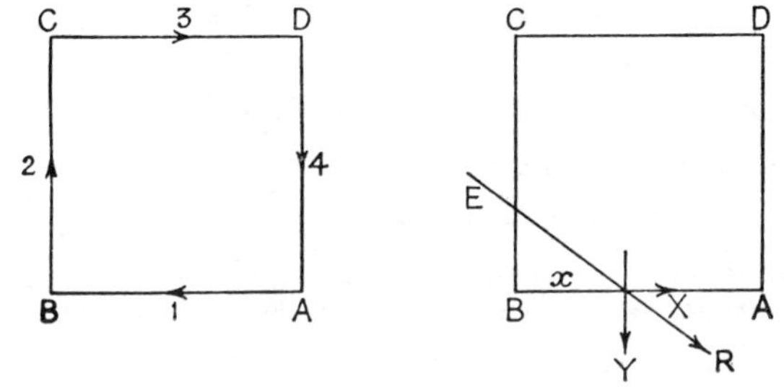

Let the resultant have components X and Y and let it cut BA at a distance x from B.

The two diagrams show *equivalent* sets of forces.

Resolving parallel to BC, $Y = 4 - 2 = 2$.

Resolving parallel to BA, $X = 3 - 1 = 2$.

$\therefore$ the resultant $R = \sqrt{(2^2 + 2^2)} = 2\sqrt{2}$ at an angle whose tangent is $\dfrac{Y}{X} = \dfrac{2}{2} = 1$ with BA.

i.e. at 45° to BA downwards

Taking moments about B, $3a + 4a = Yx = 2x$

$$\therefore\ x = 3\tfrac{1}{2}a$$

Hence if the resultant cuts BC at E, $\dfrac{\text{BE}}{x} = \tan 45°$

$$\therefore\ \text{BE} = 3\tfrac{1}{2}a$$

Example 7.—Forces of 1, 2, 3, 4, 5, 6 N act round a regular hexagon along the sides AB, BC, etc., taken in order. Reduce the system to (i) a force through the centre of the hexagon and a couple, and (ii) to a single force. The side of the hexagon is $2a$ m.

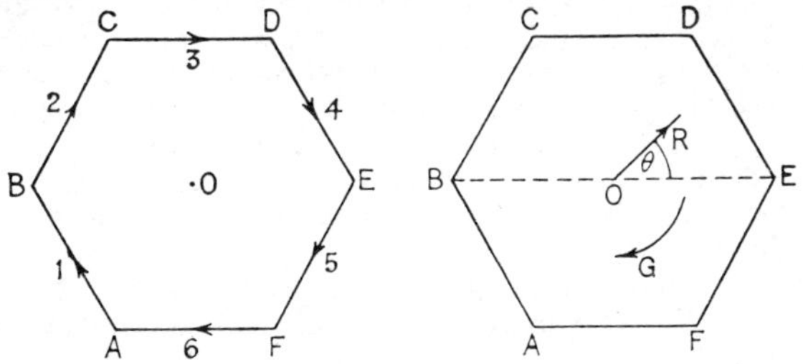

Suppose the given system is equivalent to a force R making angle θ with BE and a couple of moment G.

Resolving parallel to BE,

$$R\cos\theta = -1\cos 60° + 2\cos 60° + 3 + 4\cos 60° - 5\cos 60° - 6$$

Resolving parallel to AC, i.e. perpendicular to BE,

$$R \sin \theta = 1 \cos 30° + 2 \cos 30° - 4 \cos 30° - 5 \cos 30°$$

$$\therefore\ R \cos \theta = -3 \quad \text{and} \quad R \sin \theta = -6\sqrt{3}/2 = -3\sqrt{3}$$

Squaring and adding, $R^2 = 9 + 27 = 36$

$$\therefore\ R = 6 \text{ N}, \cos \theta = -\tfrac{1}{2} \text{ and } \sin \theta = -\sqrt{3}/2$$

$$\therefore\ \theta = 240°$$

R is 6 N at an angle of 60° with EB, i.e. parallel to EF

Taking moments about the centre $(1 + 2 + 3 + 4 + 5 + 6)\sqrt{3}a = G$

$$\therefore\ G = 21\sqrt{3}a \text{ N m}$$

G may be replaced by a force of 6 N through O at 60° with BE and a parallel force in the opposite direction through a point whose distance from O is

$$\frac{21\sqrt{3}a}{6} = \frac{7\sqrt{3}a}{2}$$

i.e. the single force is 6 N parallel to EF at a distance $\frac{7\sqrt{3}}{2}a$ from O.

EXERCISE 12A

1. ABC is an equilateral triangle with each side 2 cm long. Forces of 4 N, 8 N, and 6 N act along AB, BC, and CA respectively in directions indicated by the order of the letters. Calculate the size of the resultant force, the angle at which it cuts BC, and the distance from B of the point in which the resultant cuts BC.
2. A trapezium ABCD has parallel sides AB and DC perpendicular to AD. AB = 20 cm, DC = 60 cm, and AD = 30 cm. Forces of magnitudes 4, 10, and 12 N act along AB, BC, and CD respectively in the directions indicated by the order of the letters. Find the magnitude and direction of the resultant force and the distance of its line of action from D.
3. Forces of 1, 2, 3, 4 N act along the sides AB, BC, CD, DA of a square ABCD of side a in the direction indicated by the letters. Find the size and direction of the resultant of the forces and the distance from A at which the resultant cuts AD produced.
4. Forces of magnitude 1, 2, 3 N act along the sides AB, BC, CD respectively of a square ABCD in directions indicated by the order of the letters. A variable force P acts along DA. Prove that the resultant force passes through a fixed point. Find P if the resultant passes through the centre of the square.
5. Forces p, q, r act along the sides BC, CD, DA respectively of a square of side $2a$ whose diagonals meet at O. Find the magnitudes and senses of the three forces acting along the sides of the triangle AOB which are equivalent to the system p, q, r.
6. ABCDEF is a regular hexagon of side 1 cm. Forces of 1, 2, 3, 4 N act along AB, CD, DE, and FA respectively. Find the magnitude, direction, and position of the resultant force.
7. Ox and Oy are rectangular axes in the plane of a set of coplanar forces. The set of forces can be reduced to a force X along Ox and a force Y along Oy and a couple of moment G. Show that if they are reduced to a single force its line of action is $G - xY + yX = 0$.

8. Forces of magnitude P, $2P$, $3P$, $4P$, $5P$, and $6P$ act along the sides of a regular hexagon AB, BC, CD, DE, EF, FA. Show that their resultant is a force $6P$ acting at a distance $7d/2$ from the centre of the hexagon, where d is the distance of a side from the centre.

9. Forces represented by $2\overline{BC}$, $\overline{CA}$, and $\overline{BA}$ act along the sides BC, CA, and AB of a triangle ABC. Show that their resultant is $6\overline{DE}$, where D is the midpoint of BC and E is on CA so that CE = $\frac{1}{3}$CA.

10. ABCD is a quadrilateral and O is a point in its plane. E, F, G, H are the midpoints of AB, BC, CD, DA. Prove $\overline{OA} + \overline{OB} + \overline{OC} + \overline{OD} = 4\overline{OK}$, where K is the midpoint of EG.

11. ABCD is a quadrilateral. H and K are the midpoints of AC and BD. Prove $\overline{AB} + \overline{CB} + \overline{CD} + \overline{AD} = 4\overline{HK}$.

EXERCISE 12B

1. The sides of a triangle ABC are each 2 cm long. Forces of 1 N and 2 N act along AB and BC respectively, and a force of 3 N acts at B parallel to CA. Find the size of the resultant force and its inclination to BA. Find also the point where its line of action cuts CA produced.

2. Forces of 3, 4, 5 N act along the sides AB, BC, CA of an equilateral triangle ABC of side $2a$. Find the magnitude and direction of the resultant. Find also where the resultant cuts AB.

3. Forces of 3, 4, 5, 6 N act along AB, BC, CD, DA, the sides of a square of side a in directions indicated by the order of the letters. Find the magnitude, direction, and position of the resultant.

4. A couple of moment 20 N m acts on a square board ABCD whose side is 2 m Replace the couple by forces acting along AB, BD, and CA.

5. Forces of 1, 2, 3, 4, 5, 6 N act along the sides of a regular hexagon of side 1 m, all in the same sense, and a force P acts through the centre of the hexagon. If this system is equivalent to a couple, find the moment of the couple and the size and direction of P.

6. ABC is a right-angled triangle in which angle $B = 90°$, AB = 3 m, and BC = 4 m. Replace a couple of 30 N m in the plane of the triangle by forces p, q, r in the sides AB, BC, CA.

7. ABC is a triangular lamina in which AB = 50 cm, BC = 40 cm, and CA is 30 cm. A force of 20 N acts along AB, a force P acts at the midpoint of AC perpendicular to AC, and a force Q acts at the midpoint of BC perpendicular to BC. Show that if P and Q have certain magnitudes these three forces will keep the lamina in equilibrium, and find these values of P and Q.

8. Forces of magnitude $3F$, $4F$, F, and $5F$ act along the sides AB, CB, CD, DA of a square ABCD of side a. Find the size and direction of the resultant and the distance from D at which it cuts AD produced.

9. O is any point in the plane of a triangle ABC. Show that $\overline{OA} + \overline{OB} + \overline{OC} = \overline{OD} + \overline{OE} + \overline{OF}$, where D, E, F are the midpoints of the sides of ABC.

10. ABC is a triangle. Show that if D, E, F are the midpoints of BC, CA, AB respectively then $\overline{DA} + \overline{EB} + \overline{FC} = 0$.

11. ABC is a triangle whose medians intersect at G. O is any point in the plane of ABC. Prove that $\overline{OA} + \overline{OB} + \overline{OC} = 3\overline{OG}$.

8. Some notes on the solution of problems

We have seen that in each case the conditions governing the equilibrium of a rigid body acted upon by coplanar forces are three in number, namely two resolutions and one moment equation, or three moment equations, or two moment equations and one resolution. It follows that from these three equations we can find only three unknown quantities: three unknown forces or two forces and say one angle. It is of no use to write down a fourth equation. If this is done it will be found to be obtainable from the three already written and so does not give any new information. Hence if a problem contains four unknown quantities, with no information connecting them, the problem is insoluble. This is the case with a rod resting on rough ground and against a rough wall, when the equilibrium is not limiting, for we have two unknown normal reactions and two unknown friction forces, one of each at each contact.

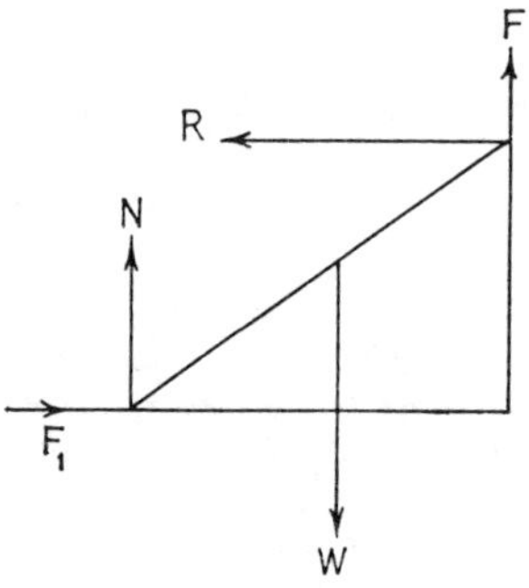

Example 8.—A uniform rod AB of length $2a$ and weight W is hinged to a vertical wall at A and is held inclined to the horizontal at 30° by a string attached to B which makes 60° with the horizontal. Find the tension in the string and the reaction of the hinge on the rod.

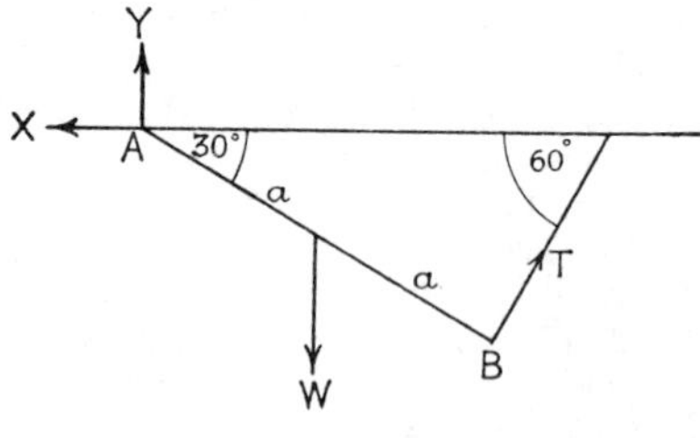

Let the components of the action of the hinge on the rod be X and Y as shown. Take moments about A,

$$2aT = Wa\cos 30°, \quad \therefore \; \underline{T = \tfrac{1}{4}W\sqrt{3}}$$

Resolve vertically,

$$Y = W - T\cos 30° = W - \tfrac{1}{4}W\sqrt{3}\,.\,\tfrac{1}{2}\sqrt{3} = \tfrac{5}{8}W$$

Resolve horizontally,

$$X = T\cos 60^\circ = \tfrac{1}{8}W\sqrt{3}$$

∴ The resultant cf X and Y is given by

$$R^2 = X^2 + Y^2 = W^2(\tfrac{3}{64} + \tfrac{25}{64}) = \tfrac{28}{64}W^2$$

$$\therefore\ \underline{R = \tfrac{1}{4}\sqrt{7}W}$$

The resultant R makes an angle with the horizontal given by

$$\tan\theta = \frac{Y}{X} = \frac{5}{\sqrt{3}} = \frac{5\sqrt{3}}{3}$$

$$\text{i.e. } \underline{\theta = 70^\circ\ 54'}$$

∴ The reaction on the rod at A is $\frac{1}{4}\sqrt{7}W$ at 70° 54′ to the horizontal.

Example 9.—Two equal uniform rods AB and BC each of length $2l$, are rigidly connected at B so that angle ABC = 90°. They rest in limiting equilibrium in contact with a fixed circular hoop of radius a with AB horizontal and BC a vertical tangent. Prove $2a(1 - \mu) = l(1 + \mu^2)$.

The forces are as shown.

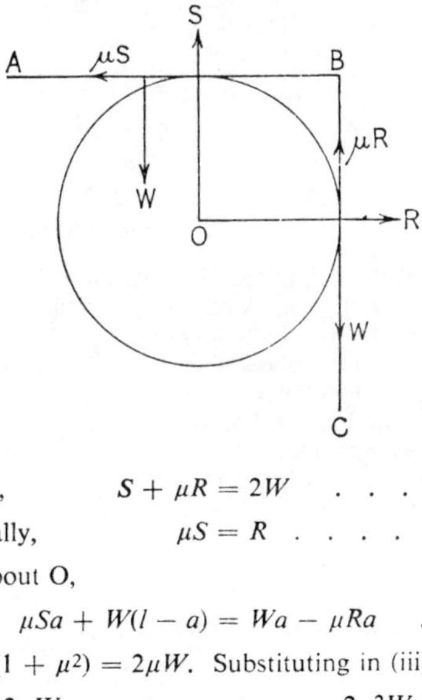

Resolving vertically, $S + \mu R = 2W$ . . . . (i)

Resolving horizontally, $\mu S = R$ . . . . (ii)

Taking moments about O,

$$\mu Sa + W(l - a) = Wa - \mu Ra \quad \text{. . . . . . . (iii)}$$

From (i) and (ii) $R(1 + \mu^2) = 2\mu W$. Substituting in (iii) for R and S,

$$\frac{2\mu Wa}{1 + \mu^2} + W(l - a) = Wa - \frac{2\mu^2 Wa}{1 + \mu^2}$$

$$2\mu Wa + 2\mu^2 Wa = W(2a - l)(1 + \mu^2)$$

$$2\mu a + 2\mu^2 a = 2a + 2a\mu^2 - l(1 + \mu^2)$$

$$l(1 + \mu^2) = 2a(1 - \mu)$$

Example 10.—Two smooth spheres of radii b and weight w are placed inside a hollow cylinder of radius a, less than $2b$, open at each end, which rests on a smooth horizontal plane; find the least weight of the cylinder in order that it shall not overturn.

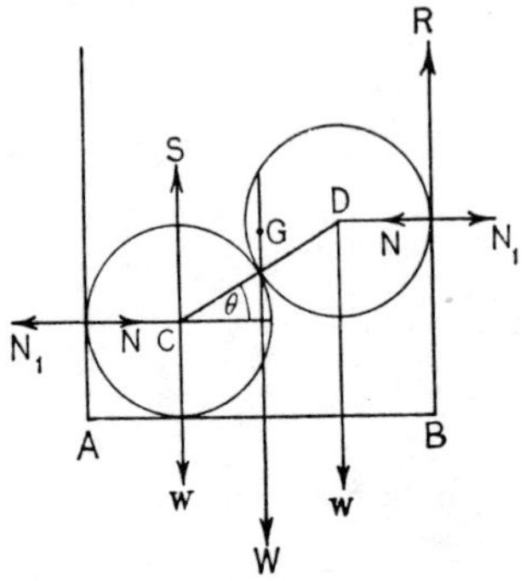

When the cylinder is about to overturn the reaction of the plane on it will act at B. This reaction R must be equal to W as these are the only vertical forces on the cylinder. These two forces R and W constitute a couple of moment Wa. The other forces on the cylinder are the two forces N_1. These must be equal and constitute a couple of moment $N_1 \times 2b \sin\theta$. For the cylinder not to overturn $Wa \geqslant N_1 . 2b \sin\theta$.

Next consider the two spheres. The lower is acted on by a force S which is equal to $2w$, leaving w up at C and w down at D, i.e. a couple of moment $w . 2b \cos\theta$, and this couple must be balanced by a couple formed by the two forces N whose moment is $N . 2b \sin\theta$. Therefore we have

$$Wa \geqslant 2bN_1 \sin\theta \quad\text{and}\quad 2b\, w \cos\theta = 2Nb \sin\theta$$

But $N = N_1$,

$$\therefore\ Wa \geqslant 2bw \cos\theta$$

$$\therefore\ W \geqslant \frac{2bw}{a} \cdot \frac{2a - 2b}{2b}$$

$$W \geqslant \frac{2(a-b)w}{a}$$

The least value of W is $\dfrac{2(a-b)w}{a}$

Example 11.—A cubical block of side $2a$ rests on a rough plane of inclination α and coefficient of limiting friction μ. A gradually increasing force is applied parallel to a line of greatest slope at the middle point of the highest edge of the cube. Show that the block will slide and not topple if $\tan\alpha < 1 - 2\mu$.

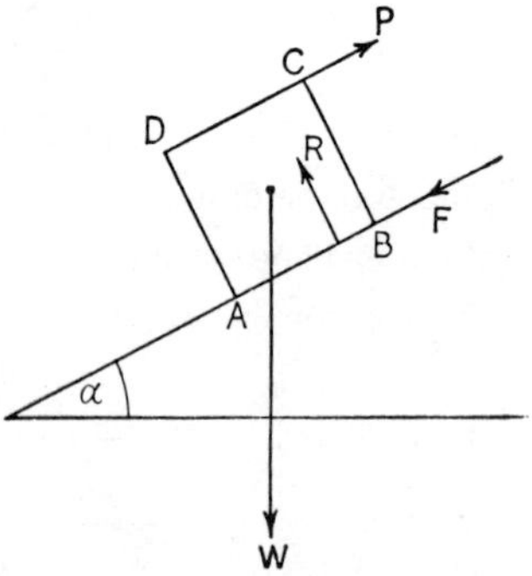

The forces acting on the block are as shown. R acts somewhere between A and B.

The block will turn about B if the moment of P about B exceeds the moment of W about B, i.e.

$$P . 2a > (W \cos\alpha + W \sin\alpha)a$$

To find F and R resolve along and normal to the plane.

$$F = P - W \sin\alpha \qquad R = W \cos\alpha$$

If the block slides $F > \mu R$, i.e.

$$P - W \sin\alpha > \mu W \cos\alpha$$

i.e.

$$P > W \sin\alpha + \mu W \cos\alpha$$

and for it to topple

$$P > \tfrac{1}{2}W(\cos\alpha + \sin\alpha)$$

Hence for it to slide before toppling

$$W \sin \alpha + \mu W \cos \alpha < \tfrac{1}{2}W(\cos \alpha + \sin \alpha)$$

$$\sin \alpha + 2\mu \cos \alpha < \cos \alpha$$

$$\tan \alpha + 2\mu < 1$$

$$\tan \alpha < 1 - 2\mu$$

EXERCISE 13A

1. A uniform rod AB 80 cm long and weight 24 N can turn freely about the end A which is fixed. The rod is held inclined to the vertical at 60° by a force P at right angles to the rod at B. Calculate P and the horizontal and vertical components of the reaction at A on the rod.
2. A uniform ladder PQ of weight 192 N and length 12·5 m rests with P on a smooth horizontal plane and Q against a smooth vertical wall. The end P is attached by a string of length 3·5 m to the junction of the wall and the floor. Find the tension in the string. If the tension is not to exceed 196 N, find how far a man of weight 600 N can ascend the ladder.
3. A uniform rod AB of weight 100 N and length 2 m rests in a vertical plane with the end A in contact with a smooth vertical wall. The end B is below A and AB is inclined at 60° to the vertical. The rod is held in equilibrium by a light string attached to B and to a point C in the wall vertically above A. Calculate the inclination of the string to the vertical, the tension in the string, and the distance CA.
4. A rectangular gate whose weight is 600 N is supported by two hinges so that the lower hinge carries the whole weight of the gate and the action of the upper hinge is entirely horizontal. The vertical distance between the hinges is 90 cm, and the centre of gravity of the gate is 120 cm from the line of the hinges. Calculate the horizontal force exerted by the upper hinge and the magnitude and direction of the force exerted by the lower hinge.
5. A man pulls a smooth cylindrical roller of radius r m and weight W over an obstacle h m high. Find the best direction in which he should pull to use the least force and the size of this least force.
6. A uniform rod of length $2a$ rests over a cylinder of radius r which is fixed in contact with a vertical wall, with one end of the rod touching the wall. If the rod is inclined at $\theta°$ to the horizontal and all contacts are smooth, prove that $a \cos \theta\,(1 + \sin \theta) = r$.
7. A uniform rod of length a rests in a vertical plane inside a fixed hemispherical bowl of radius a. If the rod rests in limiting equilibrium, inclined at $\theta°$ to the horizontal and the coefficient of limiting friction is μ, show that $\tan \theta = \dfrac{4\mu}{3 - \mu^2}$
8. A uniform rod AB of weight W and length $2a$ rests with the end A against a vertical wall. The rod is held inclined at an angle θ to the wall by a string of length $2a$ in the same vertical plane as the rod attached to B and to a point C in the wall vertically above A. Find the tension in the string and the horizontal and vertical components of the action of the wall on the rod. Show that the least angle, other than zero, the rod can make with the wall is $\cot^{-1} \tfrac{1}{3}\mu$ where μ is the coefficient of limiting friction at A.

9. A uniform rod AB of length $2a$ and weight W rests with the end A on a smooth horizontal plane and a point of the rod on a smooth peg C at a height h above the horizontal plane. If the rod is held in equilibrium inclined at θ to the horizontal by a horizontal force F applied at A, find F.

10. Two inclined planes each inclined at α to the horizontal meet in a horizontal line. A uniform rod AB rests with A on one plane and B on the other plane. If the planes are equally rough, with angle of limiting friction λ, show that if the greatest inclination of the rod to the horizontal is θ then $\tan\theta = \dfrac{\sin 2\lambda}{\cos 2\lambda - \cos 2\alpha}$

11. A uniform heavy rod of weight W and length $2a$ rests with one end against a rough vertical wall and a point of the rod over a rough peg which is a distance c from the wall. If λ is the angle of friction at both wall and peg and if the rod makes an acute angle θ with the downward vertical when in limiting equilibrium, prove $\sin^3\theta = \dfrac{c}{a}\cos^2\lambda$.

12. A uniform rod of length l passes over a rough peg A and under a rough peg B. The pegs are equally rough, the coefficient of limiting friction at each is μ, and they are c apart. Show that the least value of l for equilibrium to be possible is $c\left(1 + \dfrac{\tan\alpha}{\mu}\right)$, where α is the inclination of the line AB to the horizontal.

13. A uniform smooth sphere of weight W and radius a is suspended from a point O by a string OP of length l which is attached to P, a point on the surface of the sphere. Another body of weight W is suspended from O by a light string sufficiently long for the body to hang below the sphere. Prove that in the equilibrium position the line OP passes through Q, the centre of the sphere. Find the inclination of OP to the vertical.

14. A rod of length $a\sqrt{3}$ and weight W is placed inside a fixed smooth spherical cavity of radius a. The centre of gravity of the rod is at one of its points of trisection. Prove that the rod can rest in equilibrium inclined at an angle of 30° to the horizontal and that in this position the reactions at its lower and upper ends are $2W/\sqrt{3}$ and $W/\sqrt{3}$.

15. A uniform rod of length l rests inside a sphere of diameter $l\sqrt{2}$ in a vertical plane through the centre of the sphere. If λ is the angle of friction between the sphere and each end of the rod, show that when the rod is in limiting equilibrium the angle it makes with the horizontal is 2λ.

16. The diagram shows a vertical section of a uniform cylinder of weight W resting on two fixed inclined planes each at α to the horizontal. The cylinder is acted

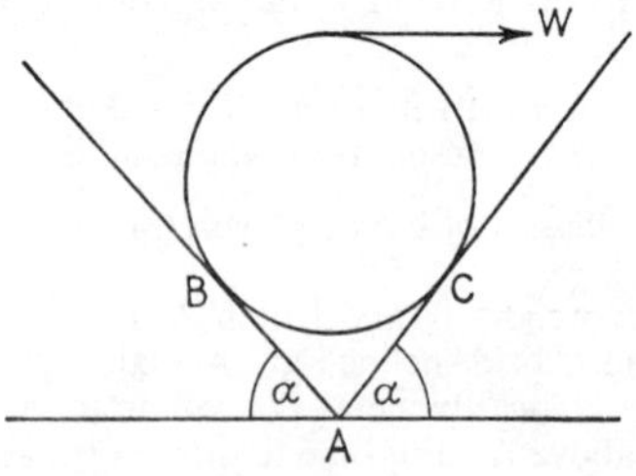

upon by a force W as shown. The plane AC is smooth, AB is rough, coefficient of friction μ. Show that equilibrium is impossible unless $\alpha > 45°$. If $\alpha = \tan^{-1}\frac{15}{8}$ show that there will be equilibrium provided μ is not less than $\frac{6}{7}$.

17. A uniform rod of weight w and length a is free to turn about a horizontal hinge at one end. The other end is attached to a point at a vertical distance a above the hinge by means of a light elastic string of natural length b. The elasticity of the string is such that when stretched by a force w its length is a. In the position of equilibrium show that the tension in the string is $\dfrac{wb}{a+b}$

18. A uniform ladder of weight W rests at an angle α with the vertical with one end on rough horizontal ground and the other end against an equally rough vertical wall. The angles of friction in each case are λ. Prove that the greatest possible value of α is 2λ.

19. A uniform ladder of length $2a$ and weight w with the one end on a rough horizontal plane and the other in contact with a rough vertical wall is in a vertical plane. The coefficient of friction is the same at both contacts. If θ is the least inclination to the horizontal at which the ladder can rest in equilibrium, show that the least inclination ϕ for safe ascent to the top of the ladder by a man of weight W is given by $\tan\phi = \tan\theta + \dfrac{W}{W+w}\sec\theta$.

20. The distance between the axles of a railway wagon is $2l$ and the centre of gravity is midway between them and at a height h above the rails. Show that if the wagon rests on an inclined plane with the upper wheels locked, the angle of inclination of the plane to the horizontal must not exceed $\tan^{-1}\dfrac{\mu l}{2l+\mu h}$, where μ is the coefficient of friction between the locked wheels and the rails.

21. A heavy uniform rod of length $2a$ rests partly inside and partly outside a smooth fixed hemispherical bowl of internal diameter d $(d > a)$ whose edge is smooth and horizontal. Prove that if the inclination of the rod to the horizontal is θ, then $4d\cos\theta = a + \sqrt{(a^2 + 8d^2)}$.

22. A solid uniform cone whose semi-vertical angle is α rests on a rough inclined plane of inclination α. The cone is pulled by an increasing force parallel to a line of greatest slope of the plane at its vertex. Prove that the cone will topple before it slides if $\tan\alpha < 4\mu$.

23. A uniform cylinder of radius a and height h stands with its base on a rough plane whose inclination to the horizontal is gradually increased. If μ is the coefficient of limiting friction show that the cylinder will topple before it slides if $2a < \mu h$.

24. A uniform cubical block of weight W stands on a rough horizontal floor. A gradually increasing force is applied at the middle point of a top edge, pulling upwards at a fixed acute angle θ with the top face. If the block begins to turn about a lower edge show that the force of friction between the floor and the block at the instant when equilibrium is broken is $\dfrac{W}{2(1+\tan\theta)}$. Find also the condition that the block shall tilt before it slides in terms of θ and μ, the coefficient of friction.

EXERCISE 13B

1. A uniform rod AB of length 60 cm and weight 4 N has a weight of 2 N attached at B. The rod is freely pivoted at A and is kept in equilibrium with B above A and AB inclined at 60° to the vertical by a horizontal string attached to B. Calculate the tension in the string and the horizontal and vertical components of the reaction of the pivot A on the rod.

2. A heavy uniform rod AB rests with the end A in contact with a smooth vertical wall, and with a light string attached to A and B and passing through a smooth ring C vertically above A. If the rod AB and the part of the string BC are inclined to the vertical at θ and α respectively, prove that $\tan\theta = 2\cot\frac{1}{2}\alpha$ and find the tension in the string in terms of W and α.

3. A uniform rod AB of length 60 cm and weight 20 N is freely hinged at A and has a weight of 10 N attached to it at B. It is held in equilibrium in a horizontal position by a string attached to B. If the string makes 30° with the horizontal, calculate the tension in the string, the horizontal and vertical components of the reaction of the hinge on the rod, and the magnitude of the resultant reaction of the hinge on the rod.

4. A rectangular door ABCD is hinged to a vertical wall at E and F two points in AD with E above F. If AB = 90 cm, BC = 2·4 m, AF = DE = 30 cm, and the door is a uniform rectangle of weight 80 N, calculate the force on each hinge if half the weight of the door is carried by each hinge.

5. A ladder whose centre of gravity is at G rests on level ground at A and against a vertical wall at B, the coefficients of friction at A and B being $\frac{1}{2}$ and $\frac{7}{24}$ respectively. If AG = 3 m and GB = 3·6 m, find the inclination of the ladder to the ground when it is on the point of slipping.

6. A uniform ladder rests with one end on rough horizontal ground and the other end against a smooth vertical wall. If the coefficient of friction between the ladder and the ground is μ and the weight of the ladder is W, find the inclination of the ladder to the ground when it is about to slip and the reactions between the ladder and the wall and the ladder and the ground.

7. Repeat question 6 when the wall also is rough with the coefficient of friction there μ_1. Find the inclination of the ladder to the ground and the normal reactions with the wall and the ground.

8. A uniform hemisphere rests in limiting equilibrium on rough ground and against an equally rough vertical wall, both coefficients of friction being μ. Show that the inclination of the plane base to the horizontal is

$$\sin^{-1}\frac{8\mu(1+\mu)}{3(1+\mu^2)}$$

9. A solid hemisphere of weight W rests in limiting equilibrium with its curved surface on a rough inclined plane. The plane face is kept horizontal by a weight w attached to a point on its rim. Show that the coefficient of friction is

$$\frac{w}{\sqrt{[W(W+2w)]}}$$

10. A uniform heavy rod of length l has one end on rough horizontal ground and the rod rests in limiting equilibrium touching a fixed coplanar semicircular hoop of radius a. If the coefficient of friction at both contacts is μ, show that θ, the inclination of the rod to the horizontal, is given by $l\sin^2\theta = \dfrac{2a\mu}{1+\mu^2}$

11. The distance between the axles of a railway truck is a and the centre of gravity is halfway between them and h above the rails. If the lower wheels are locked, the greatest slope on which the truck can rest is α. Prove that the coefficient of friction is $\dfrac{2a}{a\cot\alpha + 2h}$

12. A sphere rests on a rough inclined plane. The highest point of the sphere is tied to a point of the plane by a horizontal string. Prove that when the equilibrium is limiting $\mu = \tan \frac{1}{2}\alpha$, where α is the inclination of the plane to the horizontal. Calculate the tension in the string in terms of μ and the weight $\boldsymbol{W}$ of the sphere.

13. A drawer of depth a is jammed by pulling one handle at a distance b from the centre of the front. Prove that the least value of the coefficient of friction is $\dfrac{a}{2b}$

14. One end of a uniform rod AB of weight W is freely pivoted at A. At the end B is fastened a light elastic string of modulus λ and unstretched length l, which is fastened to a point O vertically above A so that OA $= h$. Show that in the position of equilibrium, with AB inclined to the vertical, the extension of the string is $\dfrac{Wl^2}{2\lambda h - Wl}$

15. A straight rod of weight W is suspended by four vertical elastic strings from points A, B, C, D in a horizontal line, so that AB = BC = CD. The centre of gravity of the rod is vertically below C. The strings have the same unstretched length and the same modulus of elasticity. Prove that the tensions in the string are $\dfrac{W}{10}, \dfrac{2W}{10}, \dfrac{3W}{10}, \dfrac{4W}{10}$

16. A uniform solid cone of radius a and height h rests on a rough plane whose inclination to the horizontal is gradually increased. Prove that if μ is the coefficient of friction the cone will topple before it slides if $4a < \mu h$.

17. A uniform cube rests on a rough inclined plane of inclination $\alpha(<45°)$ with two of its faces parallel to a line of greatest slope. Prove that it will not be possible to pull the cube up the plane with a force through the midpoint of the upper edge parallel to a line of greatest slope if $\mu > \frac{1}{2}(1 - \tan \alpha)$, where μ is the coefficient of friction.

9. Jointed bodies

Two bodies are often described as smoothly jointed or smoothly hinged. The hinge usually consists of a circular hole bored through each of the bodies with a loosely fitting pin through the two holes so that the bodies can turn without friction about the pin. When the joint is smooth and connects two bodies A and B the action of the joint sets up equal and opposite forces on the two bodies A and B.

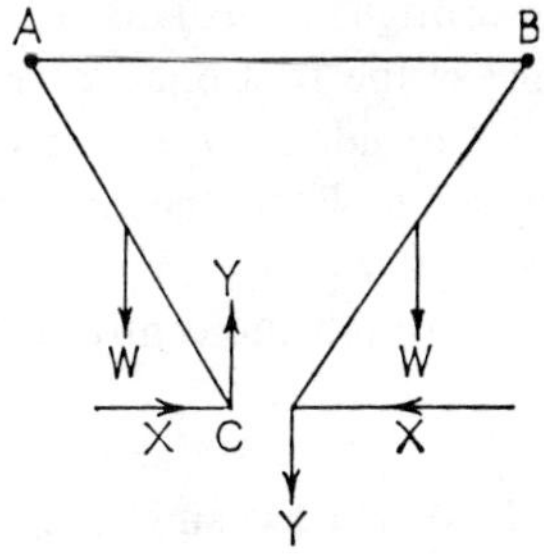

Two equal rods are hung from A and B. They are smoothly hinged at C. The rod BC is acted on by a force whose horizontal and vertical components are X and Y due to AC, and AC is acted on by an equal and opposite force whose components are also X and Y in the directions shown.

10. Symmetry

If the figure is symmetrical about the vertical through C, then the forces at C on the two rods must be symmetrical. The two forces X are symmetrical, the forces Y are not. They must be equal and opposite and they must both be in the same direction, and this is only possible if $Y = 0$. Considerations of this kind can often ease the solution of problems.

11. Several rigid bodies in contact

When dealing with several rigid bodies in contact with each other we may write down three equations for each of the separate bodies provided we include in these the reactions between each pair of bodies at their points of contact. These reactions will be equal and opposite,

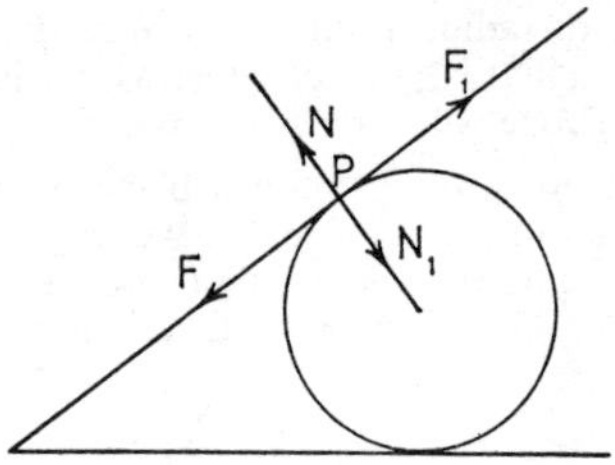

for if A presses B with a force P then B will press A with an opposite force P. If the contacts are smooth, these reactions will be perpendicular to the surfaces in contact at the point of contact. If the contacts are rough, we usually use a normal reaction and a tangential one. This last is the frictional force.

Consider, for example, a rod resting against a circular hoop. The forces at P on the rod are N and F, and the forces on the hoop are N_1 and F_1. But $N = N_1$ and $F = F_1$ in size, and both N and N_1 will go through the centre of the hoop.

12. Worked examples

Example 12.—A framework consists of three equal uniform rods each of length $2a$ and weight W freely jointed at their ends to form an equilateral triangle ABC. The framework is suspended from the middle point of AB. Calculate the reactions at the hinges.

Figure (*a*) shows the most general possible arrangement of forces on the rods. We have used only the fact that the forces on two rods at a hinge are equal and opposite. The figure is symmetrical and so $Y = Y_1$, $X = X_1$, and $M = 0$. So we may use figure (*b*).

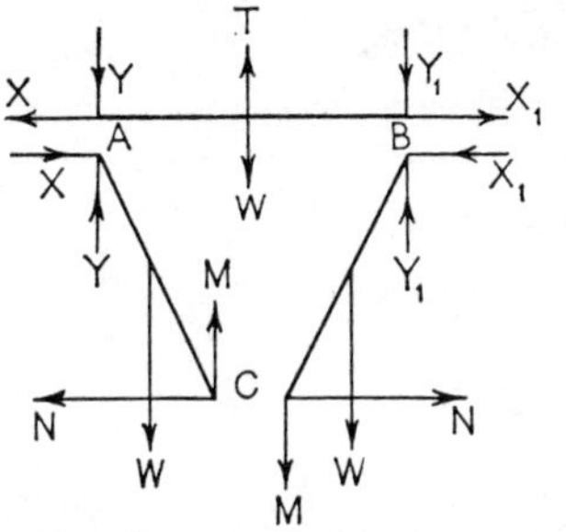

(a)

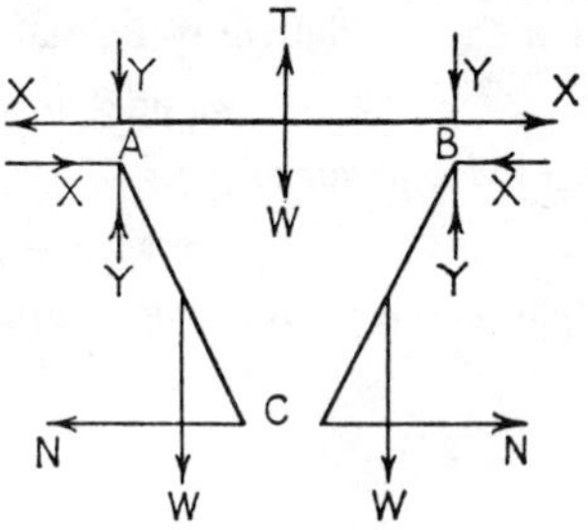

(b)

For the equilibrium of the rod AC, moments about A give

$$W \, . \, \tfrac{1}{2}a = N \, . \, a\sqrt{3}, \qquad \therefore \; N = \frac{W}{2\sqrt{3}}$$

Resolving vertically and horizontally for AC, $Y = W$ and $X = N = \dfrac{W}{2\sqrt{3}}$

$\therefore$ The action at the hinge A is $\sqrt{(X^2 + Y^2)} = \dfrac{W}{6}\sqrt{39}$

The action at B is the same and at C the reaction is $\dfrac{W}{2\sqrt{3}}$

Example 13.—Two uniform ladders AB and BC of weight W and w are smoothly jointed at B and stand in a vertical plane with A and C on a rough horizontal plane. If the length of each ladder is $2a$ and the coefficients of limiting friction at A and C are the same, show that the lighter ladder will slip first. If μ is the coefficient of friction, find the inclination of BC to the vertical when slipping takes place.

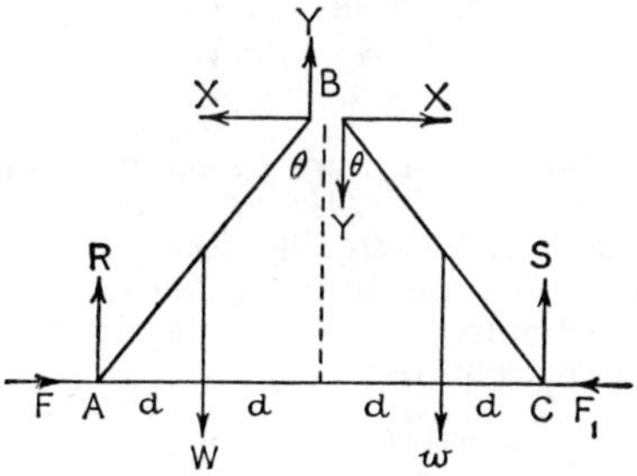

Let angle ABC $= 2\theta$. The forces are as shown. Take $W > w$. Consider the equilibrium of both ladders.

Resolving vertically and horizontally $R + S = W + w$ and $F = F_1$.

Taking moments about A,

$$4dS = w \times 3d + Wd, \qquad \therefore \; S = \frac{3w + W}{4}$$

Hence
$$R = \frac{3W + w}{4}$$

Since $W > w$, $S < R$, and since $F = F_1$, $\frac{F_1}{S} > \frac{F}{R}$

Slipping occurs when one of these reaches the value μ. Hence, since $\frac{F_1}{S} > \frac{F}{R}$ C will slip first.

For the equilibrium of BC, taking moments about B,

$$wa \sin \theta + F_1 \,.\, 2a \cos \theta = S \,.\, 2a \sin \theta$$

When BC is about to slip,

$$w \sin \theta + \mu S \,.\, 2 \cos \theta = 2S \sin \theta$$

Substituting for S, $\quad (W + w) \sin \theta = \mu(W + 3w) \cos \theta$

$$\therefore \tan \theta = \frac{\mu(W + 3w)}{W + w}$$

Example 14.—Inside a fixed hollow cylinder of radius a are placed two equal smooth cylinders of radius b with another equal cylinder placed symmetrically on the first two. If the axes of all 4 cylinders are parallel and horizontal, prove that the upper cylinder will force the lower two apart if $a > b(1 + 2\sqrt{7})$.

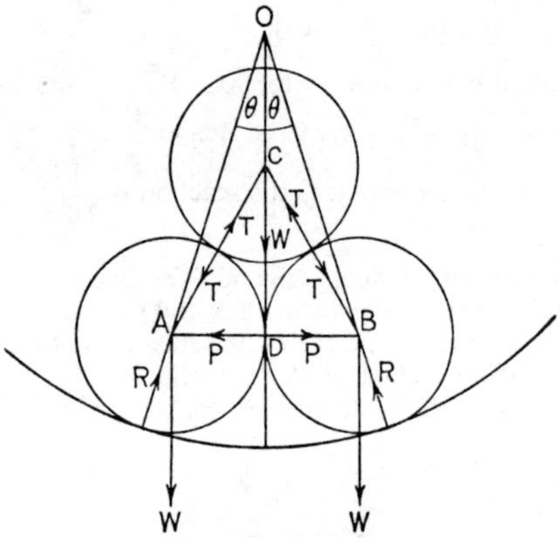

The forces are as shown, for the figure is symmetrical about OCD. Let angle AOB $= 2\theta$.

For the three cylinders resolve vertically, then $3W = 2R \cos \theta$. (Notice that the other equations we could get for the three cylinders give the $2R$s equal which we have assumed from symmetry.)

For the upper cylinder resolve vertically:

$$2T \cos 30° = W$$

For the cylinder centre A resolve horizontally:

$$P + T \cos 60° = R \sin \theta$$

The two lower cylinders will be forced apart when P is negative, i.e. when

$$T \cos 60° > R \sin \theta$$

Substituting for T and R, $\quad \frac{W}{2\sqrt{3}} > \frac{3}{2} W \tan \theta$

But $$\tan\theta = \frac{b}{\sqrt{[(a-b)^2 - b^2]}} = \frac{b}{\sqrt{(a^2 - 2ab)}}$$

$$\therefore \frac{1}{3\sqrt{3}} > \frac{b}{\sqrt{(a^2 - 2ab)}}$$

$$a^2 - 2ab > 27b^2$$

$$(a-b)^2 > 28b^2$$

$$(a-b) > 2\sqrt{7}b$$

$$a > b(1 + 2\sqrt{7})$$

EXERCISE 14A

1. Two uniform rods of weight W and W_1 rest, each in a vertical plane, with one end of each rod in contact on a smooth horizontal floor and the other ends of the rods against one of two smooth vertical parallel walls. If the inclinations of the rods to the vertical are θ and θ_1, prove $W\tan\theta = W_1\tan\theta_1$.

2. Three equal uniform rods AB, BC, CD, each of weight W, are freely jointed at B and C. The ends A and D are freely pivoted to points in a horizontal line so that when the system hangs in equilibrium angle ABC = BCD = 120°. Calculate the action of the rod AB on (i) the pivot A, and (ii) the rod BC.

3. Two equal uniform rods AB and AC, each of weight W, are freely joined at A. They stand with B and C on a smooth horizontal plane so that angle BAC is 60°, being held in this position by a light cord BC. Calculate the tension in the cord.

4. Two uniform rods OA and OB of weights W and P respectively are freely hinged at O and hang in equilibrium with A and B freely hinged to fixed pegs at the same horizontal level. If angle OAB = 60° and angle OBA = 30°, find the horizontal and vertical components of the reaction at O.

5. Two uniform rods AB and CD, each of weight W and length a, are smoothly jointed at O so that OB = OD = b. The rods rest in a vertical plane with A and C on a smooth horizontal table and B and D connected by a light string. If the inclination of each rod to the vertical is α, prove that the reaction at the joint O is $\frac{aW}{2b}\tan\alpha$.

6. Four equal uniform rods, each of weight W, are freely pivoted at their ends to form a square. The system is suspended from A and the square shape is maintained by a light rod joining A to C. Calculate the tension in the rod AC and the reaction at the joint B.

7. Two hinged ladders of equal length and weights W and w are hinged together to form a step ladder. The step ladder stands on a rough horizontal plane with the angle between the legs 2θ. If the coefficient of friction is the same at both points of contact with the ground, find its least value in terms of W, w, and α. [$W > w$ and slipping first occurs when $\theta = \alpha$.]

8. Two equal uniform rods AB and BC are smoothly jointed at B. They stand in a vertical plane with A and C on a rough horizontal plane. If AB is 3 times as heavy as BC, and if the coefficient of friction at both A and C is $\frac{2}{3}$, show that the rod BC will slip first and find the angle ABC when this happens.

9. Two equal uniform beams AB and BC, each of weight W, are connected by a hinge at B and stand in a vertical plane with A and C on a horizontal smooth plane. They are kept from falling by strings joining A and C to the middle points BC and AB respectively. Show that the tension in each string is $\frac{1}{8}W\sqrt{10}$ when the angle ABC is 90°.

10. Two smooth spheres each of radius a and weight W lie touching each other in a smooth spherical bowl of radius b. Find the reaction between the two spheres.

11. Three equal uniform rods AB, BC, CD, of equal weight are freely jointed at B and C, and the rods AB and CD rest on smooth pegs in the same horizontal line. Each of the rods AB and CD is inclined at θ to the horizontal. Prove that if the reaction at the hinge B makes an angle ϕ with the horizontal then $\cot \phi = 3 \tan \theta$. Find also the height of BC above the pegs in terms of AB and the angle θ.

12. Two equal uniform rods OA and OB, each of length $4a$, are freely jointed at O. The rods rest symmetrically on a smooth circular hoop of radius a. Show that the angle AOB is a right angle. If the weight of each rod is W, calculate the magnitude and direction of the reaction at O.

13. Two equal uniform rods AB and AC are freely jointed at A. The ends B and C rest symmetrically on two smooth inclined planes, each inclined at α to the horizontal, with the rods in a vertical plane. The ends B and C are connected by a light string. If 2θ is the angle BAC and W the weight of each rod, prove the tension in the string is $\frac{1}{2}W(\tan \theta - 2 \tan \alpha)$.

14. Two equal uniform rods AB and BC are smoothly jointed at B and are in equilibrium with C resting on a rough horizontal plane and A held above the plane. If α and β are the inclinations of CB and BA to the horizontal, prove that the coefficient of friction at C must be greater than $\dfrac{2}{\tan \beta - 3 \tan \alpha}$

15. A uniform cylinder rests on horizontal ground against a vertical wall with its axis horizontal and parallel to the wall. A uniform plank AB of weight W rests with one end A on the ground and the other end B on the cylinder so that AB is a tangent and B the point of contact. The plank makes an angle α with the ground. If the contacts of the cylinder with the plank and with the wall are smooth, find the reaction between the cylinder and the plank and between the cylinder and the wall. Find also the least value of the coefficient of friction between the plank and the ground in order that the plank may not slip.

16. A rough uniform cylinder, of radius a and weight W rests on a rough horizontal table with its axis parallel to the table. A uniform rod of length $2l$ and weight W_1 rests in a vertical plane, perpendicular to the axis of the cylinder, with one end A on the table and a point B of its length in contact with the cylinder. The inclination of the rod to the horizontal is 2α. Show that the forces of friction at the three points of contact are equal and find their magnitude. Find also the normal reaction at each point of contact.

EXAMPLES 14B

1. Two uniform beams AB and AC, each 2 m long, are smoothly jointed at A and stand in a vertical plane on a smooth horizontal plane. A string 1 m long joins the midpoints of AB and AC. If each beam weighs 30 N, calculate the tension in the string and the reaction at A.

2. Three equal uniform rods AB, BC, CA, each 30 cm long and weight 6N, are freely pivoted at A, B, and C to form a triangle. The framework is placed over a smooth peg at the middle point of BC, so that A is below BC. Find the reactions at the joints.

3. Four equal uniform rods, each of weight W, are freely jointed at their ends to form a square ABCD. The square is hung from D and is held in the form of a square by a light rod joining the midpoints of AB and BC. Calculate the thrust in the light rod.

4. Two light rods AB and AC, each 2 m long, are hinged at A and placed in a vertical plane with B and C on a smooth horizontal plane. A weight of 12 N is attached to a point D on AC 1 m above the horizontal plane and B and C are joined by a light string 2 m long. Calculate the tension in the string.

5. Two equal uniform rods are freely jointed at A. Their ends B and C rest on a smooth horizontal plane and angle CAB = 60°. The weights of the rods AB and AC are each 48 N. Another rod of weight 12 N joins the midpoints of AB and AC and is smoothly hinged to each. Calculate the magnitudes of the reactions at each joint.

6. A string is tied tightly round two smooth cylinders of radii a and b. Prove that if T is the tension in the string the force each cylinder exerts on the other is

$$\frac{4T\sqrt{(ab)}}{a+b}$$

7. Three uniform rods of lengths a, a, and b, each of weight w per unit length, are freely jointed at their ends to form an isosceles triangle. The triangle is suspended from the vertex opposite the side b. Show that the direction of the force between the rods at the lower hinges is inclined to the vertical at

$$\tan^{-1}\frac{a+b}{\sqrt{(4a^2-b^2)}}$$

8. ABCD is a quadrilateral formed by 4 rods of equal weight but different lengths freely jointed at their ends. If the system is in equilibrium in a vertical plane with AB supported in a horizontal position, prove that if CD is inclined at $\theta°$ to the horizontal then $2\tan\theta = \tan A - \tan B$.

9. AB and BC are two uniform rods each of length $2a$ and of equal weight freely hinged at B. They rest in equilibrium on two smooth pegs at the same horizontal level which are $2d$ apart, with the rods each inclined to the vertical at $\theta°$ and B below A and C. Prove $d = a\sin^3\theta$.

10. Two uniform ladders AB and AC, each of length $2a$ and weights W and $2W$, are smoothly hinged at A and rest in equilibrium in a vertical plane with B and C on a rough horizontal plane with angle BAC $= 2\alpha$. Show that the friction forces at B and C are each $\frac{3}{4}W\tan\alpha$. If the coefficients of friction at B and C are each μ, and if α is gradually increased, show that one ladder slips when α is $\tan^{-1}\frac{5}{3}\mu$ and that the resultant reaction then at the point of contact of the other ladder with the ground is of magnitude $\frac{1}{4}W\sqrt{(49+25\mu^2)}$.

11. A uniform cylinder of weight w and radius c rests on a rough table with its axis horizontal. A uniform rod ABC of weight W and length $2a$ is smoothly pivoted to the table at A and rests against the curved surface of the cylinder at B. The rod is inclined at α to the horizontal and lies in a plane through the centre of gravity of the cylinder normal to its axis. If the components normal to the cylinder of the reactions between it and the rod and the table are R and N, prove $N = w + R$. If the coefficient of friction at both these contacts is μ, show that for equilibrium to be possible $\mu \geqslant \tan\frac{1}{2}\alpha$.

12. The distance between the axles of a railway truck is $2a$ and the centre of gravity is midway between them and a distance h above the rails. With the lower wheels locked the greatest incline on which the truck can rest is α to the horizontal. Show that the coefficient of friction between the wheels and the rails is

$$\frac{2a \tan \alpha}{a + h \tan \alpha}$$

13. Two equal uniform ladders are hinged to form a step ladder which stands in a vertical plane on a rough horizontal plane with the angle between the two legs 2θ. A man whose weight is n times the weight of each ladder stands on one ladder at distance x from the top, and slipping is about to occur. Show that if the length of each leg is l and the coefficient of friction at each contact with the plane is μ, then $(nl - nx + l) \tan \theta = \mu(nl - nx + 2l)$.

14. Two uniform rods AB, BC of length $2a$, $2b$ and weight $2W$ and $3W$ respectively are smoothly hinged at B. They are in equilibrium in a vertical plane with A on a rough horizontal plane and C resting against a smooth vertical guide. A is farther from the guide than B, α is the acute angle between AB and the horizontal plane, and β the angle between CB and the downward vertical at C. Prove $3 \tan \alpha \tan \beta = 8$. If μ is the coefficient of friction at A, show that $\tan \beta \leqslant \frac{10}{3}\mu$.

15. A uniform rod of length l rests partly inside and partly outside a smooth hemispherical bowl of radius a whose rim is horizontal. If θ is the inclination of the rod to the horizontal show that $l \cos \theta = 4a \cos 2\theta$. Show also that l must lie between $\dfrac{2\sqrt{6}a}{3}$ and $4a$.

16. Two uniform rods AB and BC freely jointed at B have equal length and weight. They stand in equilibrium in a vertical plane as shown against a rough vertical wall and on a rough horizontal plane. The angles BAO and BCO are each α

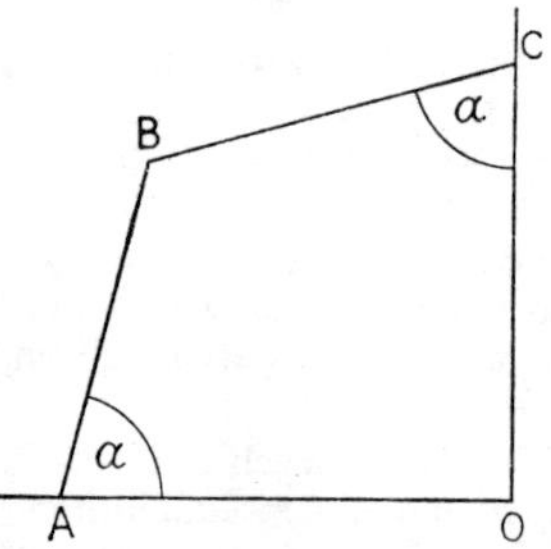

as shown, and the weight of each rod is W. Find the normal and frictional reactions on the rods at A and C. Prove that α must lie between $\frac{1}{4}\pi$ and $\frac{1}{2}\pi$, and that the frictional force at C acts upwards or downwards according as α is greater or less than $\frac{1}{3}\pi$.

17. An open cylindrical can of radius $2a$ and height $3a$ stands with its base on a horizontal table and contains, projecting from it, a uniform rod ABC of length ma. The rod is in contact at A with both the bottom and the curved surface of the can and leans against the rim at B so that AB passes through the axis of the can. All contacts are smooth. The weight of the can is nW and its centre of gravity is on its axis. The weight of the rod is W. Show that whatever the value of n the rod will slip if $m > \frac{125}{8}$. Find in terms of m the minimum value of n necessary to prevent the can from tipping.

7

Graphical Statics

1. Funicular polygon

In the previous chapter we have shown how to calculate the resultant of any number of coplanar forces. We now do similar work by drawing.

To find graphically the resultant in magnitude, direction, and position of any number of coplanar forces.

Consider four forces P, Q, S, T as shown in fig. 1. Four forces only are taken but the method is the same for any number acting on a rigid body.

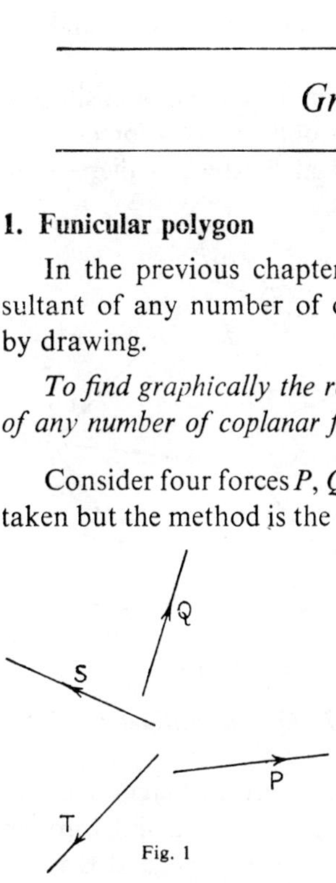

Fig. 1

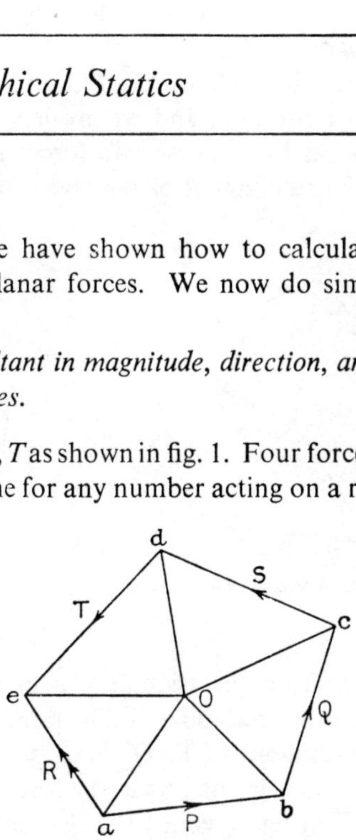

Fig. 2.—Force polygon

Draw ab parallel and proportional to P, bc parallel and proportional to Q, cd parallel and proportional to S and de parallel and proportional to T. Join ea. The figure $abcde$ is called the *force polygon* and we shall show that ae represents the resultant, R, of the forces P, Q, S, T in magnitude and direction, but not position.

Take any point o inside $abcde$ and join oa, ob, oc, od, and oe.

The force P may be replaced by forces $\overline{ao}+\overline{ob}$.

The force Q may be replaced by forces $\overline{bo}+\overline{oc}$, and so on for S and T.

Now take a point A (fig. 3) on the line of action of P and draw AE and AB parallel to ao and ob respectively, and let AB meet the force Q

at B. From B draw BC parallel to *oc* to meet the force *S* at C. From C draw CD parallel to *od* to meet the force *T* at D, and from D draw DE parallel to *oe* to meet AE at E. The point E will be shown to lie on the resultant of the forces *P*, *Q*, *S*, *T*.

In fig. 3 *P* is replaced by $\overline{ao}+\overline{ob}$, *Q* by $\overline{bo}+\overline{oc}$, *S* by $\overline{co}$ and $\overline{od}$, and *T* by $\overline{do}$ and $\overline{oe}$.

The forces in the lines AB, BC, and CD are equal and opposite and so are in equilibrium, and we are left with only the two forces $\overline{oe}$ in ED and $\overline{ao}$ in EA. These two forces meet at E which is therefore a point on the resultant *R* of the given forces.

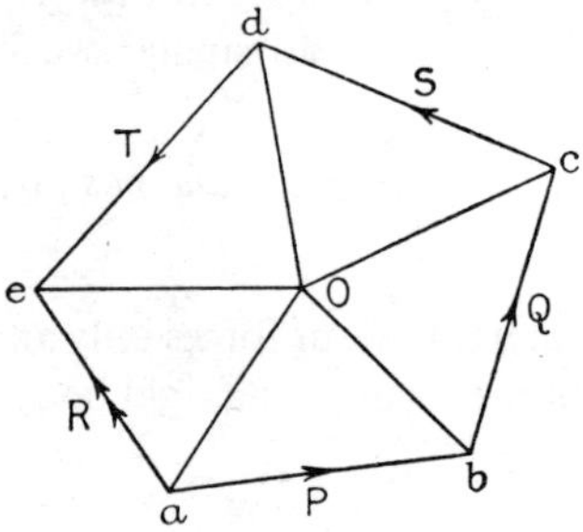

Fig. 2.—Force polygon

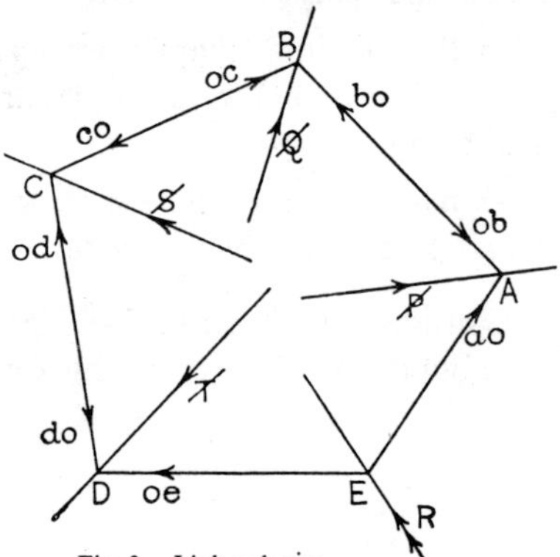

Fig. 3.—Link polygon

Further, from the triangle *aoe* in fig. 2 $\overline{ao}+\overline{oe} = \overline{ae}$, i.e. $\overline{ae}$ represents *R* in magnitude and direction.

So the resultant of *P*, *Q*, *S*, *T* is a force given in magnitude and direction by the closing line in the force polygon (fig. 2) acting through the point E in fig. 3, called the *link* or *funicular polygon*, and E is the point at which the funicular polygon closes.

If the link polygon were made of strings it would be kept in equilibrium by the forces *P*, *Q*, *S*, *T*, and *R* reversed, hence the name ' funicular ' polygon.

2. Conditions for equilibrium

If the points *a* and *e* in the force polygon coincide, then clearly $\overline{ae} = R =$ zero. This means there is no resultant force; it does not necessarily mean the forces are in equilibrium. It does mean that *ao* and *oe* coincide, and therefore in the funicular polygon (fig. 3) AE and DE will be parallel and the forces $\overline{ao}$ and $\overline{oe}$ along them will be equal and opposite, i.e. they may form a couple. They will unless

DEA is a straight line, i.e. unless the funicular polygon closes. Hence for the given forces P, Q, S, T to be in equilibrium *both the force polygon and link polygon must close.*

3. Stresses in the members of a framework

If we have a closed figure of rods, freely hinged at their ends and acted upon by a set of forces in equilibrium which act at the joints, we can adapt the previous paragraph to find the forces in the rods.

Suppose P, Q, S, T, R are a set of forces which keep the polygon of jointed rods shown in equilibrium. Let T_1, T_2, T_3, T_4, T_5 be the actions in the rods.

Draw the force polygon *abcde* for the forces P, Q, S, T, R. This must close for they are in equilibrium.

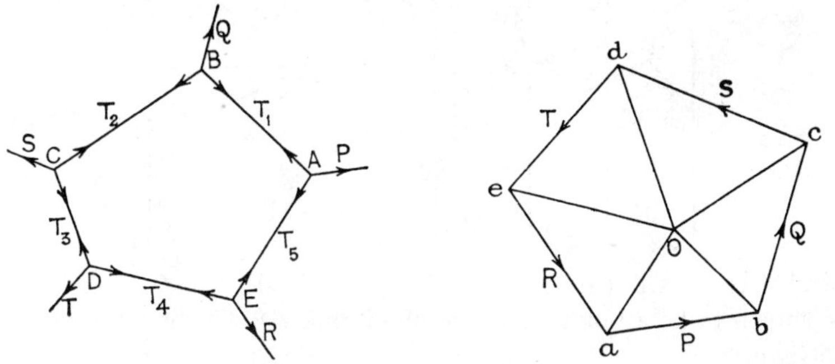

Through a draw ao parallel to AE and through b draw bo parallel to AB. Let ao and bo meet at o. Join oc, od, oe.

The triangle boa has its sides parallel to the forces P, T_1, and T_5 which act at A, and so bo represents T_1 and ao represents T_5 on the scale in which P is represented by ab. In the triangle obc the sides ob and bc represent T_1 and Q, two of the three forces which act at B, and hence co represents the third force T_2 acting at B.

In the same way do represents T_3 and eo represents T_4. Hence oa, ob, oc, od, oe represent the forces along AE, AB, BC, CD, DE in magnitude and direction.

The senses of the forces can be found from the triangles of forces, e.g. consider the triangle aob.

$$P+\overline{bo}+\overline{oa} = 0$$

These forces act at A. Hence T_1 at A acts in the direction AB and T_5 at A acts in the direction AE and similarly for the senses of T_2, T_3, T_4.

4. Parallel forces

When the forces are all parallel the method is essentially the same as that just described.

Consider four parallel forces P, Q, S, T shown on the left. The force polygon is a straight line. P is represented by $\overline{ab}$, Q by $\overline{bc}$, S by $\overline{cd}$, and T by $\overline{de}$. We shall show that the resultant R is represented by $\overline{ae}$.

Take any point o and join o to a, b, c, d, e. Choose any point A on the force P and draw AE parallel to oa, AB parallel to ob, BC

Link polygon

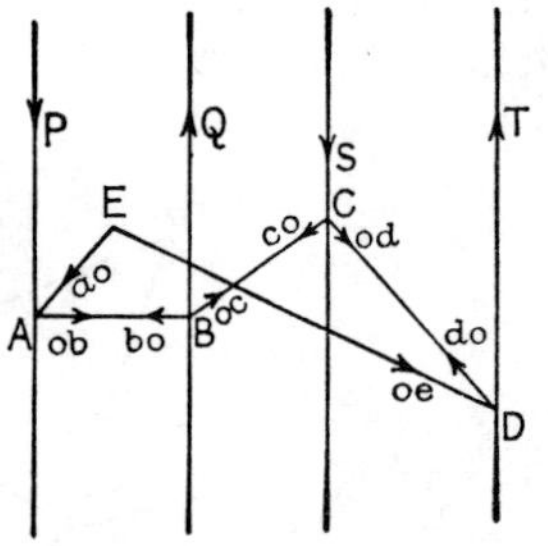

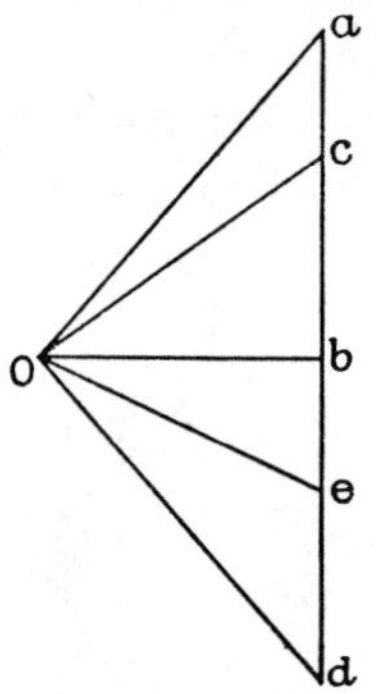

Force polygon

parallel to oc, CD parallel to od, and DE parallel to oe. Then E will be shown to be a point on the resultant and ABCDE is the funicular polygon.

From the triangles of forces such as oab, etc., P may be replaced by $\overline{ao}$ along EA$+\overline{ob}$ along AB. Q may be replaced by $\overline{bo}$ along BA$+\overline{oc}$ along BC. S may be replaced by $\overline{co}$ along CB$+\overline{od}$ along CD. T may be replaced by $\overline{do}$ along DC$+\overline{oe}$ along DE. Ignoring the equal and opposite pairs of forces in AB, BC, and CD, we are left with two forces only, namely $\overline{ao}$ along EA and $\overline{oe}$ along DE. The lines of action of these forces meet at E and so E is a point on the resultant of the given forces. Also from the triangle aoe, $\overline{ao}+\overline{oe} = \overline{ae}$. Therefore $\overline{ae}$ gives the magnitude and direction of the resultant.

The conditions for equilibrium are as before:

(i) a and e must coincide, i.e. *the force polygon closes.*

(ii) AED must be a straight line. i.e. *the link polygon closes.*

The previous paragraph clearly applies to finding the resultant of a number of weights. We can adapt it to find the force at the supports required to hold a loaded beam or structure.

Consider a beam, loaded as shown and supported at its ends. We have supposed the beam to be 40 cm long and that it has a load of 4 N at the centre—this can include its weight—and loads of 2 N and 3 N placed 8 cm and 10 cm from the ends.

Draw *abcd* a vertical straight line so that $ab = 2$ units, *bc* represents 4 units, and *cd* 3 units. Take any point *o* and join *oa*, *ob*, *oc*, and *od*.

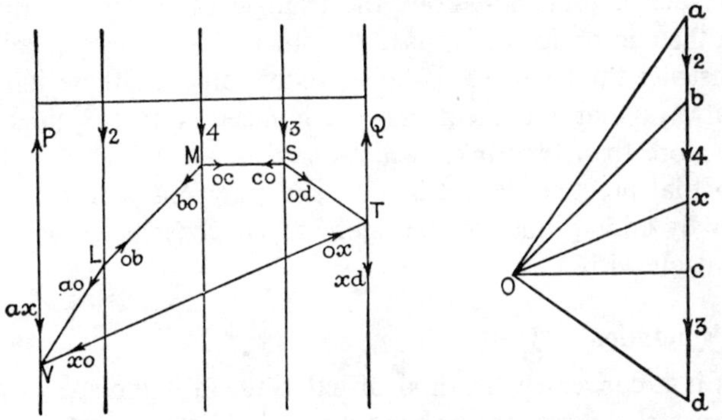

Choose a point L on the line of action of the 2-cwt force, and draw LV parallel to *oa* and LM parallel to *ob* to meet the lines of action of *P* and 4 at V and M. Draw MS parallel to *oc* to meet the line of action of 3 at S and ST parallel to *od* to meet *Q* at T. Join VT and draw a line through *o* parallel to VT to cut *ad* at *x*. Then *ax* and *xd* are proportional to *P* and *Q*.

From the triangles of force *oab*, *obc*, and *ocd* the weights 2, 3, 4 may be replaced as shown by *ao* along LV, *ob* and *bo* along LM, *oc* and *co* along MS, and *od* along OT. These reduce to *ao* along LV and *od* along ST.

From the triangle *oax*, $\overline{ao} = \overline{ax} + \overline{xo}$

From the triangle *odx*, $\overline{od} = \overline{xd} + \overline{ox}$

Add these and hence $\overline{ao} + \overline{od} = \overline{ax} + \overline{xd}$ both vertically downwards. And so $\underline{P = \overline{xa}}$ and $\underline{Q = \overline{dx}}$ for we have replaced the given system of forces by forces in vertical lines through the points of support.

5. Frameworks

By a framework we mean a number of rods jointed together at their ends. We shall assume that the joints are all smooth, that the rods are all light, and that any external forces acting on the framework act only at joints. We shall also assume that the external forces are in equilibrium and that the framework is stiff (i.e. it is not deformed by the external forces acting upon it), and that it is not over-stiff, which means that it has just sufficient bars to prevent deformity.

It is also necessary for this method that at one of the joints where an external force is applied only two bars are jointed together. We start at such a joint and draw the triangle of forces for this joint. We can then go on to the joints at the other ends of these members and can construct the triangles or polygons of force for these joints, provided always that the members which meet at these joints do not involve more than two unknown stresses.

In actual practice these triangles and polygons of force are fitted together by adding lines to the figure so far constructed as each new joint is dealt with.

6. Bow's notation

This is a convenient method of lettering. In it the spaces between any two external forces are each given a capital letter, as are the triangles or polygons between all members of the framework. If a line separates two areas A and B, then the corresponding line in the force diagram is labelled AB.

7. Struts and ties

Members of a framework which are in a state of thrust, i.e. those members which push the joint are called *struts*. Such members must be rods. Members of a framework which are in a state of tension are called *ties*. Such members could be strings. We shall indicate struts with a + sign and ties with a − sign.

These last points should be clear from a few easy examples.

Example 1.—The figure on the left shows a framework in the shape of a kite with a diagonal to keep it rigid. It is acted upon by equal and opposite forces of 10 N as shown.

The spaces between the external forces are lettered A and B, and the spaces between the members of the frame are lettered C and D. The external force diagram is the straight line AB, the forces being $\overline{AB}$ and $\overline{BA}$.

Starting at the vertex 1 we can draw the triangle of forces for the three forces there. It is the triangle ABC. The forces at the joint 1 are given by $\overline{AB}$, $\overline{BC}$, and $\overline{CA}$ in the figure on the left and so act in the directions shown in the figure on the right at the joint 1.

Next for the joint 2, the triangle of forces is BAD. This triangle gives the size and directions of the forces at joint 2. The forces there act in the directions $\overline{BA}$, $\overline{AD}$, $\overline{BD}$ as shown.

Next for joint 3, the triangle of forces is BCD. Two lines of this triangle, namely BC and BD, are already drawn, and so we join CD. This checks our previous work for it should be parallel to 3, 4, i.e. horizontal. In the triangle CBD we know that at joint 3 $\overline{BC}$ is up; it is opposite to the force at joint 1 in this line, and so the other two forces there are $\overline{CD}$ and $\overline{BD}$ as shown.

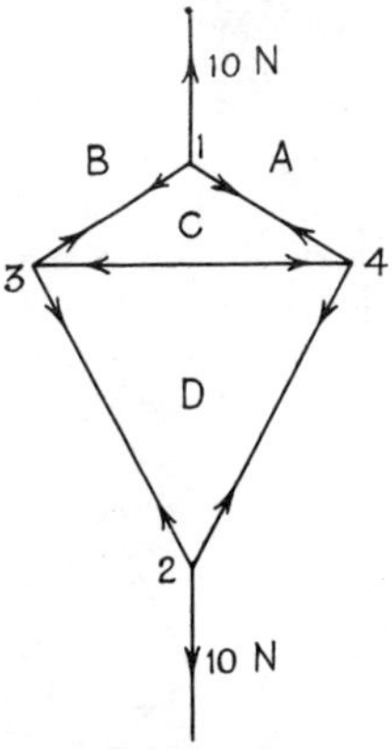

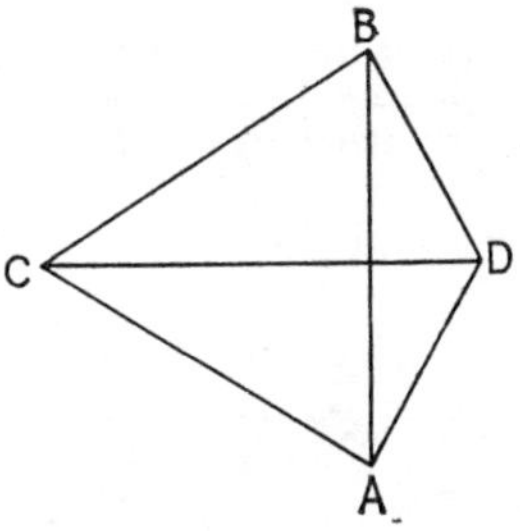

The triangle of forces for joint 4 is already drawn; it is ACD. Starting with the direction of $\overline{AC}$ the other two are $\overline{CD}$ and $\overline{DA}$, and these should be opposite to the arrows at the other ends of these members, which gives a further check.

By measurement:

1, 3 and 1, 4 are ties. The tension in them is 5 N.
2, 3 and 2, 4 are ties. The tension in them is 2·9 N.
3, 4 is a strut. The thrust in it is 5·8 N.

It is not advisable to put arrows on the force diagram for each line will have two arrows and this makes checking very difficult.

Example 2.—The framework shown is loaded with 1000 N and 500 N and supported by vertical supports *P* and *Q*. To find the actions in the members of the framework.

Letter the spaces ABCDEF. Draw the triangle of forces ABF for the corner *l*. The forces must act round this triangle in the directions AB, BF, FA, hence the arrows at *l*. Next go to the corner *m*, where there is now only one unknown letter E, for BC is vertical and equal to ½ AB. Draw CE parallel to *nm* and FE parallel to *qm*. This gives E. The directions of the forces at *m* are BC, CE, EF, FB round the quadrilateral BCEFB, hence the arrows at *m*. Next to the corner *n*, where D is the only unknown. Draw ED horizontal and the triangle of forces for *n* is ECD; the directions of the forces there are EC, CD, and DE, hence the arrows at *n*. Finally the quadrilateral of forces at *q* is EDAFE, which gives the directions and hence the arrows at *q*.

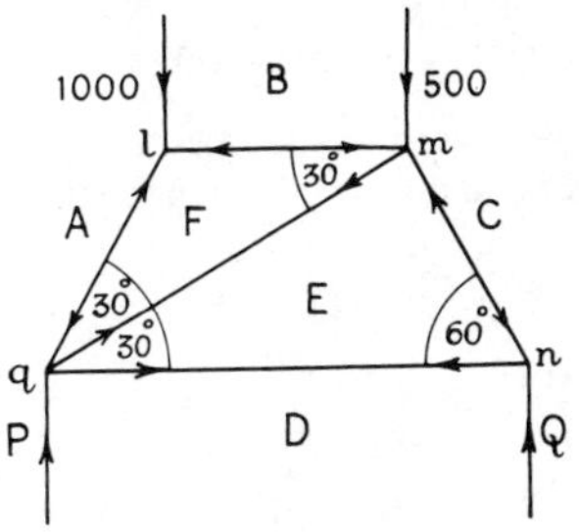

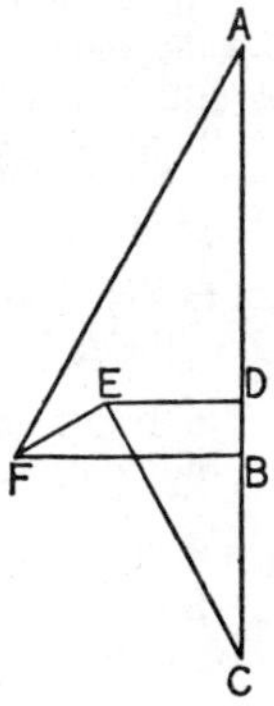

Notice that in this example we have found the external forces *P* and *Q* in the course of the work. Often it is necessary to find them first, either by taking moments or by drawing as we did for weights on a beam.

By measurement:

$P = 900$ N,	$Q = 600$ N,	$lm + 550$ N,
$mn + 700$ N,	$qn - 370$ N,	$qm - 230$ N,
$ql + 1200$ N	(+ indicates strut,	− indicates tie).

Example 3.—The figure shows a Warren girder in which all the angles are 60° loaded with 1 kN and 2 kN and supported by vertical forces *P* and *Q* as shown. To find the forces in the members. (1 kN = 1 kilonewton = 1000 N.)

Draw ABC, AB = 1 cm, BC = 2 cm. Take any point O and join OA, OB, OC. Take a point X on the line of action of the 1kN. Draw XW parallel to OA, XY parallel

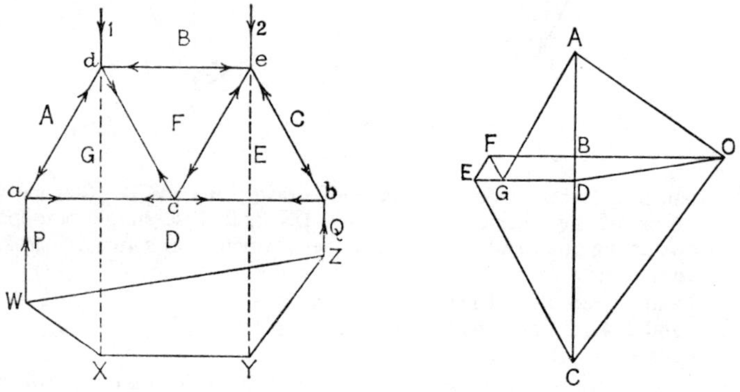

to OB, YZ parallel to OC. Join WZ and draw OD parallel to ZW. This gives the point D and so the values of *P* and *Q*. $P = 1\frac{1}{4}$, $Q = 1\frac{3}{4}$ kN. They could have been found by taking moments.

The force triangles or polygons are drawn for the joints in the order *a, b, c, d, e.* For the joint *d* all the points will be already marked; a check on the work is obtained by noting that FB is horizontal.

By measurement:

$$ac - \cdot 7,\ ad + 1\cdot 44,\ dc - \cdot 3,\ ce + \cdot 3,\ eb + 2\cdot 0,\ de + \cdot 9,\ cb + 1\cdot 0$$

all in kN. The calculation of these forces is straightforward trigonometry.

In the above we could have found the external forces *P* and *Q* by taking moments:

Moments about *a* give $Q \times 4 = 2 \times 3 + 1 \times 1, \quad \therefore\ Q = 1\frac{3}{4}$ kN
Moments about *b* give $P \times 4 = 1 \times 3 + 2 \times 1, \quad \therefore\ P = 1\frac{1}{4}$ kN

These check for $P + Q = 3$ kN by resolving vertically.

Example 4.—In the frame shown, ABD and CBD are equilateral triangles, and AB = AE. The frame is hinged to a vertical wall at A, E. BC is horizontal, and a mass of weight 100 N hangs from C.

To find the forces at A, E and those in the bars.

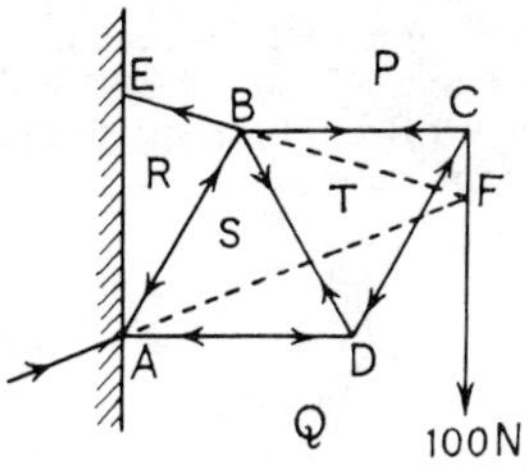

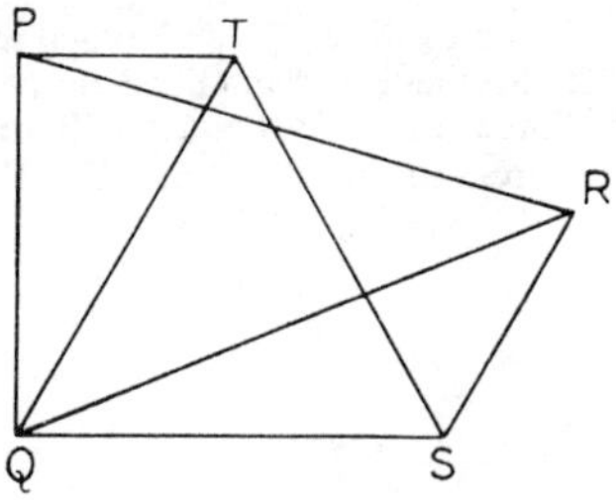

We may regard the force in EB as an external force applied to the frame ABCD, and we begin by considering the external forces at A, B and C. Being three in number and not parallel, they must be concurrent, and we produce EB to meet the vertical through C at F. Then the line of the force at A passes through F, and this line is continued outside the frame.

Using Bow's notation we place the letters P, Q, R outside the frame, and S, T inside. Note that the dotted lines are not bars of the frame, and that the line of the force at A is drawn outside the frame.

The triangle of forces, PQR, can now be drawn, giving the forces at A, B, in the senses shown by the arrows, and the force diagram is completed, considering first the triangle QRS or PQT.

It is found that the reaction at A is 161 N, at 21° 45′ to the horizontal, and the forces in the rods are: EB, − 155; AB, + 69; BC, − 58; CD, + 115; DA, + 115; BD, − 115, all in newtons.

EXERCISE 15A

1. ABCDEF is a regular hexagon; forces of 1, 2, 3, 4, 5 N act along AB, AC, DA, EA, and AF respectively. Find graphically the magnitude and direction of their resultant.

2. ABC is an equilateral triangle; forces of 2, 3, 5 N act along AB, BC, and CA respectively. Find the magnitude, direction, and position of their resultant graphically. If the side of the equilateral triangle is 1 m, how far from C does the resultant cut AC produced?

3. ABCD are four consecutive vertices of a regular hexagon. Forces of 6, 4, and 7 N act along AB, BC, and DC. If the side of the hexagon is 20 cm, find graphically the magnitude, direction, and position of the resultant of the three forces. Give the distance from B at which the resultant cuts BC.

4. Three light strings are knotted together to form an equilateral triangle ABC. A weight of 10 N is hung from A, and the triangle is hung by strings BX and CY so that angle CBX = angle BCY = 135°. Calculate the tensions in AB, BC, and BX. Give also a graphical solution.

5. Five light rods are jointed together to form a kite ABCD with DB the other bar. The joints are smooth, AD = AB and CD = CB. Angle ADB = 45° and angle BDC = 30°. The kite is hung from A and from C a weight of 200 N is hung. Find graphically the forces in the rods AB, BC, and BD.

6. Five equal light rods are freely jointed at their ends to form a regular pentagon ABCDE. Stiffness is maintained by two other light rods AD and AC. Two opposite forces each of 10 N act at B and E parallel to EB and BE respectively. Find the forces in the rods AB, BC, CD, and AC graphically.

7. The five rods AB, BC, CA, AD, DB are jointed freely together. Forces of 100 N each act at C and D as shown. Find graphically the forces in the rods AB, BC, and CA.

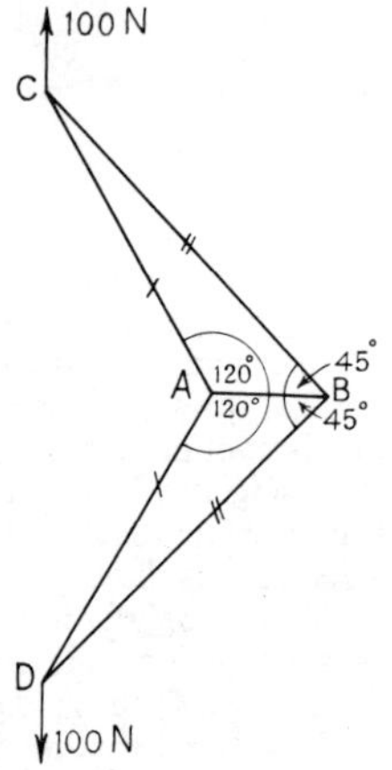

8. ABCDEF is a regular hexagon. Forces of p, q, r and 20 N act along AC, AF, DE, and EC. If the forces are in equilibrium, find graphically p, q and r.

9. Five light rods are freely jointed together as shown in the figure. A vertical force of 10 N acts at c, a horizontal force p acts at b and a force q acts at a. Find the forces in the rods.

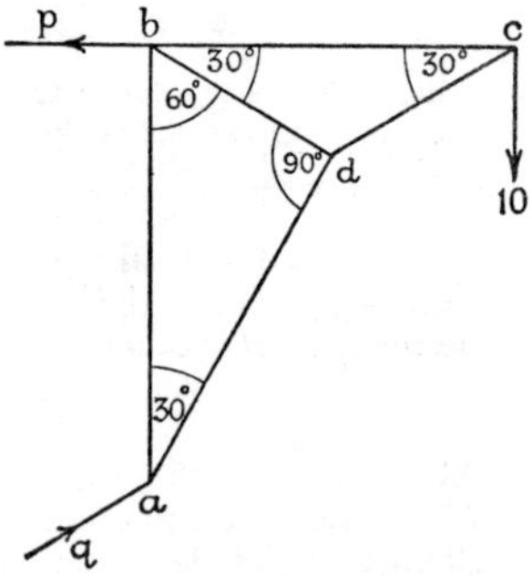

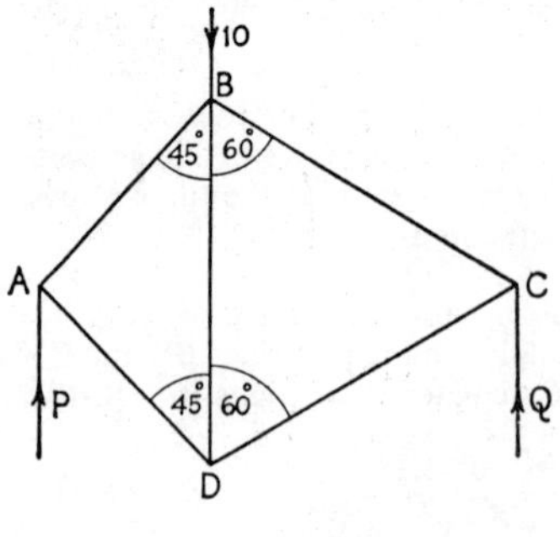

10. Five light rods are jointed together as shown in the figure. A vertical force of 10 N acts at B and vertical forces P and Q act at A and C. If the external forces are in equilibrium calculate P and Q and the forces in the rods.

11. Parallel forces of 1, 2, 3, 4, 5 N are each 30 cm apart and in the same sense; find graphically the distance of the resultant from the 1 N.

12. A uniform beam AB of length 1·2 m and weight 40 N has loads of 20 N at A, 40 N 20 cm from A, and 80 N 20 cm from B. It is supported at B and C, and C is 40 cm from A. Find graphically the reactions at the supports.

13. A uniform plank AB of length 10 m weighs 200 N and has a weight of 100 N attached to it at B. The plank passes under a smooth peg C and over a smooth peg D. The line CD is horizontal, AC is 3 m, and CD is 1 m. Calculate the reaction between each of the pegs and the plank. Check graphically.

Find also the least weight to be attached to the plank for equilibrium to be broken. Where should this weight be attached?

14. A uniform plank of weight 160 N and length 2·4 m carries a load of 48 N at a point 90 cm from one end. The plank is supported at its ends. Find graphically the reactions at the supports.

15. ABCDE is a light rod supported at B and D. AB = DE = 20 cm, BC = 30 cm, and CD = 50 cm. It carries loads of 16, 12, 8 N at A, C, and E respectively. Find graphically the reactions at the supports.

16. The figure opposite shows a framework of 4 light rods freely hinged and hinged to a vertical wall at *a* and *b*, At *c* is a load of 1 kN. Find graphically the reactions at *a* and *b* and the forces in the rods.

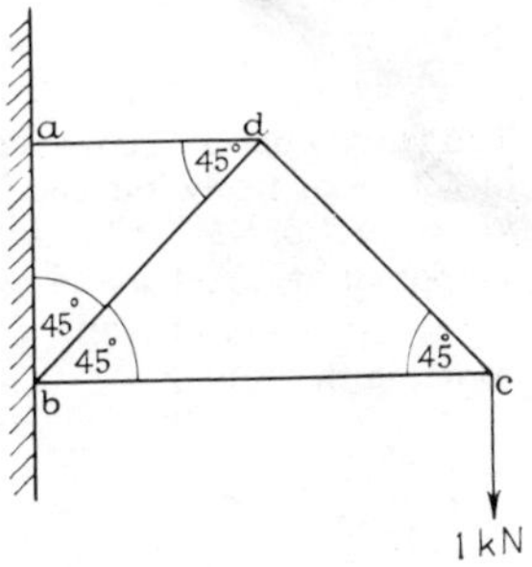

17. In the figure opposite the angles are all 45° or 90°. The framework loaded as shown is fastened to a vertical wall by smooth pivots at A and B. Find the reactions at A and B and the forces in the bars.

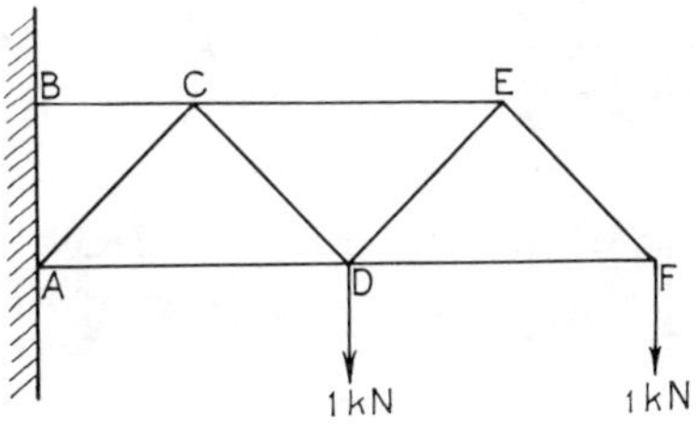

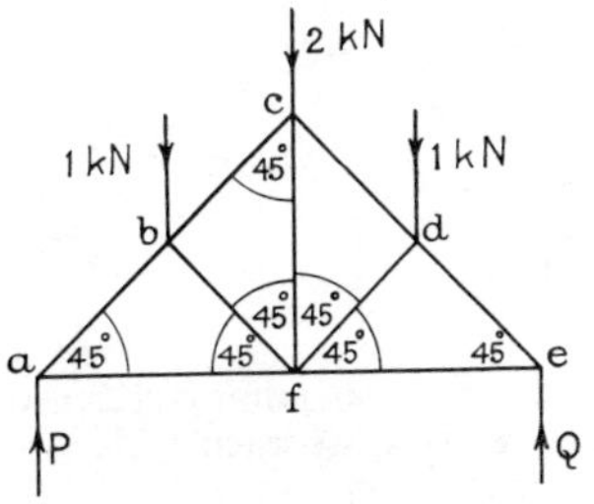

18. In the smoothly jointed framework opposite, loaded and supported as shown, calculate *P* and *Q* and find graphically the forces in the rods.

19. The symmetrical framework of light rods freely hinged is loaded as shown opposite and is supported at *a* and *b*. Find graphically the forces in the rods.

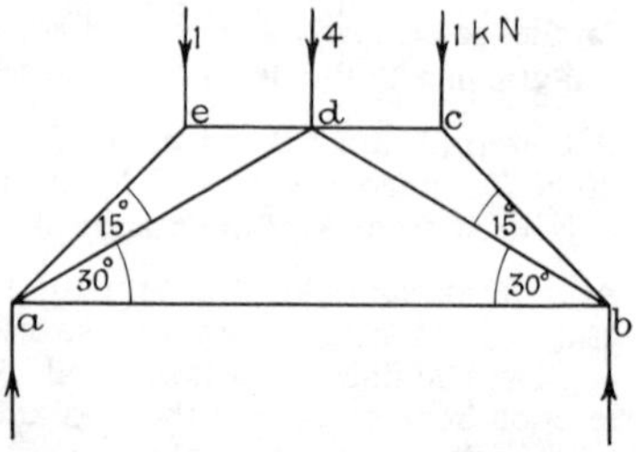

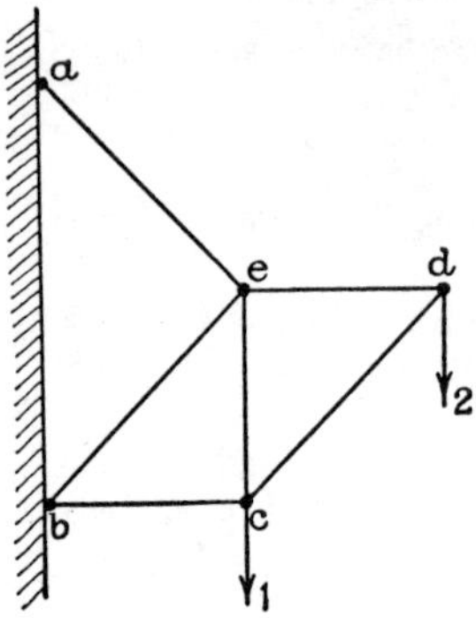

20. In the figure opposite all the angles are 90° or 45°. It represents a pin-jointed framework fastened to a vertical wall at *a* and *b* and loaded with 1 kN and 2 kN at *c* and *d*. Find the forces in the rods.

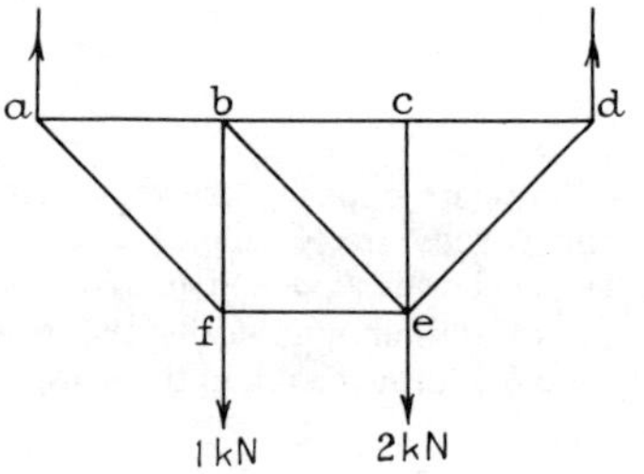

21. The figure opposite shows a framework of light rods freely jointed. The angles are all 90° or 45°. The framework is supported at *a* and *d* and loaded with 1 kN and 2 kN at *f* and *e*. Find the forces in the bars.

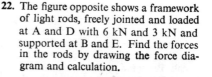

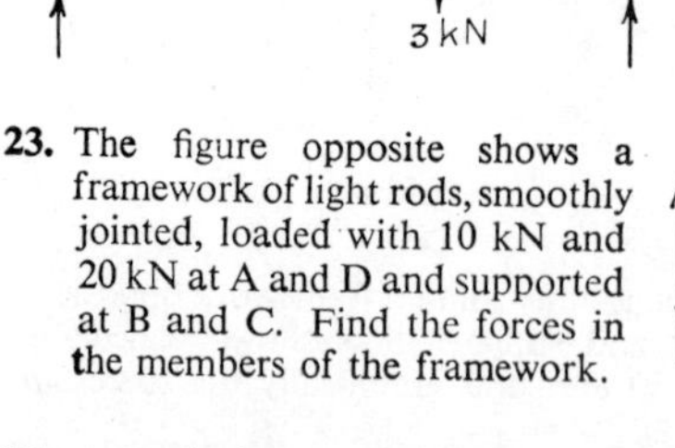

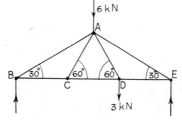

22. The figure opposite shows a framework of light rods, freely jointed and loaded at A and D with 6 kN and 3 kN and supported at B and E. Find the forces in the rods by drawing the force diagram and calculation.

23. The figure opposite shows a framework of light rods, smoothly jointed, loaded with 10 kN and 20 kN at A and D and supported at B and C. Find the forces in the members of the framework.

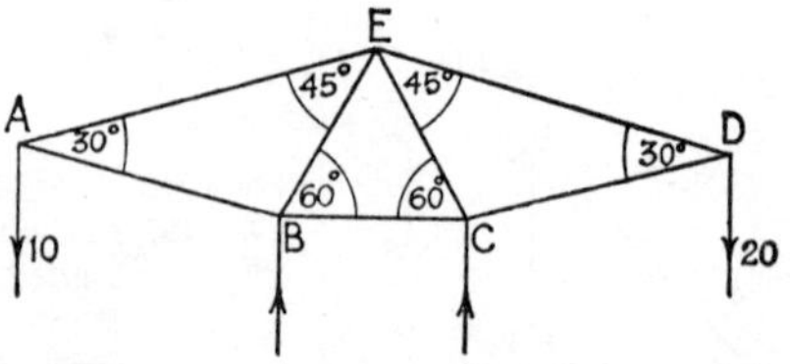

24. A regular hexagon of light rods ABCDEF connected by smooth joints is stiffened by light rods FB, FC, FD, and is suspended from A. Weights, each 1 kN, are attached to B, C, D, E, F. Find graphically the force in each rod.

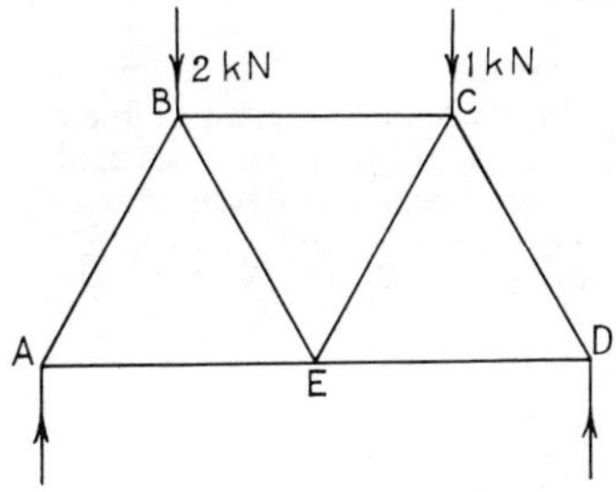

25. The figure opposite shows a Warren girder of light smoothly jointed bars loaded and supported as shown. All the angles are 60°. Find the forces in the bars.

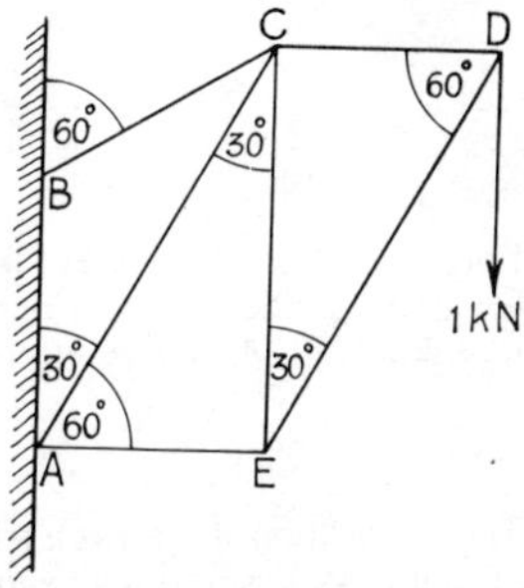

26. The diagram opposite shows a pin-jointed framework of light rods joined to a vertical wall at A and B and loaded with 1 kN at D. Find graphically the forces in the rods.

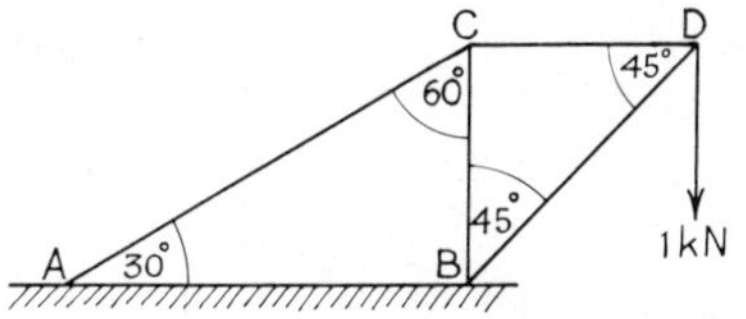

27. The figure opposite shows a crane composed of 4 freely jointed light rods AC, CD, CB, and BD. It carries a load of 1 kN at D and is fixed to horizontal ground at A and B. Sketch the force diagram and hence calculate the forces in the rods.

EXERCISE 15B

1. ABC is an equilateral triangle. Forces of 1, 3, 2 N act along AB, BC, and AC. Find graphically the magnitude and direction of their resultant. Give the angle it makes with AC.

2. ABCD is a square of side 1 m. A force of 6 N acts at A along DA, a force of $\sqrt{2}$ N at C along AC, and a force of 4 N acts at D along the side DC. Find graphically the size of the resultant, the angle at which it cuts AB, and the distance from D at which it cuts DA.

3. In the figure opposite ABC is a triangular framework of 3 light rods smoothly jointed. AB is horizontal. The framework carries a load of 100 N at C, W N at A, and is supported at B by a smooth hinge. Draw the force diagram and from it calculate the forces in the rods, the value of W, and the reaction at the hinge B.

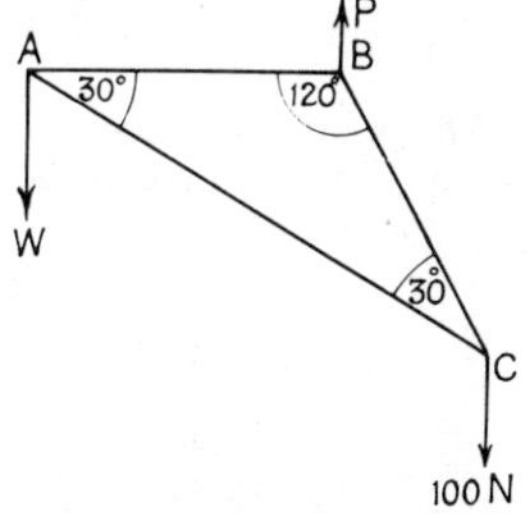

4. ABCDEF is a regular hexagon. Forces of 1, 2, 1, 2, 1, 2 N act along AB, CB, CD, ED, EF, AF. Draw the force polygon. Would this system of forces keep a rigid body in equilibrium? State the moment of the equivalent couple if AB = 2 m.

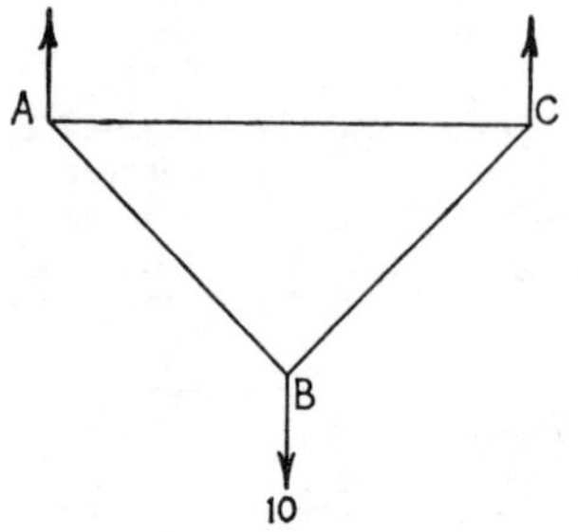

5. In the figure on the left ABC is a framework of 3 light rods freely jointed at their ends. The angles *A* and *C* are each 45° and a load of 10 N is attached to B; the framework is supported at A and C with AC horizontal. Find the forces in AB and AC.

6. Forces of 1, 2, 3, 2 N act along the sides AB, BC, CD, and DE of a regular hexagon of side 20 cm. Find graphically the size of the resultant and show that it cuts AB produced 28 cm from B.

7. The figure opposite shows a triangular framework of light rods smoothly jointed. It is loaded with 1 kN at B and is supported at A and C. Find graphically the forces in the bars and the reactions at A and C.

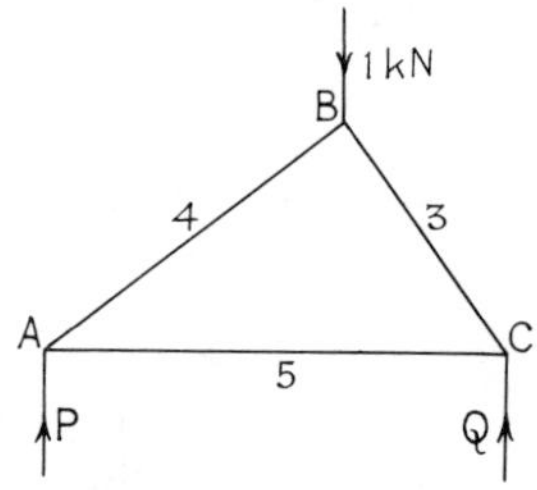

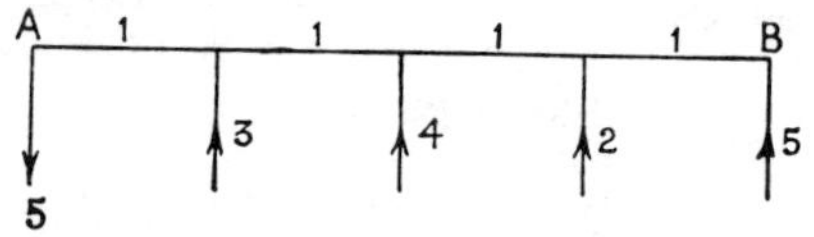

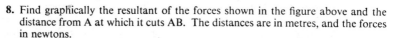

8. Find graphically the resultant of the forces shown in the figure above and the distance from A at which it cuts AB. The distances are in metres, and the forces in newtons.

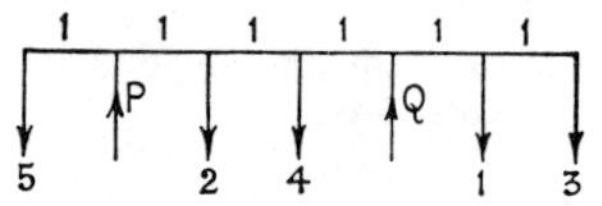

9. The beam shown above is loaded as shown. Find the reactions *P* and *Q* graphically. The loads are in kN and include the weight of the beam. The distances are in metres.

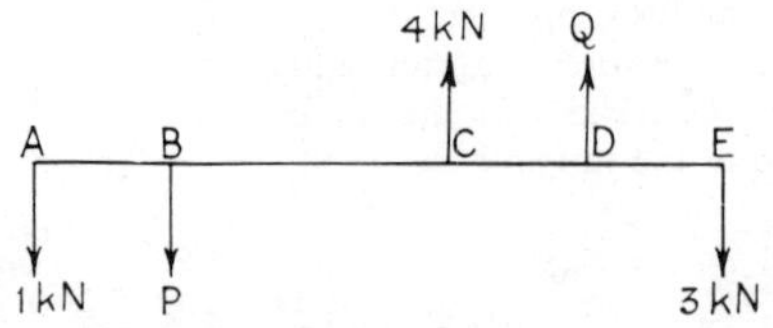

10. The light beam AE shown above passes under a smooth peg B and over a smooth peg D. It is loaded as shown. Find graphically the reactions of the pegs on the beam if AB = 1 m, BC = 2 m, CD = 1 m, and DE = 1 m.

11. In the figure on the right ABCD is a framework of light rods smoothly jointed. There is a load of 1 kN at B and the framework, which is made rigid by a light rod BD, is supported at A and C. Find the forces in the rods.

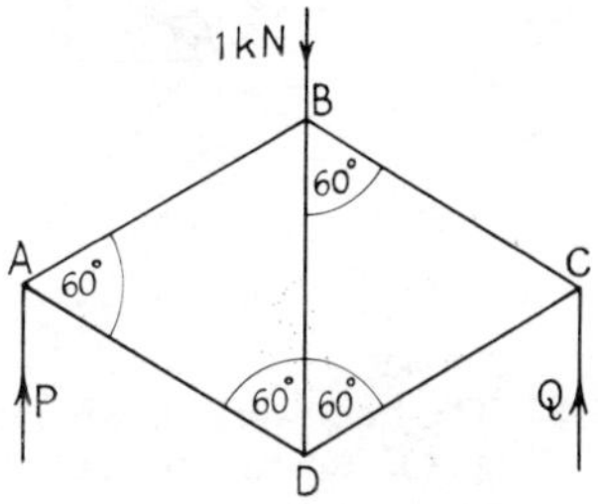

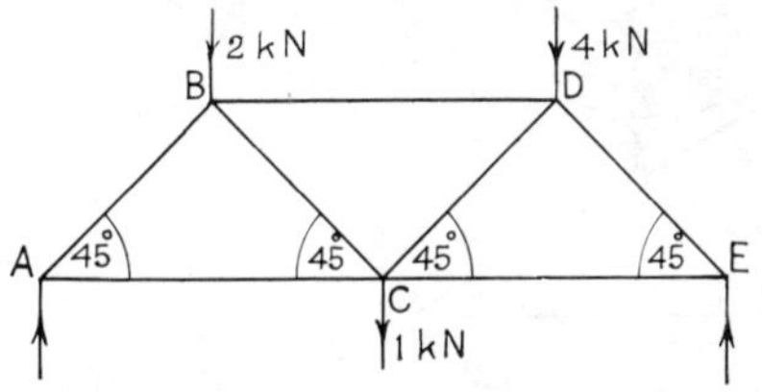

12. The figure on the left shows a Warren girder supported at A and E. Sketch the force diagram and from it calculate the forces in the bars which are all light and smoothly jointed.

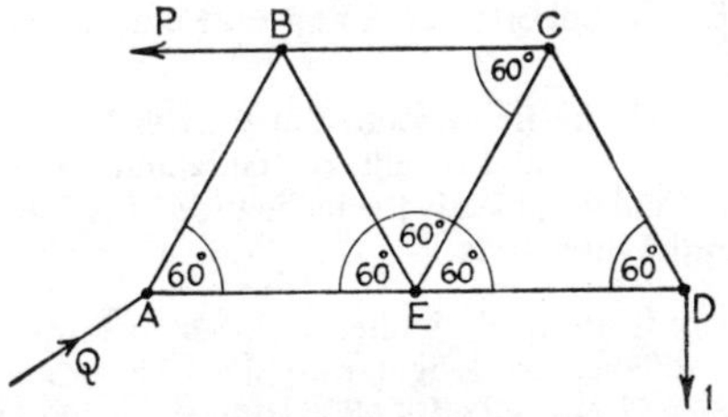

13. The framework of light rods which are freely jointed is loaded with 1 kN at D and supported at B by a horizontal rod and at A by a smooth pivot. Draw the force diagram and so find P and Q and the forces in the rods.

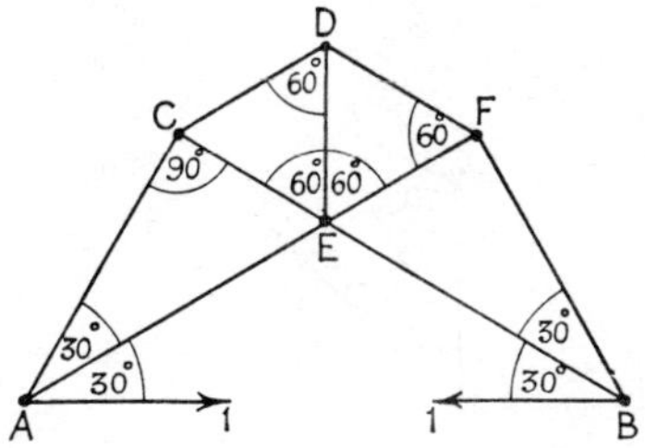

14. The framework of light freely jointed rods has external forces each 1 kN applied horizontally at A and B. Draw the force diagram and find the forces in AC CD, CE, DE, and AE.

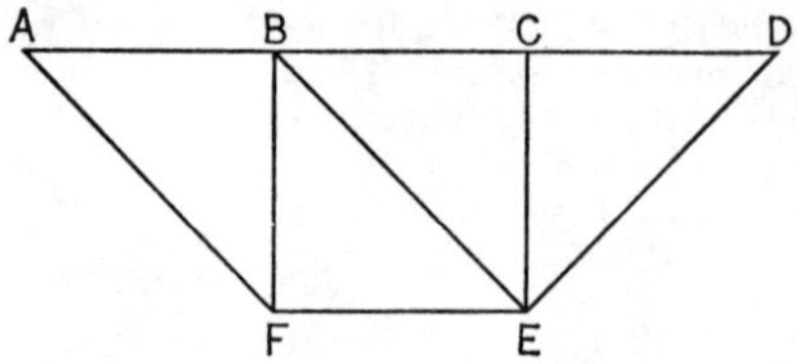

15. The framework shown above has its angles 45° or 90°. It is supported at A and D and carries loads of 2 kN and 1 kN at E and F. Find the forces in the bars.

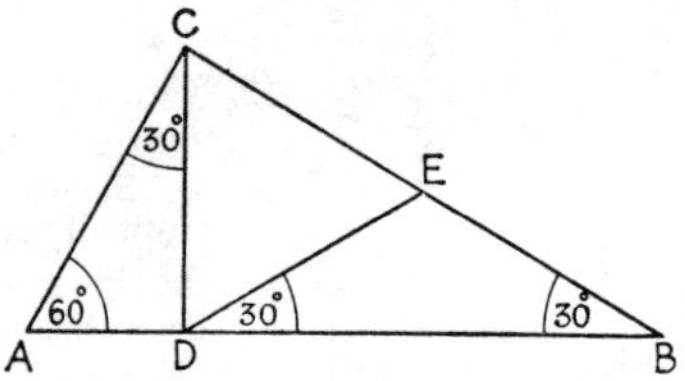

16. The framework of smoothly jointed light rods shown is loaded with 2 kN at C and 1 kN at E and supported at A and B. Find the forces in the members of the framework.

17. Three equal light rods are freely jointed at their ends. The framework is suspended from one vertex A and weights of 100 N and 200 N are attached to the vertices B and C. Find graphically the inclination of AB to the vertical and find also the forces in the three rods.

18. Four equal light rods are freely jointed together to form a square framework ABCD which is stiffened by a light rod BD. The framework is suspended from A and weights of 1, 2, 3 N are hung from B, C, and D respectively. Find graphically the inclination of AD to the vertical and the forces in the rods.

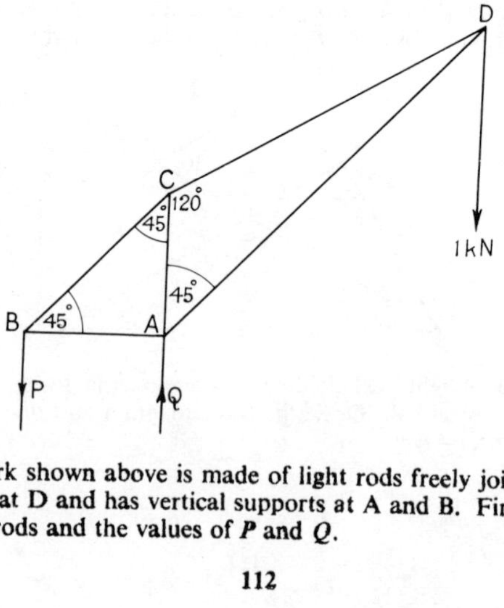

19. The framework shown above is made of light rods freely jointed. It carries a load of 1 kN at D and has vertical supports at A and B. Find graphically the forces in the rods and the values of P and Q.

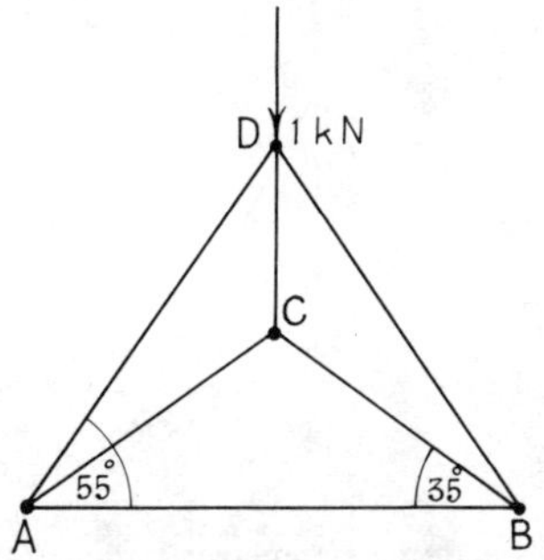

20. The figure above shows a framework of 5 light freely jointed rods AD, DB, AC, CB and DC. There is a load of 1 kN at D and vertical supports at A and B. The figure is symmetrical and angle DAB = 55° and angle CBA = 35°. Draw the force diagram and so find the forces in AD, AC, and DC.

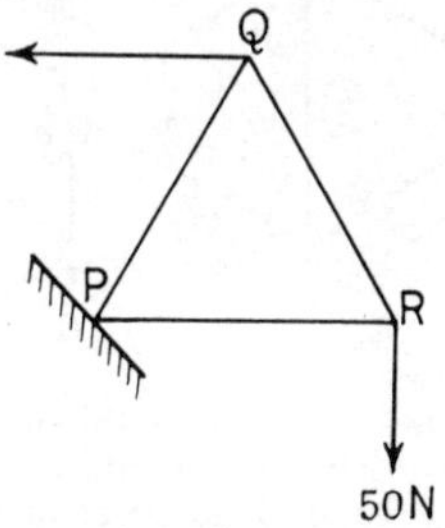

21. The frame shown is an equilateral triangle. It is hinged to a fixed support at P, PR is horizontal, a vertical load of 50 N is applied at R, and the force at Q is horizontal. Find the size and direction of the force at P, and the forces at Q and in the bars of the frame.

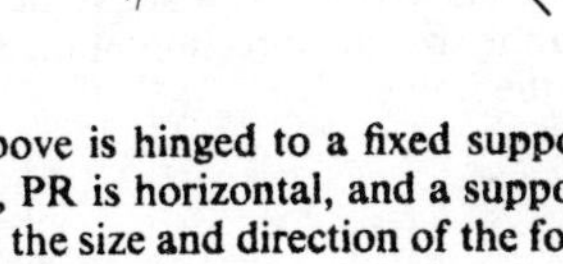

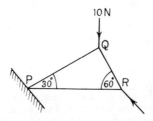

22. The frame shown above is hinged to a fixed support at P. A vertical load of 10 N is applied at Q, PR is horizontal, and a supporting force at R is at 45° to the horizontal. Find the size and direction of the force at P, and the forces at R and in the bars of the frame.

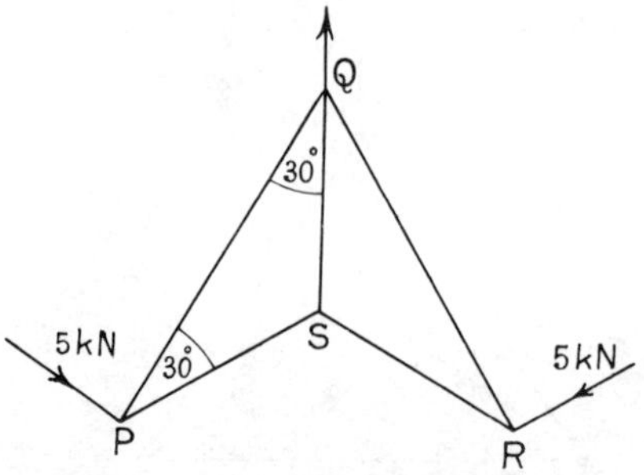

23. The figure above shows a frame symmetrical about QS, which is vertical. The external forces at P, R are at right angles to PQ, RQ, and there is a supporting force at Q. Find this force and the forces in the bars of the frame.

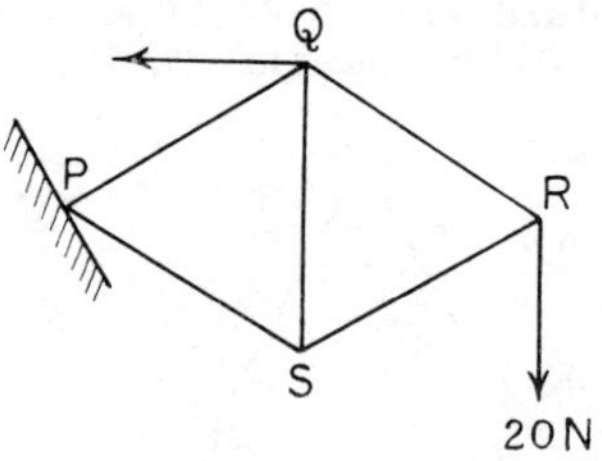

24. In the frame shown, the triangles are equilateral and QS is vertical. P is hinged to a fixed support, the load at R is vertical, and the external force at Q is horizontal. Find the size and direction of the force at P, and the forces at Q and in the bars of the frame.

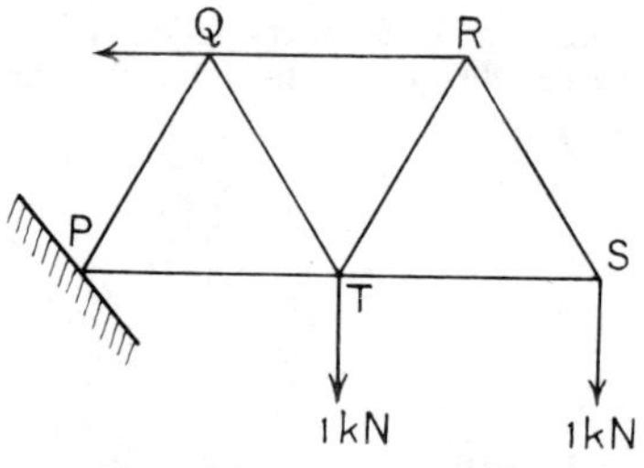

25. In the frame shown, the triangles are equilateral and PTS is horizontal. P is hinged to a fixed support, the loads at T, S are vertical, and the external force at Q is horizontal. Find the size and direction of the force at P, and the forces at Q and in the bars of the frame.

8

The Laws of Motion. Units and Dimensions

The science of Dynamics is based on Newton's three Laws of Motion. The proof of these, as of all scientific laws, is that predictions made by their use are fulfilled with extreme accuracy. Only in a few very special cases need Newton's Laws be modified in accordance with the Theory of Relativity.

1. First Law

The First Law states that

every body continues in a state of rest or uniform motion in a straight line unless it is compelled by forces to change that state.

A force may be thought of as an external agency such as the pressure of another body, the pull of a string, or the influence of the Earth's attraction, which we call *weight.* This tendency of a body to continue in a state of rest or uniform motion is called *inertia*, and its measure is the *mass* of the body. The description of mass given in chap. 4 as ' quantity of matter ', does not provide any method of comparing masses, but since the force of the Earth's attraction for a body is, by Newton's Law of Universal Gravitation, directly proportional to its mass, the masses of bodies are in the same ratio as their weights, if these are measured at the same place.

It cannot, however, be too emphatically stated that *mass and weight are fundamentally different.* The mass of a body is a scalar quantity, and absolutely constant, unless the composition of the body is changed; its weight is a vector, and slightly variable, as explained in section 3 of this chapter.

2. Second Law

From the First Law we see that the tendency of a force is to change the state of motion of a body; that is, to give it an acceleration. Newton's Second Law states that

the rate of change of momentum of a body is proportional to the applied force, and is in the direction of this force.

The *momentum* of a body is defined as the product of its mass and its velocity. It is therefore a vector, having the direction of the velocity. In all cases that we shall consider, the composition of the body will not change, so that its mass will remain constant, and the rate of change of momentum will be given by mass × rate of change of velocity, that is, mass × acceleration. Symbolically,

$$\frac{d}{dt}(mv) = m\frac{dv}{dt} = mf$$

Thus any force P acting on a body of mass m produces an acceleration f in the direction of the force, and $P = kmf$, where k is some constant. If f is the actual acceleration of the body, P must be the resultant of all the forces acting on it, but the equation, being between vectors, implies that the total component force on a body in any direction is equal to k times the component of mass-acceleration in that direction.

3. Units

Units are chosen to reduce the constant k to unity. In the International System of units (SI) the unit mass is 1 kg and the unit acceleration is 1 m/s². The force which gives unit acceleration to unit mass is called a *newton* (N). With these units, the ' equation of motion ' given by Newton's second law may be written as $P = mf$.

The weight of a unit mass is not suitable as a unit of force, because it is variable. The acceleration (g) of a body influenced only by its weight varies by about 0·5% at different points on the Earth's surface, and above that surface varies inversely as the square of the distance from the Earth's centre. It is approximately 9·8 m/s² at the surface, but results found by the use of this number are reliable only to 2 significant figures.

Note that 1000 kg = 1 tonne (t) and 1000 N = 1 kN, so that a force of 1 kN gives a mass of 1 t an acceleration of 1 m/s².

4. Third Law

Newton's Third Law states that

to every action there is an equal and opposite reaction.

This means that the force which one body exerts on another is always accompanied by an equal and opposite force exerted by the second on the first. Examples of this have already been given in statical problems about bodies in contact. Again, when a car moves forward it is because the driving wheels exert a backward force on the road surface. The equal and opposite reaction pushes the car forward, and, if this reaction exceeds the sum of the various resisting forces, the car has a forward acceleration. The weight of a body is no exception to the law, for the attraction of the Earth on a body is accompanied by an equal and opposite attraction exerted by the body on the Earth. This, of course, has no appreciable effect on the motion of such a great mass as the Earth.

In the application of the laws of motion the following points should be observed.

A *particle* is defined as *a piece of matter whose dimensions are small in comparison with other distances involved.* Thus all forces acting on it are concurrent, and the question of its rotation does not arise. Bodies of appreciable size may, however, be regarded as particles if their motion is such that they do not rotate, for it can be proved that the motion of the centre of gravity of a body is precisely found if the whole mass is supposed to be concentrated there, and if the forces on the body are supposed to act at the centre of gravity, in directions parallel to their actual lines of action. Thus no error is involved in treating a body of appreciable size as a particle, provided that it does not rotate. This matter will be more fully considered in Chapter 15.

When bodies are said to be connected by light inextensible strings, it is to be supposed that the string has no mass, and that the tension is the same at all points along it; in particular, that it exerts equal and opposite forces at its ends. Neither these conditions nor the property of being inextensible can of course be realized in practice, but for the present we are simplifying our problems, and are content with answers approximately true.

Another convention is that the tension of a string is unaltered if the string passes over a smooth peg or round a pulley. In practice, no peg is perfectly smooth, and no pulley is without mass, so that, even if its bearings could be frictionless, force would be needed to rotate it. These details also are disregarded for the present.

When two bodies are described as *smooth*, the force between them is taken to be wholly along the common normal at the point of contact. If they are not smooth, and one is moving over the other, limiting friction is supposed to act, opposing the relative motion, and the coefficient of this ‘dynamical friction’ is taken to be equal to that of ‘statical friction’ already discussed.

Example 1.—A train whose total mass, including the engine, is 125 t is travelling on a level track. The resistance to motion (due to friction and air resistance) is 40 N per tonne. The tractive force exerted by the engine is 120 kN. (i) Find the acceleration of the train. (ii) If the mass of the engine is 50 t, find the tension in the coupling between the engine and the rest of the train. (iii) If all the engine's weight is carried by the driving wheels, what must be the coefficient of friction between wheels and track to prevent skidding? (iv) What acceleration will be produced by the same force, against the same resistances, when the train is travelling up a slope of 1 in 98?

(i) Total horizontal forward force on train

$$= \text{tractive force} - \text{resistance}$$
$$= 120{,}000 \text{ N} - 5000 \text{ N}$$
$$= 115{,}000 \text{ N}$$

Mass of train $= 125 \times 1000$ kg

The equation $P = mf$ gives

$$115{,}000 = 125{,}000f$$
$$\therefore f = \frac{115}{125}$$
$$= 0{\cdot}92 \text{ m/s}^2$$

(ii) Let T N be the tension in the coupling. (Note that the unit is specified.)
Total horizontal forward force on rest of train

$$= (T - 40 \times 75) \text{ N}$$

Mass of rest of train $= 75 \times 1000$ kg
Acceleration $= 0{\cdot}92 \text{ m/s}^2$
The equation $P = mf$ now gives

$$T - 40 \times 75 = 75{,}000 \times 0{\cdot}92$$
$$T - 3000 = 69{,}000$$
$$\therefore T = 72 \text{ kN}$$

(iii) Frictional force between wheels and track

$$= \text{tractive force} = 120 \text{ kN}$$

Normal force = weight of engine (since there is no vertical motion)

$$= 50{,}000 \times 9{\cdot}8 \text{ N}$$

$$\therefore \text{Coefficient of friction} \geqslant \frac{120}{50 \times 9{\cdot}8} \text{ (both forces in same units)}$$
$$\geqslant 0{\cdot}245 \text{ (approx.)}$$

(iv) The diagram shows the forces acting on the train. A gradient of 1 in 98, on a railway or a road means a rise of 1 unit vertically for every 98 units, along the track, so that sin α in the diagram is 1/98.

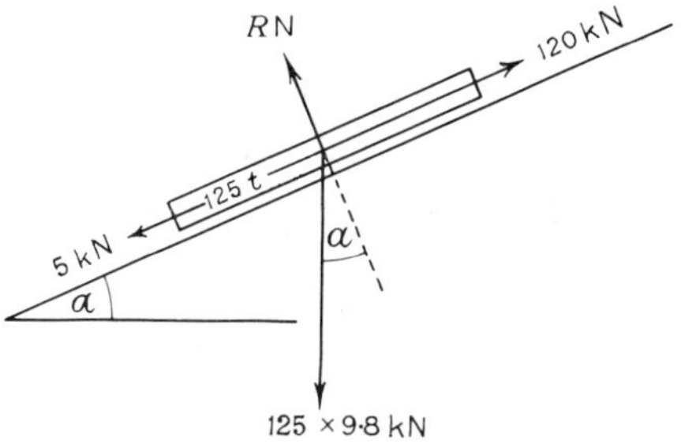

Total force parallel to the track, up the slope,

$$= (120 - 5 - 125 \times 9{\cdot}8 \sin \alpha) \text{ kN}$$
$$= (115 - 12{\cdot}5) \text{ kN}$$
$$= 102{\cdot}5 \text{ kN}$$

Total mass = 125 t and the equation of motion gives

$$102{\cdot}5 \times 1000 = 125 \times 1000f$$
$$\therefore f = \frac{102{\cdot}5}{125} = 0{\cdot}82 \text{ m/s}^2$$

Example 2.—Masses of 5 kg and 3 kg are attached to the ends of a light inextensible string which passes over a light frictionless pulley, and hang freely. The system is released from rest when the heavier particle is 19·6 m above the floor. Find (i) the tension in the string, (ii) the time taken by the heavier mass to reach the floor.

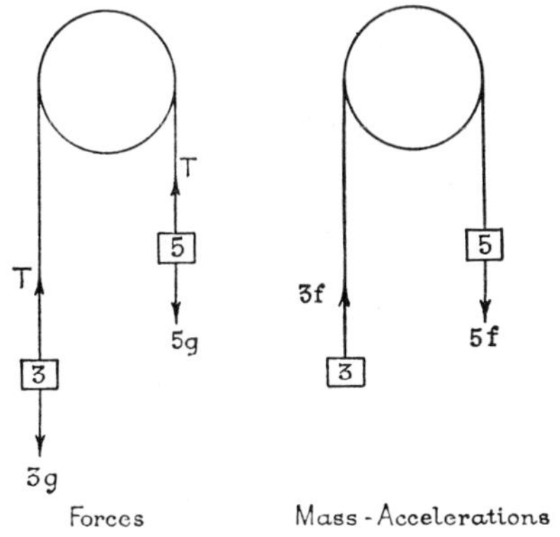

We show forces and mass-accelerations in separate diagrams. Forces are given in absolute units, i.e. newtons.

(i) For the lighter mass, $T - 3g = 3f$
For the heavier, $5g - T = 5f$
Adding, $2g = 8f$

$$\therefore f = \tfrac{1}{4}g = 2{\cdot}45 \text{ m/s}^2$$

From the first equation,

$$T = 3g + 3f$$
$$= 3\tfrac{3}{4}g \text{ N}$$

(ii) The heavier mass is to travel 19·6 m from rest with acceleration 2·45 m/s². Using the formula $s = ut + \frac{1}{2}ft^2$, we have

$$19{\cdot}6 = 1{\cdot}225t^2, \text{ and } t^2 = 16$$
$$\therefore t = 4 \quad \text{(the negative root being disregarded)}$$
$$\therefore \text{Time} = 4 \text{ sec}$$

In such systems as this, the acceleration may be more rapidly found by considering the equivalent system shown.

The acceleration is now the same in all parts of the system, which was not the case originally; upward and downward accelerations are different. We can write down a single equation of motion for the whole system, and the forces T, being a pair of equal and opposite reactions, do not come into it.

$$\text{Total force (to right)} = \text{total mass-acceleration (to right)}$$
$$2g = 8f$$

and so on as before. If the tension is required, it can be found by considering one of the masses separately.

Example 3.—A mass m kg is placed at the foot of an inclined plane 13 m long, whose upper end is 5 m higher than the foot. The mass is connected by a light inextensible string, passing over a smooth pulley at the top of the plane, to an equal mass which hangs level with the top of the plane, 5 m above the floor. The coefficient of friction between the first mass and the plane is $\frac{1}{2}$. The system is allowed to move. Describe its motion.

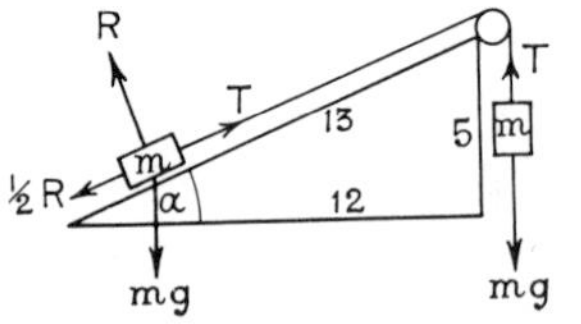

Forces

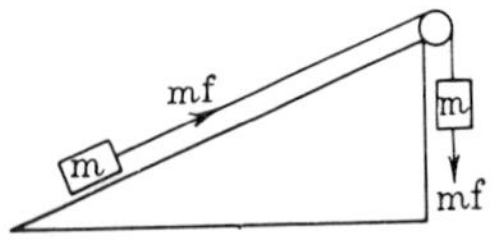

Mass - Accelerations

(i) While both masses are in motion, forces and mass-accelerations are as shown. By geometry we find the base of the triangle to be 12 m, so that $\sin \alpha = \frac{5}{13}$, $\cos \alpha = \frac{12}{13}$.
For the first mass, resolving perpendicularly to the plane,

$$R - mg \cos \alpha = 0$$
$$\therefore R = \tfrac{12}{13}mg. \qquad \text{(i)}$$

Resolving parallel to the plane,

$$T - \tfrac{1}{2}R - mg \sin \alpha = mf$$

$\therefore$ From (i), $$T - \tfrac{6}{13}\, mg - \tfrac{5}{13}\, mg = mf \quad \text{. . . . . (ii)}$$

For the second mass,

$$mg - T = mf \quad \text{. . . . . (iii)}$$

Adding (ii) and (iii) $$\tfrac{2}{13}\, mg = 2mf$$

$$\therefore f = \frac{g}{13}$$

(ii) This state of affairs continues until the hanging mass strikes the floor, when the tension in the string vanishes. At this instant, the velocities of the masses are given by $v^2 = 2f \times 5 = \tfrac{10}{13}\, g$.
The forces and mass-accelerations are now as shown:

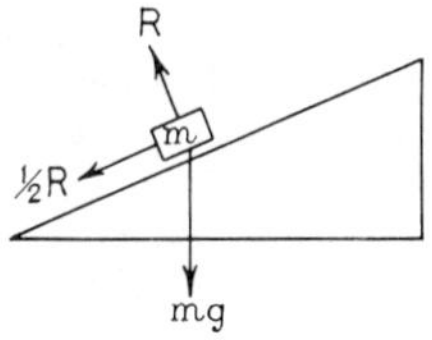

Forces

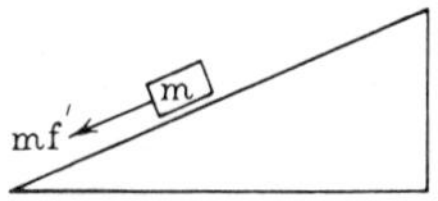

Mass - Accelerations

As before, $$R = \tfrac{12}{13}\, mg$$

Resolving parallel to the plane,

$$\tfrac{1}{2}R + mg \sin \alpha = mf'$$
$$\therefore \tfrac{6}{13}\, mg + \tfrac{5}{13}\, mg = mf'$$
$$\therefore f' = \tfrac{11}{13}\, g \text{ (down the plane)}$$

The mass is reduced to rest after travelling a distance x m given by

$$0 - v^2 = -2f'x$$
$$\therefore \tfrac{10}{13}\, g = \tfrac{22}{13}\, gx$$
$$\therefore x = \tfrac{5}{11}$$

The mass comes to rest $5\tfrac{5}{11}$ m from the foot of the plane.

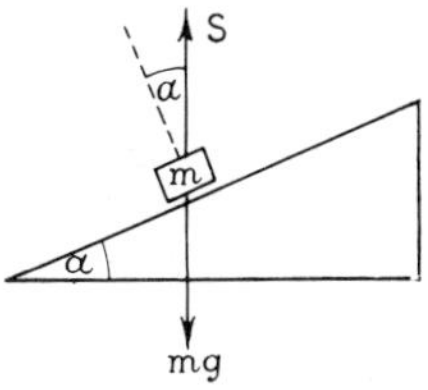

(iii) The mass now remains at rest under the forces shown. ***S*** is the total reaction of the plane, and for equilibrium must act at α to the normal. This is possible, since $\tan \alpha = \tfrac{5}{12}$, and this is less than the coefficient of friction.

EXERCISE 16A

Take g as 9·8 m/s².

1. Find the tractive force needed to move a train of mass 280 t up an incline of 1 in 196 against resistances amounting to 50 N/t (i) at a constant speed (ii) with acceleration 1 m/s².
2. A car of mass 750 kg will just move down an incline of 1 in 19·6 with the brakes off. Find the resistance to motion. If this resistance remains constant, what acceleration will be given to the car by a tractive force of 1200 N (i) on the level, (ii) up an incline of 1 in 9·8?
3. A particle of mass 5 kg on a level floor is suddenly given a velocity of 7 m/s. If the coefficient of friction with the floor is $\frac{1}{4}$, how far will it travel?
4. A train of mass 300 t is travelling uniformly at 54 km/h on the level, and the resistances to motion are 50 N/t. The rear coach, of mass 50 t, is slipped, and the engine's tractive force is unaltered. Find (i) the acceleration of the rest of the train, (ii) the distance run by the slipped coach before coming to rest, if the brakes are not applied.
5. A bullet of mass 10 g, travelling at 800 m/s, comes to rest after penetrating 20 cm into a block of wood. Find the resisting force exerted by the wood, in kN, assuming it to be constant.
6. A bullet of mass 10 g acquires a velocity of 800 m/s while travelling 1 m along a rifle barrel. What constant force, acting for what time, will give this velocity? If the diameter of the base of the bullet is 7·6 mm, find the pressure on the base in kN/m². Neglect the effect of friction.
7. A particle of mass 1 kg is at the foot of a plane 2·45 m long, inclined at 30° to the horizontal, and the coefficient of friction between particle and plane is $\frac{1}{2}$. With what speed must the particle be started if it is to travel just to the top of the plane? (Answer to 2 significant figures.)
8. Masses of 10 kg and 6 kg are attached to the ends of a light inextensible string which passes over a smooth pulley. The 6-kg mass is held at floor level, and the other mass is 4·9 m above the floor. The system is released (i) After what time will the 10-kg mass hit the floor? (ii) To what height will the 6-kg mass rise?
9. Masses m_1, m_2 $(m_1 > m_2)$ are attached to the ends of a light inextensible string passing over a smooth pulley. Find the ratio $m_1 : m_2$ so that the acceleration of the system shall be g/n, where $n > 1$. Why is the condition $n > 1$ necessary?
10. A particle of mass m on a rough horizontal table is connected by a light inextensible string passing over a smooth pulley at the edge of the table to an equal mass hanging freely. If μ is the coefficient of friction between the first particle and the table, and $\mu < 1$, find the acceleration of the system and the tension in the string.
11. A trolley of mass 1 kg runs on light wheels on a horizontal bench. It is pulled by a string which passes over a smooth pulley at the end of the bench and carries at the other end a scale-pan of mass 25 g. With no weights in the pan the trolley moves uniformly. Find the resistance to motion. Find the acceleration of the trolley when a 25-g mass is placed in the scale-pan. (Answer to 2 significant figures.)

12. Under the conditions of question 11, what mass must be placed in the pan to give the trolley an acceleration g/n, where $n > 1$?

13. Two planes inclined at 30° to the horizontal meet in a horizontal ridge. Particles of mass 3 kg and 1 kg rest, one on each plane, and are connected by a light inextensible string which passes over a smooth pulley on the ridge. Find the acceleration of the system (i) if both planes are smooth, (ii) if the coefficient of friction at each contact is $1/(3\sqrt{3})$.

14. A light inextensible string fixed at one end passes under a light movable pulley to which is attached a mass M, over a fixed pulley, and carries a mass m at its free end. Both pulleys are frictionless, and the string is vertical except where it passes round the pulleys. Find the acceleration of the mass M and the tension in the string when $M > 2m$. What happens when $M < 2m$?

15. A particle of mass m rests on a rough plank of mass M, which is supported by a smooth horizontal table. The coefficient of friction between particle and plank is μ. The particle is suddenly given a velocity u, and slides over the plank. Find the accelerations of particle and plank while the sliding continues. For how long will it continue, and what will be the final common velocity of particle and plank? How far does the particle move relatively to the plank?

EXERCISE 16B

1. An engine has mass 100 t, $\frac{4}{5}$ of this resting on the driving wheels. The rest of the train has mass 300 t. The coefficient of friction between driving wheels and rails is 0·2. Resistances to motion on engine and train amount to $\frac{1}{2}$ per cent of the weight. Find the greatest acceleration possible (i) on the level, (ii) up a gradient of 1 in 200, assuming the same tractive force and resistance as on the level.

2. (i) The brakes of a car can just hold it on a slope of 1 in 3. Assuming constant resistances to motion at all speeds, find the shortest distance in which the car can be brought to rest from a speed of 36 km/h on the level. (ii) The maximum tractive force available will just move the car up a slope of 1 in 3. Find the greatest acceleration possible on the level, assuming constant resistances.

3. A cycle and rider together have mass 80 kg. The rider can free-wheel down a slope of 1 in 20 at a uniform speed. (i) What tractive force is required if he is to go up the same incline at the same speed? (ii) If he exerts the same force when travelling at the same speed on the level, what will be his acceleration?

4. Two masses of 500 g each are connected by a light string passing over a smooth pulley. A rider of mass 50 g is placed on one mass and the system is released from rest. After travelling 2·1 m the rider is lifted off. What is then the speed of each mass?

5. A board 10 m long has one end on the floor, and the other supported 6 m above the floor. A block of mass m kg at the foot of the board is connected by a light string over a smooth pulley at the upper end of the board to an equal mass hanging freely at the level of the upper end. The coefficient of friction between block and board is $\frac{1}{4}$. The system is released from rest. Find (i) the velocity of the block when the string slackens, (ii) the whole distance which the block moves along the board, (iii) the velocity of the block just before the string again tightens. (Give velocities to 2 significant figures.)

6. A train consists of coaches, each of mass m kg. The resistance to motion is x per cent of the weight. Find the tension in the nth coupling from the rear (i) when the acceleration is f m/s^2 on the level, (ii) when the speed is constant up an incline of y in 100.

7. The train described in question 6 is moving at constant speed down an incline of z in 100 ($z < x$) when the brakes are applied (to the engine only) so as to give an increasing retardation. Show that when this reaches a certain value (to be found) the couplings all slacken simultaneously.

8. A light inextensible string connecting masses of 1 kg and 2 kg passes over two fixed pulleys and under a movable light pulley between them. A mass of 4 kg is attached to the movable pulley. The string between the pulleys is vertical. Show that if the upward accelerations of the 1-kg and 2-kg masses are f_1, f_2, the downward acceleration of the movable pulley is $\frac{1}{2}(f_1 + f_2)$. Find the tension in the string, and the values of f_1, f_2.

9. A block of mass 5 kg resting on a smooth horizontal floor is pulled with a horizontal force which increases from zero at the rate of $\frac{1}{20}\,g$ N per sec. Find its speed after 20 sec.

10. If in question 9 the coefficient of friction between block and floor is 0·2, and the force ceases after $\frac{1}{2}$ min, find the total distance travelled by the block, to two significant figures.

11. Masses m and M are connected by a light smooth string which passes along a line of greatest slope of a smooth inclined plane and through a small smooth hole O in the plane. Initially the mass m is held at a distance l from O measured along the plane and at a vertical distance h below O, whilst the mass M hangs freely below O. When motion is allowed to ensue, the mass m moves up the plane and the mass M moves vertically downwards. After the mass m has moved a distance $x(<l)$ up the plane the string is cut and m just reaches the hole. Prove that $x = \dfrac{(m + M)hl}{M(h + l)}$. If the mass m travels from its initial position to the hole in time t, show that $gh(Ml - mh)t^2 = 2M(h + l)l^2$.

12. A particle of mass m lies on a fixed rough plane inclined at an angle α to the horizontal. To the particle is attached a light inextensible string which passes over a small smooth pulley at the top of the plane. Another particle of mass m is attached to the other end of the string and hangs freely vertically below the pulley. The particles and string all lie in the vertical plane through a line of greatest slope of the inclined plane. If the coefficient of friction between the particle and the plane is μ, show that motion will ensue if $\mu < (1 - \sin\alpha)/\cos\alpha$, and if this condition is satisfied find the acceleration of the sliding particle and the tension in the string.

13. A particle A of mass m_1 and a particle B of mass m are connected by a light inextensible string. A second light inextensible string connects B to another particle C of mass $m_2(<m_1)$. The particle B is placed on a rough horizontal table, the coefficient of friction between B and the table being μ. The strings pass over opposite parallel edges of the table and the particles A and C hang freely. The system is released from rest with A, B, C in a vertical plane perpendicular to the parallel edges of the table. Prove that the particles will move if $m_1 > m_2 + \mu m$, and in this case find the acceleration of the system and the tensions of the two strings.

14. The weight of a train is W, the maximum tractive force which can be exerted by the locomotive is P, and the maximum braking force is B; there is in addition a constant resistance R. Show that the least possible time t which the train can take to cover a distance s from rest to rest is $\left\{\dfrac{2sW(P+B)}{g(P-R)(B+R)}\right\}^{1/2}$. If v is the greatest velocity attained under the above conditions, and an additional condition is imposed that the velocity of the train shall not exceed $V(<v)$, show that the least time which may be taken exceeds t by $\dfrac{s(v-V)^2}{Vv^2}$

15. A car of mass 600 kg is pulled along a straight level road by a resultant forward force which decreases uniformly from 600 N at the start to zero in 20 sec. The car is initially at rest. Obtain a formula for the acceleration of the car at time t sec after the start. Deduce an expression for the velocity at time t sec, and find the total distance covered in the 20 sec.

5. Units and dimensions

All physical quantities are specified by comparison with some standard. When a length is described as 4 m, the number 4 is called the measure of the quantity, and the metre is the unit. A number is an abstraction whose nature we shall not discuss; it is sufficient to be able to manipulate numbers by the rules of arithmetic. The *metre*, which is the unit of length in the SI system, was originally defined by reference to a material standard in the custody of the French Government, but is now defined in terms of the wavelength of a particular radiation. In practice, copies of the material standard are used.

No such material standard is needed for units of area, since they are derived from units of length. A square metre is the area of a square whose side is one metre. In finding an area, it is sometimes said that we multiply one length by another, but this is not an accurate statement. It is not possible to multiply 3 m by 2 m. What we do is to multiply together the two measures of length—the numbers 3 and 2—to find the measure of the area. An area is always found by multiplying together two measures of length, and possibly multiplying the result by a pure number, as in finding the area of a circle. Accordingly we say that area is of two dimensions in length, and express this by using the 'dimension symbol' L^2 for areas.

The SI unit of mass is the *kilogramme*, which, as previously mentioned, is referred to a material standard. The *second*, which is the unit of time, was formerly defined by reference to the rotation of the Earth, but is now defined in terms of the frequency of a certain atomic radiation. Further details of these standards are available in official publications.

Just as the unit of area is derived from that of length, so by the use of the fundamental units of length, mass, and time all units required in mechanics can be defined. For instance, in finding a velocity we divide a measure of length by a measure of time: the dimension symbol of velocity is LT^{-1}. Similarly it can be seen that the symbol for acceleration is LT^{-2}. Since a force is equivalent to a mass-acceleration the symbol for force is LMT^{-2}. An angle in circular measure, being given by the ratio of two lengths, has dimension zero, and an angular velocity has the dimension symbol T^{-1}. Other quantities which have been, or will be, encountered can be readily dealt with on the same lines.

A glance at the dimensions of an expression is often sufficient to detect a slip in working. No sum or difference of terms, and no equation, can have any meaning unless all the terms are of the same dimensions. Thus, if v, g, a are measures of velocity, acceleration, and distance respectively, $v^2 = ga$ is a possible statement, since both sides of the equation have dimensions L^2T^{-2}, but $v^2 = g/a$ is not possible.

This check cannot be used when any of the quantities involved have purely numerical measures. With a numerical value for g the equation above might be $v^2 = 9{\cdot}8a$, which cannot be checked dimensionally.

The dimension symbol is also useful when converting from one system of units to another.

Before the introduction of SI (sometimes known as MKS, or metre-kilogramme-second) units, the systems most in use were the FPS, or foot-pound-second, and the CGS, or centimetre-gramme-second, systems. It should be mentioned that the unit of force in the FPS system is the *poundal*, which gives a mass of 1 lb an acceleration of 1 ft/s², and in the CGS system the *dyne*, which gives a mass of 1 *g* an acceleration of 1 cm/s². Thus the weight of 1 lb (1 lb wt or 1 lbf), which was sometimes used as a measure of force, is *g* poundals, approximately 32 pdl, and the weight of 1 gramme is *g* dynes, approximately 980 dynes.

Example 4.—Convert 10^6 dynes to poundals.

The dimension symbol for a force is LMT^{-2}. Writing CGS and FPS for the fundamental units in the two systems, we have

$$x\mathrm{FPS}^{-2} = 10^6\mathrm{CGS}^{-2} \qquad \therefore\ x = 10^6\,\frac{\mathrm{C}}{\mathrm{F}}\cdot\frac{\mathrm{G}}{\mathrm{P}}$$

The ratios C/F and G/P are known to be approximately 1/30·5 and 1/454. Hence

$$\underline{x \simeq 72{\cdot}2}$$

Example 5.—Convert 14·6 lb wt/sq in to megadynes/sq cm, using the same equivalents as in example 4, and taking g as 32·2 ft/s². [The prefix ‘mega-’ means ‘one million times’.]

The lb wt is not the FPS unit of force, and must be expressed in poundals. We must also change force per sq in to force per sq ft, and megadynes to dynes.

Let x megadynes/sq cm = 14·6 lb wt/sq in.

Then 10^6x dynes/sq cm = 14·6 × 32·2 × 144 pdl/sq ft.

The dimension symbol for pressure is $\frac{\text{ML}}{\text{T}^2} \cdot \frac{1}{\text{L}^2} = \text{ML}^{-1}\text{T}^{-2}$.

We have, as before in example 4,

$$10^6 x\text{GC}^{-1}\text{S}^{-2} = 14{\cdot}6 \times 32{\cdot}2 \times 144\ \text{PF}^{-1}\text{S}^{-2}$$

$$\therefore\ x = \frac{14{\cdot}6 \times 32{\cdot}2 \times 144}{10^6} \cdot \frac{\text{P}}{\text{G}} \cdot \frac{\text{C}}{\text{F}}$$

$$= \frac{14{\cdot}6 \times 32{\cdot}2 \times 144 \times 454}{10^6 \times 30{\cdot}5}$$

$$\simeq 1{\cdot}01$$

EXERCISE 17

1. What are the dimension symbols for momentum, pressure, density, angular acceleration, weight per unit length, moment of a force?

2. In this question m is a measure of mass, a or x of length, v of velocity, g of acceleration, P of force, t of time, ω of angular velocity. Find the dimensions of:

   (i) mv^2, $\frac{m_1 v}{m_1 + m_2}$, $m_1 x_1^2 + m_2 x_2^2$

   (ii) $a\omega^2$, $a \sin \omega t$, $a\omega \cos \omega t$

   (iii) Px, Pt, Pv

   (iv) v^2/x, $v\frac{dv}{dx}$, $\sqrt{(g/a)}$

3. With the notation of question 2, (i) if $\frac{km_1 m_2}{a^2}$ measures a force, what are the dimensions of k? (ii) if $g^p a^q$ measures a time, find the values of p, q.

4. Convert 70,000 pdl per sq ft to dynes per sq cm.

5. (i) Convert 1 newton to dynes.
   (ii) Convert 1 kN to poundals.
   (iii) Express 10^6 dynes per sq cm in SI units.

6. Express the momentum of a mass of 10 lb moving at 6 ft/s in SI units.

7. Convert 30 m.p.h. to cm/s.

8. Express a density of 8·3 g/cm³ in lb/cu ft.

9. A force of 20 pdl acts at a distance of 6 in from an axis. Find its moment about the axis in N m.

10. A square metre of sheet metal has mass 10 kg. Find the surface density in lb/sq ft.

9

Projectiles

It must be stated at the outset that the results found in this chapter have no practical application. The effect of air resistance, which is in practice very important, has been neglected, because it is too difficult to deal with at present.

We take the x-axis horizontally, and the y-axis vertically upwards, and suppose that a particle is projected from the origin with speed V at an elevation α, that is, at an angle α with the x-axis in the xy-plane.

Since there is no horizontal force,

$$\ddot{x} = 0 \quad \ldots \quad (1)$$

Integrating, and using the initial conditions,

$$\dot{x} = V \cos \alpha \quad \ldots \quad (2)$$

Integrating again, $$x = V \cos \alpha \,.\, t \quad \ldots \quad (3)$$

Since the vertical force is that of gravity,

$$\ddot{y} = -g \quad \ldots \quad (4)$$

Integrating, $$\dot{y} = V \sin \alpha - gt \quad \ldots \quad (5)$$

Integrating again, $$y = V \sin \alpha \,.\, t - \tfrac{1}{2}gt^2 \quad \ldots \quad (6)$$

These six equations are the basis of our work.

1. Standard results

To find the range on a horizontal plane through the point of projection.

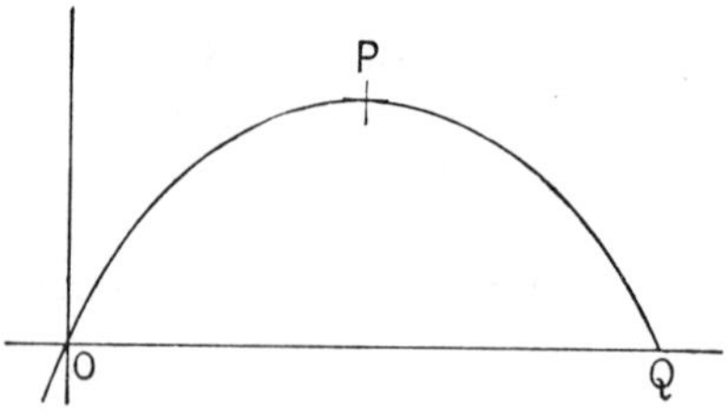

At Q, $$y = 0$$

$\therefore$ From (6)

$$t(V \sin \alpha - \tfrac{1}{2}gt) = 0$$

The solution $t = 0$ corresponds to point O.

$$\therefore\ t = \frac{2V \sin \alpha}{g}$$

giving the time of flight.

Substituting in (3),

$$x = \frac{2V^2 \sin \alpha \cos \alpha}{g}$$
$$= \frac{V^2 \sin 2\alpha}{g}$$

the required range.

Hence the maximum range is attained when $\sin 2\alpha = 1$, that is, when $\alpha = 45°$, and this maximum range is V^2/g.

For a range less than the maximum, $\sin 2\alpha < 1$ and there are two values of 2α, which are supplementary. Thus there are two values of α, which are complementary; in other words, equally inclined to the direction 45° which gives maximum range.

To find the greatest height reached.

If P is the highest point of the path, then at P, $\dot{y} = 0$.

$\therefore$ From (5)

$$V \sin \alpha = gt$$
$$t = \frac{V \sin \alpha}{g}$$

which is half the time from O to Q.

Substituting in (6) we have

$$y = \frac{V^2 \sin^2 \alpha}{g} - \tfrac{1}{2}g \frac{V^2 \sin^2 \alpha}{g^2}$$
$$= \frac{V^2 \sin^2 \alpha}{2g}$$

2. Trajectory

To find the equation of the trajectory or path of the projectile. For any point on the path, equations (3) and (6) are true. Eliminating t, we find

$$y = V \sin \alpha \frac{x}{V \cos \alpha} - \tfrac{1}{2}g \frac{x^2}{V^2 \cos^2 \alpha}$$
$$= x \tan \alpha - \frac{g x^2 \sec^2 \alpha}{2V^2}$$

This can be recognized as the equation of a parabola, since the only second-degree term is that containing x^2, but the equation of the trajectory is simpler if we take horizontal and vertical axes through its highest point. Using (X, Y) for the new coordinates we have $X = V\cos\alpha\,.\,t$ as before, and $Y = -\frac{1}{2}gt^2$, since the vertical velocity at the highest point is zero. Eliminating t gives

$$Y = -\tfrac{1}{2}g\frac{X^2}{V^2\cos^2\alpha}, \quad \text{or} \quad X^2 = \frac{-2V^2\cos^2\alpha}{g}Y$$

By comparing this with the standard equation of a parabola we see that the latus rectum is $2V^2\cos^2\alpha/g$, and that the directrix is a horizontal line at a distance $V^2\cos^2\alpha/(2g)$ above the highest point of the trajectory. Since the height of this point above the level of projection is $V^2\sin^2\alpha/(2g)$, we find that the directrix is at a height $V^2/(2g)$ above the point of projection. This height is independent of α, so that all trajectories from the same point with the same initial speed have the same directrix. This fact sometimes enables us to solve a problem graphically. (See example 5, later in this chapter.)

Note that the height of the directrix, $V^2/(2g)$, is the greatest height reached if a particle is projected vertically upwards with speed V.

Note also that at any point of the path, $\dot{x} = V\cos\alpha$, and $\dot{y} = V\sin\alpha - gt$. Hence, if v is the speed at this point,

$$\begin{aligned} v^2 &= V^2\cos^2\alpha + (V\sin\alpha - gt)^2 \\ &= V^2 - 2gtV\sin\alpha + g^2t^2 \end{aligned}$$

Now if d is the distance of this point below the directrix,

$$\begin{aligned} d &= V^2/(2g) - (V\sin\alpha\,.\,t - \tfrac{1}{2}gt^2) \\ \therefore\ 2gd &= V^2 - 2gt\,.\,V\sin\alpha + g^2t^2 \\ &= v^2 \end{aligned}$$

showing that the speed at any point of the path is that which would be attained by a particle falling freely to that point from the directrix.

3. Enveloping parabola

We now find the equation of the curve which bounds the region within range from a given point with a given speed of projection. With the point of projection as origin, the equation already found for the trajectory may be written

$$y = x\tan\alpha - \frac{gx^2}{2V^2}(1+\tan^2\alpha)$$

$$\therefore\ 2V^2y - 2V^2x\tan\alpha + gx^2(1+\tan^2\alpha) = 0$$

$$\therefore\ gx^2\tan^2\alpha - 2V^2x\tan\alpha + (2V^2y+gx^2) = 0$$

In this form the equation may be considered as a quadratic in $\tan\alpha$. Given V, and a point (x, y) through which the trajectory must pass, the equation gives the necessary elevation α. Three cases arise:

(i) The equation may have two real and different roots. When a point is in range, there are two possible elevations at which the particle may be projected. (See example 5 later in the chapter.)

(ii) The equation may have no real root. In this case the point (x, y) is out of range.

(iii) The equation may have two coincident roots. This corresponds to the border-line case when the point (x, y) is just within range.

The condition for (iii) in the standard quadratic $ax^2+bx+c = 0$ is $b^2 = 4ac$. In this case it becomes $4V^4x^2 = 4gx^2(2V^2y+gx^2)$. The solution $x = 0$ is algebraically possible, and in fact a point on the line $x = 0$ above the origin can be reached with only one elevation, namely 90°, but $x = 0$ is not part of the boundary of the region within range. Rejecting this solution, we have

$$V^4 = g(2V^2y+gx^2)$$

This equation represents a parabola, called the *enveloping parabola* because it touches all the trajectories from the origin with given velocity V. It is convex upwards, its axis is vertical, its latus rectum is $2V^2/g$, and its vertex is at $(0, V^2/2g)$ which is the highest point that can be reached by a particle projected from the origin with speed V.

4. Vectors

A vector diagram of velocities is instructive, and this method is sometimes useful in solving problems.

Let $\overline{\text{OP}}$ represent the initial velocity in magnitude and direction.

Since the acceleration is vertically downwards, of constant magnitude g, the change of velocity in time t is gt vertically downwards, and this is shown as $\overline{PQ}$.

$\therefore$ $\overline{OQ}$ represents the velocity at time t.

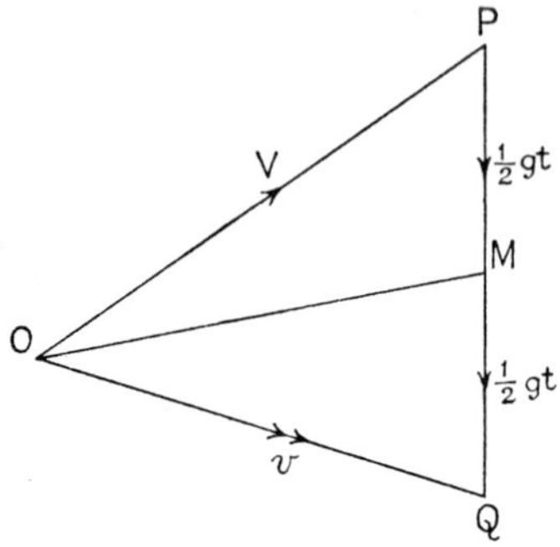

What we have done is to integrate a vector equation. In considering straight-line motion with uniform acceleration, we start with $\frac{dv}{dt} = f$, and derive by integration the equations $v = u+ft$, and $s = ut+\frac{1}{2}ft^2$. So here we start with $\frac{d\bar{v}}{dt} = \bar{g}$. the symbols $\bar{v}$, $\bar{g}$ representing vectors. Integrating, and remembering that the initial velocity is $\bar{V}$, we have, since $\bar{g}$ is constant, $\bar{v} = \bar{V}+\bar{g}t$, which the diagram shows as $\overline{OQ} = \overline{OP} + \overline{PQ}$.

A second integration gives the displacement $\bar{s}$, since $\bar{v} = \frac{d\bar{s}}{dt}$. We find $\bar{s} = \bar{V}t+\frac{1}{2}\bar{g}t^2$. A diagram showing displacements as vectors could be drawn to illustrate this, and such a diagram is occasionally useful.

In this connection, it should be noted that if M in the diagram above is the midpoint of PQ, then

$$\overline{OM} = \bar{V}+\tfrac{1}{2}\bar{g}t$$

and

$$t\,.\,\overline{OM} = \bar{V}t+\tfrac{1}{2}\bar{g}t^2 = \bar{s}$$

so that OM gives the direction from O to the position of the projectile at time t.

Example 1.—A particle is projected with speed 39·2 m/s at an elevation of 30°. Find its greatest height, the range on a horizontal plane through the point of projection, the time of flight, and the velocity (in magnitude and direction) after $\frac{1}{2}$ sec.

Using standard formulæ, we have:

$$\text{Greatest height} = \frac{V^2 \sin^2 \alpha}{2g} = \frac{39{\cdot}2 \times 39{\cdot}2}{19{\cdot}6 \times 4} = \underline{19{\cdot}6 \text{ m}}$$

$$\text{Range} = \frac{V^2 \sin 2\alpha}{g} = \frac{39{\cdot}2 \times 39{\cdot}2 \times \sqrt{3}}{9{\cdot}8 \times 2} = 78{\cdot}4\sqrt{3} \simeq \underline{136 \text{ m}}$$

$$\text{Time of flight} = \frac{2V \sin \alpha}{g} = \frac{2 \times 39{\cdot}2}{9{\cdot}8 \times 2} = \underline{4 \text{ sec}}$$

After ½ sec, $\dot{x} = V \cos \alpha = \underline{19{\cdot}6\sqrt{3} \text{ m/s}}$

$$\begin{aligned} \dot{y} &= V \sin \alpha - gt \\ &= 19{\cdot}6 - 4{\cdot}9 = \underline{14{\cdot}7 \text{ m/s}} \end{aligned}$$

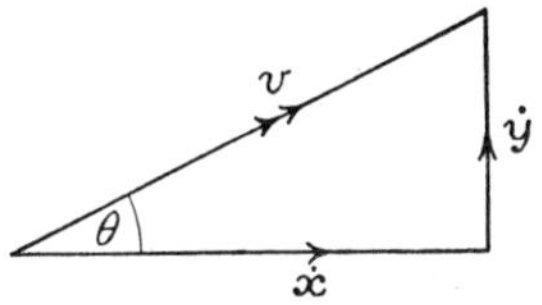

If the velocity is v ft/s, at angle θ above the horizontal,

$$\begin{aligned} v^2 &= (19{\cdot}6\sqrt{3})^2 + 14{\cdot}7^2 \\ &\simeq 1369 \\ \therefore\ v &\simeq \underline{37 \text{ m/s}} \end{aligned}$$

$$\tan \theta = \frac{\dot{y}}{\dot{x}} = \frac{14{\cdot}7}{19{\cdot}6\sqrt{3}} = \frac{7}{8\sqrt{3}}$$

$$\therefore \theta \simeq \underline{26^\circ\ 48'}$$

Example 2.—After time t a projectile is moving at right angles to its initial direction. Prove that its speed v at that time is given by $v^2 = g^2t^2 - V^2$, where V is the speed of projection.

Let α be the initial elevation. Then since v is at right angles to the original direction of motion, the angle between $\dot{y}$ and v is α.

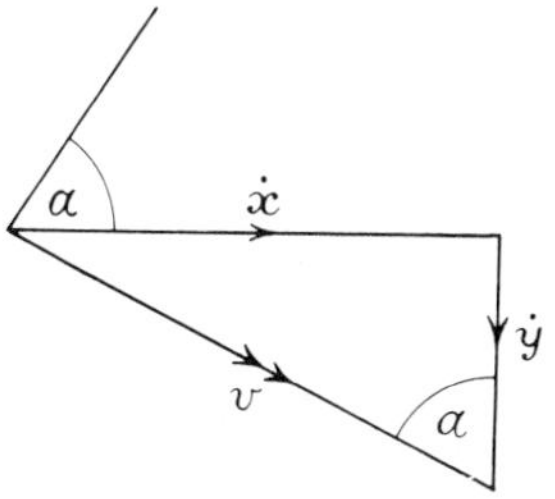

In the diagram, $\dot{y}$ is a downward speed, whereas $\dot{y}$ in our standard equation (5) is upward. So here

$$\dot{y} = gt - V \sin \alpha$$

$$\dot{x} = V \cos \alpha$$

Squaring and adding, we have

$$v^2 = g^2t^2 - 2gtV \sin \alpha + V^2 \qquad \ldots\ldots \quad \text{(A)}$$

But we have also

$$\dot{y} = \dot{x} \cot \alpha$$

$$\therefore\ gt - V \sin \alpha = \frac{V \cos^2 \alpha}{\sin \alpha}$$

$$\therefore\ gt \sin \alpha = V \cos^2 \alpha + V \sin^2 \alpha = V$$

and substituting in equation (A) gives $\underline{v^2 = g^2t^2 - V^2}$

The vector diagram of velocities gives a much neater solution.

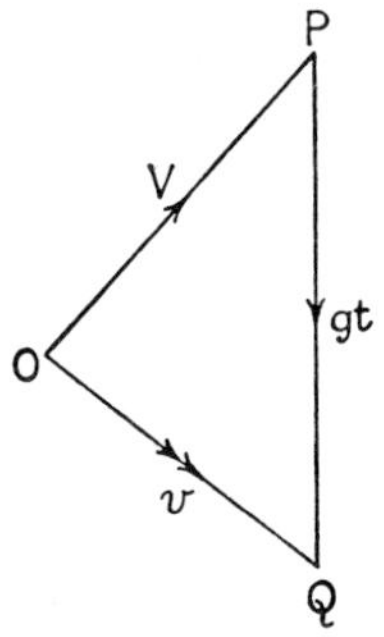

Since angle POQ = 90°, we have at once, by Pythagoras' Theorem,

$$v^2 = g^2t^2 - V^2$$

Example 3.—**A particle is projected with speed** ***kg*** **m/s. Find the least possible value of** ***k*** **for it to pass through a point 2*g* m horizontally and $1\frac{1}{2}g$ m vertically from the point of projection.**

The equation of the enveloping parabola, which passes through all the points which are just within range for speed of projection V, is

$$V^4 = 2V^2gy + g^2x^2$$

Substituting given values, we get

$$k^4g^4 = 2k^2g^3 \,.\, 1\tfrac{1}{2}g + g^2 \,.\, 4g^2$$

$$\therefore\ k^4 = 3k^2 + 4$$

$$\therefore\ k^4 - 3k^2 - 4 = 0$$

$$\therefore\ (k^2 - 4)(k^2 + 1) = 0$$

$$\therefore\ \underline{k^2 = 4 \text{ or } -1}$$

The second solution gives no real value of k. The first gives $k = 2$; we reject the negative solution.

Example 4.—**A particle A is projected horizontally at 50 m/s from the top of a tower 100 m high. At the same instant another particle B is projected from the bottom of the tower, in the same vertical plane, at 100 m/s with elevation 60°. Prove that the particles will collide, and find where they do so.**

Since the particles have the same acceleration, their relative acceleration is zero, and their relative velocity remains constant.

The velocity of B can be resolved into 50 m/s horizontally, and $50\sqrt{3}$ m/s vertically upwards.

$\therefore$ The constant velocity of B relative to A is $50\sqrt{3}$ m/s vertically upward, and since B starts 100 m vertically below A, B and A will collide after $2/\sqrt{3}$ sec.

At this time, the position of B relative to the foot of the tower is given by

$$x = 50 \times 2/\sqrt{3} \simeq 58 \text{ m}$$

$$y = 50\sqrt{3} \times 2/\sqrt{3} - 4{\cdot}9 \times \tfrac{4}{3} \simeq 93 \text{ m}$$

Example 5.—From a given point A a particle is to be projected with speed V so as to pass through a given point B. Find the direction of projection.

Draw LM at a height above A representing $V^2/(2g)$ on the scale chosen for the diagram. LM is the directrix of the trajectory. Draw AP, BQ perpendicular to LM, and let circles with centres A, B and radii AP, BQ intersect in S_1, S_2. Then either S_1 or S_2 is a possible focus for the trajectory.

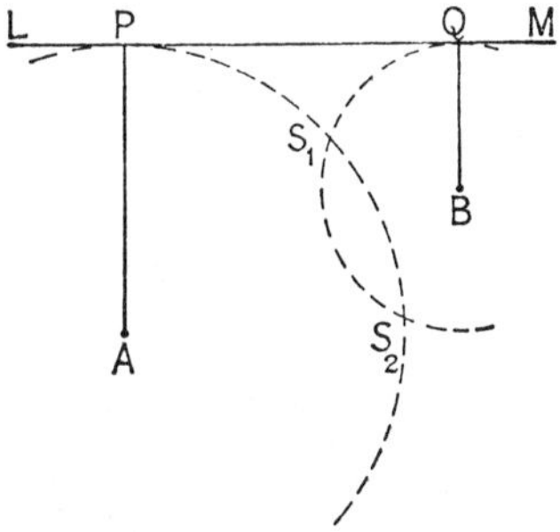

In either case SA = AP and SB = BQ, so that a parabola with focus S and directrix LM will pass through A, B.

From the geometry of the parabola it is known that the tangent at A bisects angle SAP, giving the direction of projection.

A sketch of the trajectory may be made, using the further facts that the tangent at B bisects angle SBQ and that the vertex of the parabola is midway between S and LM.

Note that if B is within range for a particle projected from A with speed V, there will be two possible trajectories. If B is on the boundary of the region that can be reached, the circles will touch, and there will be only one trajectory, with its focus on AB. If B is out of range from A, the circles will not meet.

EXERCISE 18A

1. A particle is projected with speed 49 m/s at an elevation $\tan^{-1} \frac{3}{4}$. Find its greatest height, horizontal range, and time of flight.

2. A particle is projected as in question 1. Find its horizontal and vertical distances from the point of projection, and its velocity (in magnitude and direction) after 2 sec.

3. The maximum horizontal range of a projectile is 980 m. Find the speed of projection. With this speed, find the elevation necessary for a range of 490 m.
4. Prove that the speed of a projectile when at a given height is the same on the upward and downward parts of the trajectory, and that these velocities are equally inclined to the horizontal.
5. A particle is projected at elevation 30° with speed 49 m/s. Find the time required to reach a level (i) 19·6 m above, (ii) 29·4 m below, the point of projection.
6. A particle is projected at 19·6 m/s. Taking the usual axes, and 1 m as unit of distance, find the elevations necessary to pass through (i) (19·6, 9·8), (ii) (19·6. 14·7), (iii) (19·6, 19·6).
7. A bullet is fired with speed 735 m/s at elevation 5°. (i) Find the horizontal range in metres. (ii) If the elevation is changed by 10′, what is the change in range? Answer to the nearest 100 m in each case.
[*Note.*—This question well illustrates the effect of neglecting air resistance. The actual range of a rifle bullet under the given conditions is about 2000 m.]
8. A boy can throw a cricket ball 80 metres and no more. With what speed in metres per second must it be thrown? Take $g = 9{\cdot}8$ m/s².
9. The initial velocity of a projectile has horizontal component u and vertical component v. Find expressions for the horizontal range and the greatest height reached.
10. A projectile has initial horizontal velocity 12·25 m/s and vertical velocity 14·7 m/s. (i) Find its horizontal range. (ii) Over what fraction of this distance is its height greater than three-quarters of the greatest height?
11. A stone is thrown with a velocity of 24·5 m/s at an elevation of 30° to the horizontal from the top of a tower 29·4 m high. Calculate (i) the time the stone takes to reach the horizontal plane through the foot of the tower, (ii) the horizontal displacement of the stone from the foot of the tower when it strikes this plane, (iii) the magnitude of the velocity with which it strikes the plane.
12. A cricket ball is thrown from a point 1·625 m above the ground so that after $1\frac{1}{2}$ sec it just clears the top of a tree 20 m high and 30 m away measured horizontally. Assuming that the ground is level, calculate (i) the horizontal and vertical components of the velocity with which the ball was thrown, (ii) the greatest height reached, (iii) the horizontal distance travelled by the time the ball is again at a height of 1·625 m.
13. A particle is projected from a point O with a velocity of which the horizontal and vertical components are u, v respectively. Prove that its range on the horizontal plane through O is $2uv/g$.

A ball projected from a point O on level ground strikes the ground again at a distance 100 m from O after a time 5 sec. Find the horizontal and vertical components of its initial velocity. The ball just passes over a tree standing 20 m from O. Find the height of the tree, and the speed of the ball at the moment when it is passing the tree.
14. A particle is projected from a point O with a velocity V inclined at angle α to the horizontal. Obtain expressions defining the position and the direction of motion of the particle at a time t after projection.

If, after a time t, the angle of elevation from O is ϕ and the angle of inclination to the horizontal of the velocity of the particle is θ, prove that $\tan\alpha$, $\tan\phi$, and $\tan\theta$ are in arithmetical progression.

Hence, or otherwise, show that $R = 4H\cot\alpha$, where R is the range on a horizontal plane through O and H is the greatest height above O.

15. Prove that when a particle projected with speed V is again moving with speed V it must be on a level with the point of projection.

16. A particle is projected from a point O, and after time t its distance from O is s. Prove that when $s = \frac{1}{2}gt^2$ the particle is moving at right angles to its direction of projection.

17. A projectile is fired horizontally with speed V from a point A on the top of a cliff, and is observed to strike the sea at a point B where AB makes an angle α with the horizontal. Show that A is at a height $2V^2 \tan^2 \alpha/g$ above sea-level. Show also that at B the direction of motion of the projectile makes an angle β with the horizontal, where $\tan \beta = 2 \tan \alpha$.

Prove that at any point P of the path the speed of the projectile is the same as that of a body which has fallen vertically to P from rest at a height $V^2/(2g)$ above the level of A.

18. Prove that, referred to a horizontal axis Ox and a vertically downward axis Oy through the highest point O of the path of a particle projected with velocity V inclined at an angle α to the horizontal, the equation of the path is

$$2V^2 = gx^2 \sec^2 \alpha$$

A particle is projected from a point A with velocity V inclined at 45° to the horizontal. If the particle passes through two points, each at a height h above the horizontal plane through A and at a horizontal distance a apart, prove that $V^2 = 2gh + g\sqrt{(a^2 + 4h^2)}$.

EXERCISE 18B

1. A stone is thrown upwards at 60° to the horizontal. If the greatest height it reaches is 14·7 m, calculate (i) the speed at which it was thrown, (ii) the time taken to reach the greatest height, and (iii) the horizontal distance between the thrower and the highest point of the path.

2. From a point 61·25 m above horizontal ground a stone is thrown at 60° to the downward vertical at a speed of 24·5 m/s. Calculate (i) the time taken to reach the ground, (ii) the horizontal distance travelled by the stone before it hits the ground, (iii) the speed with which it hits the ground.

3. A ball is thrown at 10 m/s from a point 1 m above floor level in a corridor 4 m high. Find the greatest elevation at which it can be thrown so as not to hit the ceiling.

4. A projectile has a horizontal range of 3500 m, and its greatest height is 125 m. Find its velocity of projection in magnitude and direction.

5. A projectile has initially a speed of 36·75 m/s. After 1 sec its speed is 31·85 m/s. Find the horizontal and vertical components of its initial velocity.

6. Particles are projected with different elevations at the same speed, of 49 m/s, so that the range is in each case 235·2 m. Find the difference between (i) the greatest heights reached, (ii) the times of flight.

7. A particle is to be projected from ground level at 45° to the horizontal. Find the least speed of projection that will enable it to clear a wall 5·2 m high and 15 m away from the point of projection.

8. A particle is projected at 35 m/s to pass through a point 50 m away horizontally, and 50 m higher than the point of projection. Find the necessary elevation.

9. A projectile starts with a speed of 245 m/s at 60° above the horizontal. Find its speed, and its height above the starting point, when its direction of motion is (i) 45°, (ii) 30° above the horizontal.

10. A particle projected with speed V from a point O arrives at a point P with speed v after t seconds. OP $= s$. Prove that $v^2 = \dfrac{2s^2}{t^2} + \frac{1}{4}g^2t^2 - V^2$.

11. Under the conditions of question 10, if the velocity at P is perpendicular to OP, prove that $3v^2 = g^2t^2 - V^2$.

12. A particle is projected from a point O in a horizontal plane with a velocity V at an angle α to the horizontal. Find expressions for the time of flight, the greatest height attained, and the range OR on the horizontal plane. Find also the equation of the path referred to horizontal and vertical axes through O.

Another particle is moving in the horizontal plane in a straight line through O, in the same vertical plane as the path of the first particle, with constant speed U directly towards O. If the particle is initially beyond R and its distance from O at the instant of projection is d, show that, for the first particle to hit the second,
$$\sin 2\alpha + \frac{2U}{V}\sin\alpha - \frac{gd}{V^2} = 0.$$

13. A particle is projected from a point O with speed V m/s at an angle α above the horizontal. Find expressions for its coordinates (x, y) referred to horizontal and vertical axes through O in the plane of motion, after time t sec.

If $\alpha = 60°$, it is observed that when $t = 2$ the elevation of the particle as viewed from O is 45°. Find the initial speed V, and the range on the horizontal plane through O.

14. A particle is projected under gravity from a point A to a point B, where AB $= l$, and the line AB is inclined at an angle α to the horizontal. Show that the minimum speed of projection for which this is possible is $\sqrt{\{gl(1 + \sin\alpha)\}}$ and find the corresponding angle of projection. Show also that in this trajectory the directions of motion at A and B are perpendicular.

15. If a particle is projected horizontally with speed V from a point A at a height $V^2/(2g)$ above another point O, show that it will describe the enveloping parabola of the family of trajectories obtained by projecting particles from O with speed V in varying directions in the same vertical plane.

A particle is projected horizontally from A with speed V, and another is projected from O with the same speed at any angle to the horizontal, and in the same vertical plane as the first trajectory. Prove that the difference of the squares of their speeds at the common point of their paths is V^2.

16. A number of particles are projected at the same instant from the same point with the same speed, but in different directions. Prove that at any subsequent instant they lie on the surface of a sphere. [Use the vector diagram of displacements.]

5. Projectile on an inclined plane

When the motion of a projectile is related to an inclined plane it is sometimes convenient to take axes parallel and perpendicular to the plane, as shown.

Let the plane be inclined at an angle θ to the horizontal, and let the particle be projected from a point O on the plane with speed V at an angle β to the plane.

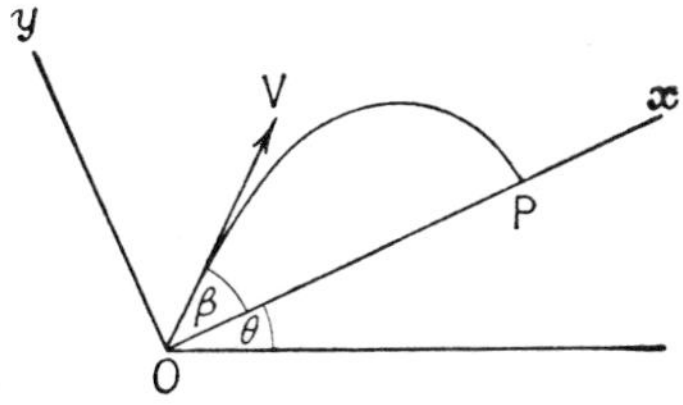

The components of acceleration due to gravity are $-g \sin\theta$ and $-g \cos\theta$ parallel to the axes of x and y respectively. Integrating, and using the initial conditions as in the early part of this chapter, we have:

$$\ddot{x} = -g \sin\theta$$
$$\dot{x} = V\cos\beta - g\sin\theta \,.\, t$$
$$x = V\cos\beta \,.\, t - \tfrac{1}{2}g\sin\theta \,.\, t^2$$
$$\ddot{y} = -g\cos\theta$$
$$\dot{y} = V\sin\beta - g\cos\theta \,.\, t$$
$$y = V\sin\beta \,.\, t - \tfrac{1}{2}g\cos\theta \,.\, t^2$$

To find the range on the plane, in the upward direction.

At P, $\qquad y = 0$

$$\therefore\ V\sin\beta = \tfrac{1}{2}g\cos\theta \,.\, t$$

[The solution $t = 0$ gives the point O.]

$$\therefore\ t = \frac{2V\sin\beta}{g\cos\theta}$$

Substituting in the third equation, we have

$$x = \frac{2V^2\sin\beta\cos\beta}{g\cos\theta} - \frac{g\sin\theta}{2}\cdot\frac{4V^2\sin^2\beta}{g^2\cos^2\theta}$$
$$= \frac{2V^2\sin\beta}{g\cos^2\theta}\{\cos\beta\cos\theta - \sin\beta\sin\theta\}$$
$$= \frac{2V^2\sin\beta\cos(\beta+\theta)}{g\cos^2\theta}$$

The range on the plane in the downward direction is found by changing the sign of θ in this formula to be

$$\frac{2V^2 \sin\beta \cos(\beta-\theta)}{g\cos^2\theta}$$

If the maximum range up the plane is required, we change the product in the top line of the formula to a difference of sines, giving

$$x = \frac{V^2}{g\cos^2\theta}\{\sin(2\beta+\theta)-\sin\theta\}$$

As the elevation β changes, the greatest value of $\sin(2\beta+\theta)$ is unity, which occurs when $2\beta+\theta = 90°$, or $\beta = \frac{1}{2}(90°-\theta)$, showing that the direction of projection for maximum range bisects the angle between OP and the vertical.

The maximum range is

$$\begin{aligned} R &= \frac{V^2(1-\sin\theta)}{g\cos^2\theta} \\ &= \frac{V^2(1-\sin\theta)}{g(1-\sin^2\theta)} \\ &= \frac{V^2}{g(1+\sin\theta)} \end{aligned}$$

For ranges less than the maximum, $\sin(2\beta+\theta)$ will have a value less than unity, and there will be two possible elevations β_1, β_2, such that

$$\begin{aligned} 2\beta_1+\theta+2\beta_2+\theta &= 180° \\ \therefore\ \beta_1+\beta_2 &= 90°-\theta \end{aligned}$$

showing that the two directions are equally inclined to the direction for maximum range.

As before, results applying to projection down the plane are found by changing the sign of θ.

The maximum range can also be obtained by finding where the enveloping parabola is cut by a line representing the plane. The equations, referred to horizontal and vertical axes through the point of projection, are $V^4 = 2V^2gy+g^2x^2$ and $y = x\tan\theta$.

A simpler method, however, is that used earlier in example 5. It is there shown that if a particle projected from A just reaches B the focus of the trajectory is on AB.

Let the maximum range AB $= R$

Then $\quad AS = AP = V^2/(2g)$

$\quad BS = BQ = V^2/(2g) - R\sin\theta$

But $\quad AB = AS + BS$

$$\therefore\ R = V^2/g - R\sin\theta$$

$$\therefore\ R = \frac{V^2}{g(1+\sin\theta)}$$

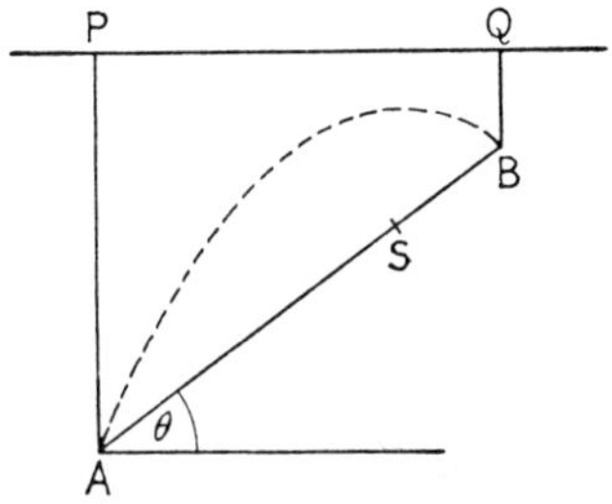

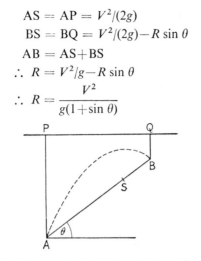

The tangents to the parabola at A and B are known to bisect the angles SAP, SBQ. Thus we have the direction of projection for maximum range, and also the result that the direction of motion at B is perpendicular to that at A.

EXERCISE 19

1. A particle is projected up a plane inclined at 30° to the horizontal, from a point on the plane, at **29·4** m/s.The range is **58·8**($\sqrt{3}$ − **1**) m. Find the two possible angles of elevation relative to the plane. Find also the range down the plane if the particle is projected with the same speed at each of these elevations relative to the plane.

2. The extreme ranges of a particle up and down a plane are in the ratio 1 : 3, and their sum is **120** m. Find the inclination of the plane and the speed of projection.

3. A particle is projected from a point on a plane inclined at an angle β to the horizontal and its initial angle of elevation is α from the horizontal. Prove that it will strike the plane at right angles if $\cot(\alpha - \beta) = 2\tan\beta$.

4. A particle is projected with speed V at an angle α to the horizontal from a point at a height h above a horizontal plane. Prove that the greatest range on the plane is attained when $\tan\alpha = \sqrt{\dfrac{V^2}{V^2+2gh}}$

5. Three particles are projected from the same point at the same time, with velocities differing in magnitude and direction. Show that the triangle which they form at any subsequent time is similar to that formed at any other time.

6. A particle is projected from the lowest point A of the inside of a hollow sphere of radius a, so as to strike the sphere at right angles at a point B such that the chord AB is of length a. Find the angle between the direction of projection and the horizontal.

7. A projectile is fired horizontally with speed V from a point A on the top of a cliff, and strikes the sea at a point B where AB makes an angle α with the horizontal. Show that A is at a height $(2V^2 \tan^2 \alpha)/g$ above sea-level. Show also that at B the direction of motion of the projectile makes an angle β with the horizontal, where $\tan \beta = 2 \tan \alpha$.

Prove that at any point P of the path the speed of the projectile is that of a body which has fallen vertically to P from rest at a height $V^2/(2g)$ above the level of A.

8. A shell is fired with speed V at inclination θ to the horizontal from a fort at sea-level. Find the horizontal and vertical distances of the shell from the fort t sec after firing. Hence find the range of the shell on the horizontal plane through the fort.

A target is towed with speed v directly away from the fort. If the target is hit by the shell, show that when the shell is fired the distance between the target and the fort is $(V^2 \sin 2\theta - 2Vv \sin \theta)/g$.

If the target had stopped at the moment the shell was fired, show that the shell would have passed over the target at a height h, where

$$gh = 2v \tan \theta (V \sin \theta - v \tan \theta)$$

9. A projectile is fired from a fixed point O at an elevation α and hits a stationary target A, where OA is horizontal. Find the speed V of projection, neglecting air resistance.

On another occasion the target starts to rise vertically from A with uniform speed U at the same instant as a projectile is fired from O with elevation α. If the moving target is to be hit, show that the speed of projection must be increased to W, where W is the positive root of the equation

$$W^2 - WU \operatorname{cosec} \alpha = V^2$$

10. Two projectiles are fired simultaneously from a point A with the same speed V. They follow different paths, and both strike a point B at a distance d from A and on the same level as A. Show that the difference in their time of arrival at B is $\dfrac{2V}{g}\left(1 - \dfrac{gd}{V^2}\right)^{1/2}$

11. It is required to fire a shell from a gun on a horizontal plane so as to pass over a wall of height h, the plane of the trajectory being perpendicular to the wall. The muzzle velocity is $\sqrt{(2ga)}$ and the angle of projection cannot exceed α, where $\alpha < \frac{1}{4}\pi$. Show that the gun may be fired from any point in a strip of the plane of width $4 \cos \alpha (a^2 \sin^2 \alpha - ah)^{1/2}$. Interpret this result if $h > a \sin^2 \alpha$.

12. A particle is projected from a point O with velocity $\sqrt{(2gH)}$ at an angle α to the horizontal. Taking O as origin, the y-axis being vertically upwards, show that the equation of the trajectory is $y = x \tan \alpha - \dfrac{x^2(1 + \tan^2 \alpha)}{4H}$

A projectile fired at elevation $\tan^{-1}(\frac{1}{3})$ from a point O 60 m above sea-level strikes the sea at a point 120 m from the vertical line through O. Find the speed of projection.

It is desired to increase the range from 120 m to 121 m without altering the speed of projection. By applying differentiation to the above equation find an approximate relation between corresponding small changes in x and α when y is constant, and deduce an approximate value for the appropriate change in elevation, giving the answer in minutes of arc.

10

Work, Power, and Energy

1. Work

When a force moves the point to which it is applied, it is said to do work. It will be convenient to consider first a constant force P.

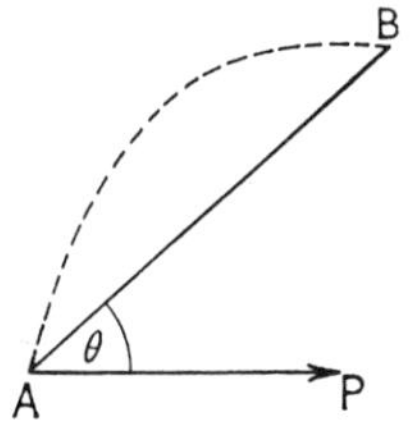

If its point of application moves from A to B by any path whatever, the work done is defined as P . AB $\cos\theta$, where θ is the angle between the directions of AB and P. When $\theta = 90°$, the work done is zero. θ is obtuse, the work done by P is negative, and this is sometimes expressed by saying that work is done *against* P. The amount of work may also be found by resolving P into components parallel and perpendicular to AB. The component parallel to AB does work $P\cos\theta$. AB; the other component does no work.
Note that work is a scalar quantity; no distinction is made between work done by a horizontal force in moving a body horizontally and work done by a vertical force in moving a body vertically.

2. Units

In the SI system the unit of work is the *joule* (J), which is the work done by a force of 1 N in moving its point of application 1 m along its line of action.

When a force varies, in magnitude or direction or both, we consider the motion as taking place in small steps, and arrive at the work done by summing the amounts done in them. We are thus led to the definition of the work done by a variable force P as its point of application moves along a path, of any form, whose length is s. If at any point the force makes an angle θ with the direction of motion, the work done is $\int P \cos \theta \, ds$, between proper limits. The value of this integral may sometimes be found as the area under a force-distance graph. (See example 1, p. 146.)

The effect of a couple is to rotate the body to which it is applied. Suppose that a body is rotating about a fixed point O and is acted on by a couple of moment G. We may consider the couple as being provided by forces G/a perpendicular to a line OA, of length a, fixed

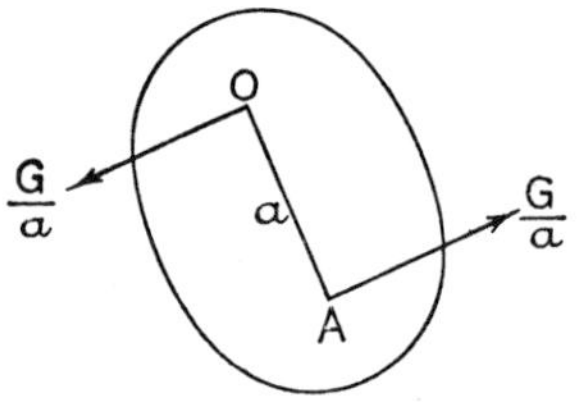

in the body. During a small rotation of amount $\delta\theta$ one of these forces does no work, since O is a fixed point, and the other does work $(G/a)a\,\delta\theta = G\,\delta\theta$. Thus if G is constant the work done in a rotation through angle θ is $G\theta$; if G is not constant the work is given by $\int G\,d\theta$ between proper limits. Note that θ must be measured in radians.

3. Power

The rate at which work is done is called the *power* of the person, animal, or machine doing it. If the component force in the direction of motion is R, the work done in a small displacement δs is $R\,\delta s$, and the power is therefore $R\,ds/dt$, or Rv. The unit of power in the SI system is 1 joule per sec, which is called 1 *watt* (W).

4. Energy

Energy is the capacity for doing work. It may exist in several forms. When coal is burned under a boiler, steam is produced, which may be made to drive an engine coupled to a dynamo, and so to generate electricity. Here we see energy first in a chemical form, then as heat, then in a mechanical form, and finally as electricity.

The *Law of Conservation of Energy*, founded on experience, stated that energy can be changed from one form to another, or converted into work, but cannot be created or destroyed. This law must be modified in view of the fact that in certain circumstances matter is transformed into energy, as for example in the interior of the sun. Considerations such as this are outside our scope. In fact, we are concerned in Mechanics with only two forms of energy: *kinetic energy* and *potential energy*.

5. Kinetic energy

The *kinetic energy* (KE) of a body is the capacity for doing work which it possesses in virtue of its motion, and is measured by the amount of work it can do before coming to rest. If the resultant force on the body in the direction of motion is R, then with the usual symbols

$$R = m\frac{dv}{dt} = mv\frac{dv}{ds}$$

The differential expressing work done is then $mv\,dv$, or $\frac{1}{2}m\,d(v^2)$. Thus the change in the quantity $\frac{1}{2}mv^2$ gives the total work done on the body, and the kinetic energy of a body is measured by $\frac{1}{2}mv^2$. Since energy is convertible into work, the same units are used for both.

The statement just made about the change in $\frac{1}{2}mv^2$ is known as the *Principle of Energy*, and is of great importance. It is best applied in the form:

Final KE—Initial KE = Work done on body by forces.

In calculating the right-hand side of this equation it must be remembered that the work done by a force is positive if the displacement is in the same sense as the component of force along the line of motion, but negative if force and displacement are in opposite senses.

Example 1.—An elastic string of natural length a and modulus λ is gradually extended, first through a distance x and then through a farther distance y. The extension is made so slowly that at any time the extending force may be taken as equal to the tension in the string. Find the amounts of work done on the string in the two stages.

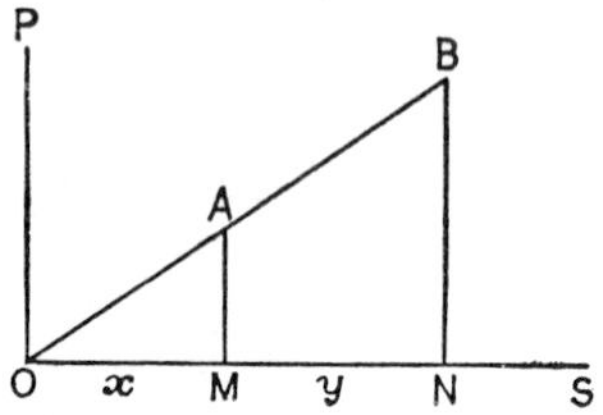

In the diagram, P is the external force, s the elongation. By Hooke's Law the tension in the string at any elongation is $\lambda s/a$ and thus the value of P at A is $\lambda x/a$, and at Q it is $\lambda(x + y)/a$.

$$\text{Work done in first stage} = \int_0^x P\,ds$$
$$= \text{area OAM}$$
$$= \tfrac{1}{2}\lambda x^2/a$$
$$\text{Work done in second stage} = \int_x^{x+y} P\,ds$$
$$= \text{area ABNM}$$
$$= \tfrac{1}{2}\lambda(2x + y)y/a$$

These results may both be expressed as

(Mean of initial and final tensions) $\times$ increase in length

Example 2.—A train of total mass 500 t is travelling uniformly at 72 km/h up an incline of 1 in 100. Resistances to motion amount to $\frac{1}{2}$ per cent of the weight. Find (i) the power exerted by the engine, (ii) the KE of the train.

(i) Since there is no acceleration, the tractive force of the engine is in equilibrium with the opposing forces, gravity and the resistances, and the work done by the engine in any interval is equal to the work done against those forces.

In 1 sec the train moves 20 m up an incline of 1 in 100.

Weight of train $= 500g$ kN vertically downwards.

Vertical displacement $= 0{\cdot}2$ m upwards.

$\therefore$ Work done against gravity $= 100g$ kJ.

Resistance $= 2\frac{1}{2}g$ kN.

Distances moved against resistance $= 20$ m.

$\therefore$ Work done against resistance $= 50g$ kJ.

$\therefore$ Rate of working by engine $= 150g$ kJ per sec

$= 150g$ kW

$\simeq 1470$ kW.

(ii) The formula $\frac{1}{2}mv^2$ gives the kinetic energy. The result is given in megajoules. (1 MJ = 10^6 J.)

$$\text{KE of train} = \tfrac{1}{2} \times 500{,}000 \times (20)^2 \text{ J}$$
$$= 100 \text{ MJ}$$

Example 3.—A particle of mass m kg, attached to a fixed point by a light inextensible string 2 m long, is hanging freely when it is suddenly given a horizontal velocity of 5 m/s. (i) Find its speed when the string is at 60° to the downward vertical. (ii) Find the greatest height to which the particle will rise, assuming that the string does not slacken.

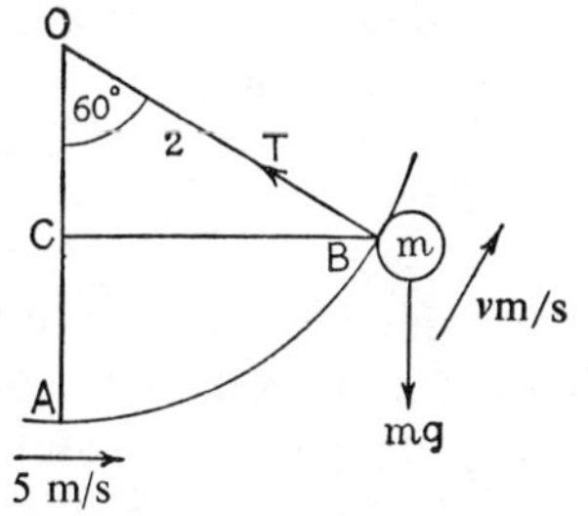

(i) Let the required speed be v m/s. During the motion from A to B the forces acting are the tension T in the string and the weight mg of the particle. Since at any point of the path the tension is at right angles to the displacement, the froce T does no work. Since the weight acts vertically, the displacement to be considered in connection with it is the vertical distance AC, which is 2 – 2 cos 60°, or 1 m. But this displacement is vertically upwards, while the force is vertically downwards, so that the work done by gravity is $-\,mg$ J.

The Principle of Energy gives $\frac{1}{2}mv^2 - \frac{1}{2}m \times 25 = -\,mg$

$$\therefore v^2 = 25 - 2g$$
$$\triangleq 5{\cdot}4$$
$$\therefore v = \pm\, 2{\cdot}3$$

and the speed is 2·3 m/s.

(ii) Let the greatest height above A be h m. Then if the string does not slacken, the velocity at the greatest height is zero. [The proviso is necessary, because if the string rises above the horizontal and slackens, the particle describes a parabola, and the velocity at the highest point is not zero.]

The Principle of Energy gives $0 - \frac{1}{2}m \times 25 = -\,mgh$

$$\therefore h = 25/2g$$
$$\triangleq 1{\cdot}3 \text{ m}$$

Thus we see that the string does not in fact rise above the horizontal. The particle will fall back along its circular path, and the negative value of v found in (i) gives its speed when again passing B.

6. Conservative forces

Conservative forces are forces such that under their influence the work done in moving a body from one point to another is independent of the path taken. Weight is an example. When a body moves under gravity the displacement to be taken into account in calculating the work done is the vertical distance covered, which is the same for all paths between two given points. Another example is a force directed to a fixed point and depending only on the distance from that point, for instance the tension in an elastic string fixed at one end. The work done by the tension depends only on the change of length of the string; if the string changes only in direction no work is done, the force being always perpendicular to the displacement.

7. Potential energy

If a body is influenced only by conservative forces the work done by them in moving it from any position to some standard position (which must be specified in any particular case) is called the *potential energy* (PE) of the body in its original position.

Forces such as friction and air resistance are not conservative, since they act always in the opposite direction to the motion of a body, and do negative work which depends on the length of the path taken. If a body is influenced by non-conservative forces the concept of potential energy cannot be used.

In a system consisting entirely of conservative forces any quantity of work done on the body decreases the potential energy by that amount, and the Principle of Energy then states that:

$$\text{Increase of KE} = \text{decrease of PE}$$

or

$$\text{KE}+\text{PE} = \text{constant}$$

This is sometimes a convenient way of applying the principle, but it must be emphasized that the term *potential energy* has no meaning except in conservative systems of forces. On the other hand, the Principle of Energy as originally stated can always be used, if account is taken of the negative work done by resisting forces.

Energy absorbed by resistances is for the most part converted into heat, which is one of the forms of energy with which we are not concerned while discussing Mechanics. The energy is, from this restricted point of view, lost.

8. Equilibrium

When a particle in a conservative field of force is free to move, the resultant force on it will do positive work. The particle will not, under the forces of the field, move so as to increase its potential energy. In a position of equilibrium the resultant force is zero, and the potential energy has a stationary value. The various cases that arise are illustrated in the following diagrams, which represent a particle under gravity and in contact with a barrier. Two-dimensional motion only is considered.

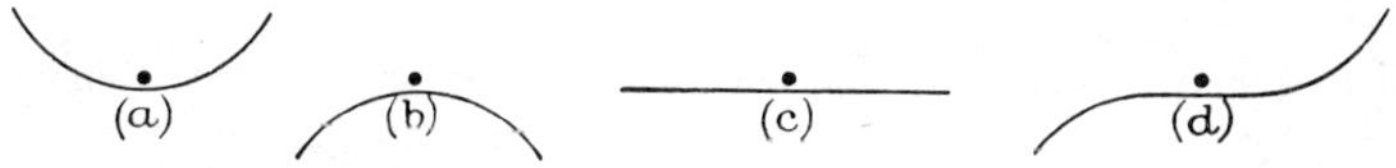

In (*a*) the potential energy is a minimum. If the particle is displaced from its equilibrium position it will return to this position under gravity. The equilibrium is said to be *stable.*

In (*b*) the potential energy is a maximum. If the particle is displaced, it will move farther away from the equilibrium position. The equilibrium is said to be *unstable.*

In (*c*) a particle displaced from an equilibrium position is still in equilibrium. The equilibrium is *neutral.*

In (*d*) equilibrium would appear to be stable for displacement to the right, and unstable for those to the left, but a particle released from a position to the right of that shown would arrive at the equilibrium position with a velocity which would carry it over to the left, so that the equilibrium is practically unstable.

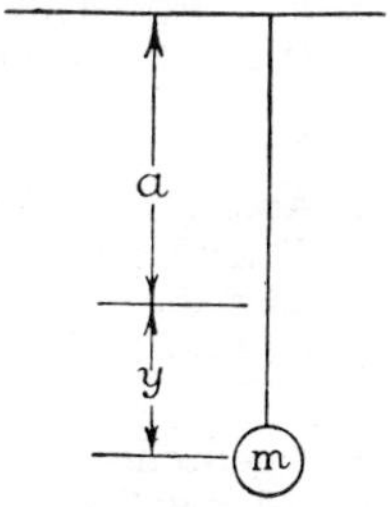

Example 4.—A particle of mass m is attached to a fixed point by an elastic string of natural length a and modulus λ, and can move in a vertical line through the fixed point. Find its potential energy in any position. Hence find the equilibrium position, and show that equilibrium is stable.

Let the 'standard position' for potential energy be at distance a below the fixed point. Then if the particle is at distance y below the standard position, the potential energy due to the string is $\frac{1}{2}\lambda y^2/a$, for this amount of work must be done against the tension of the string to give elongation y (see Example 1) and the tension will do the same amount of work in returning to its natural length.

The potential energy due to gravity is $-mgy$.

Thus the total potential energy $= \frac{1}{2}\lambda y^2/a - mgy = V$ say

$$\frac{dV}{dy} = \frac{\lambda y}{a} - mg$$

$$\frac{d^2V}{dy^2} = \frac{\lambda}{a}$$

The equilibrium position is given by $\frac{dV}{dy} = 0$

$$\frac{\lambda y}{a} = mg$$

$$y = \frac{mga}{\lambda}$$

Since d^2V/dy^2 is positive, the potential energy is a minimum, and equilibrium is stable.

Note.—The equilibrium position can be found without consideration of the potential energy by putting the tension, $\lambda y/a$, equal to the weight, mg. The use of potential energy, however, determines the stability of the equilibrium very shortly and simply.

EXERCISE 20A

1. A radiator consumes energy at the rate of 1 kW. How many megajoules does it use in 1 hour?

2. Find the kinetic energy of (i) a car of mass 750 kg travelling at 36 km/h, (ii) a rifle bullet of mass 10 g travelling at 800 m/s.

3. A mass of 5 kg hangs from a fixed point by an elastic cord of natural length 30 cm. In the equilibrium position the length of the cord is 40 cm. Find the work done (i) in pulling the mass 30 cm vertically downwards, (ii) in moving the mass 30 cm horizontally.

4. A cyclist and his cycle together have mass 50 kg. He can free-wheel at a steady 18 km/h down an incline of 1 in 20. Find the resistance to his motion. If he rides up the same incline at the same uniform speed, what power is he exerting?

5. An engine working at 200 kW draws a train of total mass 200 t at a steady 72 km/h on the level. If air resistance and friction together make a constant retarding force, find its magnitude. Find also the maximum speed of the train up an incline of 1 in 196, if the same resistance acts and the power is unchanged.

6. A horizontal rectangular channel 3 m wide and 1 m deep is full of water moving at 2 m/s. The water falls an average distance of 3 m and drives a turbine at the foot of the fall. 1 m^3 of water has mass 1 t. Find (i) the mass of water flowing per sec, (ii) its kinetic energy at the top of the fall, (iii) the energy per second available for the turbine, (iv) the effective power of the turbine, if its efficiency is 40 per cent.

7. A liner of mass 20,000 t is driven at its maximum speed of 36 km/h when the engines are working at 45,000 kW. Find the resistance to motion in N per tonne.

8. A parcel of mass 5 kg is placed at the top of a chute 40 m long, and slides to the bottom, 20 m below the top. The average frictional resistance throughout is 4 N. Find the velocity of the parcel when it arrives at the bottom.

9. A pulley of diameter d m is driven at n rev/min by a belt. The tension in the tighter side of the belt is T_1 N, and in the slacker side T_0 N. Find the power which is being supplied, assuming that the belt does not slip over the pulley.

10. A motor cyclist and his machine together have mass 200 kg. He goes up a gradient of 1 in 14 at 36 km/h, and comes down the same gradient at 72 km/h, using the same power. Assuming that the resistance to motion varies as the square of the speed, find the power used.

11. The resistance to the motion of a cyclist is kv^2 N, where v m/s is the speed. If his maximum speed on the level is u m/s, show that his greatest power is ku^3 W. If the man and machine together have mass 100 kg, and when exerting his greatest power the speed on the level is 10 m/s, and up an incline of 1 in 20 the speed is 5 m/s, show that his maximum power is about 280 W.

12. A tank contains 180 m³ of water of mass 1 t per m³. It is emptied by a pump in 2 hr at a uniform rate through a horizontal nozzle whose cross-section is 40 cm² and whose axis is 1·5 m above the initial position of the centre of gravity of the water. Find the average speed of the water as it leaves the nozzle and find the average power at which the pump is working.

13. A car of mass 3 t moves along a level road against a constant resistance of 500 N. The greatest tractive force which the engine can exert is 2000 N, and the greatest power is 12 kW. If the car starts from rest and moves so as to have always the greatest possible acceleration, find at what speed the greatest power will be developed, and the least time in which it can acquire this speed. Find also the maximum speed of which the car is capable.

14. The motion of a lift of mass 450 kg in a vertical shaft is subject to a constant frictional resistance of 90 N. An engine, working at 9 kW pulls the lift upwards at uniform speed; find the speed in m per sec. Find also the acceleration of the lift when its speed is 1 m per sec while being hauled upwards, if the engine develops 6·3 kW at this speed.

15. An elastic string of natural length 1·2 m and modulus 12 N is stretched between points 2·4 m apart on a smooth horizontal table. A particle of mass 100 g attached to the middle point of the string is pulled 30 cm from its equilibrium position, along the line of the string, and released. Find its velocity when it passes through the equilibrium position. Answer the same question when the particle is pulled 50 cm from its equilibrium position, perpendicularly to the line of the string.

16. Particles of mass m are attached, one to each end of a long inextensible string, and one to the middle point, and the string is placed symmetrically over two smooth pegs at the same level, distant $2a$ apart. The central particle moves in a vertical line midway between the pegs. Find an expression for the potential energy of the system when the string between the pegs makes an angle θ with the horizontal. Hence find the equilibrium position of the system, and show that it is stable.

17. An elastic string of natural length $2a$ and modulus λ, having a particle of mass m attached to its mid-point, is stretched between points A, B, distant $4a$ apart in a vertical line. Find the potential energy of the system when the particle is at a distance $x(<a)$ below the mid-point of AB. Find the value of x in the equilibrium position, (i) using the potential energy, (ii) by considering the tensions in the two parts of the string. Show that the equilibrium is stable.

18. A cylinder of radius a m can rotate without friction about a horizontal axis. The rotation is opposed by a spring which exerts a couple $k\theta$ N m when the cylinder has turned through θ radians. A particle of mass m kg is attached to the end of a light inextensible string wound round the cylinder, and hangs freely. Find the potential energy of the system when the particle has descended x m from the position in which the spring was unstrained. Find the value of x in the equilibrium position, (i) using the potential energy, (ii) by considering the forces acting on the cylinder. Show that the equilibrium is stable.

EXERCISE 20B

1. One horse-power lifts a mass of 550 lb against gravity at 1 ft/s. Taking 1 lb = 453·6 g and 1 ft = 0·3048 m, express this rate of working in watts, to 3 significant figures.

2. A cyclist, whose mass together with that of his cycle is 70 kg starts at a speed of 54 km/h to ride up a hill 200 m long with a gradient of 1 in 20. His speed at the top of the hill is 18 km/h. Taking the resistance to motion due to friction and air resistance as 24 N, find the amount of work he has done, in joules. Answer to 2 significant figures.

3. A flywheel has mass 5 kg, which may be regarded as concentrated in its rim, of radius 14 cm. Find in joules the kinetic energy of the wheel at 800 rev/min and the work needed to increase the speed by 10 per cent. Take π as $3\frac{1}{7}$.

4. A bullet of mass 10 g passes through a board 2·5 cm thick, and its speed is thereby reduced from 400 m/s to 200 m/s. Find the average resisting force exerted by the board, in kN, to 2 significant figures.

5. Every $12\frac{1}{2}$ hr the level of water in a dock of area 10,000 m^2 is raised 3 m by the tide. If the energy thus supplied could be used at a uniform rate, what power would it provide? Take the mass of 1 m^3 of water as 1 t, and answer to 2 significant figures.

6. A board 3 m long has one end at floor level and the other 1·5 m higher. A particle of mass 250 g is projected up the board, starting from the lower end at 8 m/s. The frictional resistance is $\frac{1}{4}$ N, and air resistance is to be neglected. Find the speed of the particle (i) when it reaches the top of the board, (ii) when it reaches the floor after leaving the board.

7. A train of total mass 300 t is found to run down a slope of 1 in 98 at constant speed with the steam off. Find what power would be needed to pull it up the same slope at a constant speed of 54 km/h, the frictional resistance remaining at the same constant value throughout. Assuming that the engine is working at this power when the train is travelling on the level at 18 km/h, with resistances as before, find the acceleration.

8. Calculate the least power required to pull a train of mass 250 t on a level track at a steady 72 km/h against resistances of 80 N/t. If the engine works at 450 kW when the train is travelling at 72 km/h, what is the acceleration?

9. A truck of mass 5 t is pulled up an incline of 1 in 70 by a rope parallel to the track. Resistances to motion amount to 160 N/t. At an instant when the speed is 18 km/h and the acceleration 0·1 m/s², both up the track, find (i) the tension in the rope, in N, (ii) the power exerted by the hauling engine.

10. A moving staircase which travels at 30 m per min has a rise of 15 m. In one minute the staircase is used by 180 people, of an average mass of 50 kg, travelling upwards. Prove that if they stand still on the staircase approximately 22 kW is required, neglecting friction, and find the power needed if they walk up, moving at 24 m per min relative to the staircase.

11. Find the average power required to pump out a dock 100 m long, 30 m wide, and 6 m deep in 4 hr, delivering the water at a uniform rate through a pipe of cross-sectional area 0·1 m², which is 10 m above the bottom of the dock. Take the mass of 1 m³ of water as 1 t, and the efficiency of the pump as 80 per cent.

12. A pump delivers m kg per sec of water having density ρ kg/m³ through an orifice of area A m², at a height h m above the intake. Neglecting friction, prove that the power required is

$$mgh + \frac{m^3}{2\rho^2 A^2} \text{ watts.}$$

Find the least power required to deliver 75 kg of water per sec at a height of 100 m, whatever the size of orifice, and find the area of orifice which involves an increase of 3 per cent in this power.

13. A ship of mass 10,000 tonnes is driven at a uniform speed of 36 km/hr by engines working at 12,000 kilowatts. The resistance to motion varies as the cube of the speed. Prove that $\frac{dv}{dt} \times 10^7 = F - 1200v^3$, where F newtons is the propulsive force, and v m/s is the speed. (1 tonne = 1000 kg.) The engines are stopped when the ship is travelling at 36 km/h. Find the time after which the speed is reduced to 18 km/h.

14. When a spiral spring has been wound up through θ radians from the unstressed position it exerts a torque of $k\theta$ N m. The rotation is opposed by a constant frictional couple of F N m. Find the work done in winding the spring through n revolutions from the unstressed position.

15. A car of total mass 2 t is driven up a slope of 1 in 98. The frictional resistance is constant, and of magnitude 400 N. The car starts from rest at a point A; when it reaches B, distant 40 m from A, its speed is 18 km/h and its acceleration is zero. Find the power at which the engine is working when the car is at B. Find also, in joules, the work done by the engine in driving the car from A to B.

16. A light rod of length a is freely hinged to a fixed point O. A particle of mass m is attached to the other end A of the rod, and connected by an elastic string of natural length a and modulus mg to a point B, at distance a vertically above O. Find an expression for the potential energy of the system when angle AOB = θ and the string is in tension. Hence find the equilibrium position of the system, and show that it is stable.

17. A light rod is pivoted freely at one end and supported by a vertical elastic string attached to a point on it distant $2a$ from the pivot. The string is so long that it may be assumed to remain vertical when the rod is displaced, and it exerts an upward force mgx/a when the point of attachment is at a distance x below the level of the pivot. A particle of mass m is attached to the rod at a distance a from the pivot. Find the potential energy of the system when the rod is at an angle θ below the horizontal, and show that in equilibrium $\theta = \frac{1}{2}\pi$ or $\theta = \sin^{-1}\frac{1}{4}$. Investigate the stability of the two positions of equilibrium.

18. A long smooth light rod is pivoted freely at one end and supported by a plunger working in frictionless guides in a vertical line at a distance b from the pivot. The plunger exerts an upward force kx when the point of contact with the rod is at a distance x below the level of the pivot. A particle of mass m is attached to the rod at a distance a from the pivot. Show by using potential energy that in equilibrium $kb^2 \sin\theta = mga\cos^4\theta$, where θ is the angle by which the rod is below the horizontal. Verify this result by taking moments about the pivot for the forces on the rod. [The reaction between the rod and the plunger needs careful consideration.] Show that the equilibrium is stable.

9. Machines

It is possible by arrangements of ropes and pulleys, gear wheels, or screws, to change the point of application of a force and at the same time to alter its magnitude. Devices for doing this are known as *machines*, and the most common form is that in which a weight is lifted. The effect is said to be produced on a *load*, by an *effort* applied to the machine.

Work is done by the effort, and in an ideal machine the whole of this would be applied to the load. In practice, work is absorbed by resistances, and the ratio

$$\frac{\text{work done on load}}{\text{work done by effort}}$$

is called the *efficiency* of the machine.

It is often expressed as a percentage.

The arrangement usually is that the effort shall be of smaller magnitude than the load, but shall move through a greater distance. The ratio

$$\frac{\text{distance travelled by effort}}{\text{distance travelled by load}}$$

is called the *velocity ratio* of the machine.

It can be found by examining the design of the machine, as will be shown.

The ratio $\dfrac{\text{load}}{\text{effort}}$ is called the *mechanical advantage* of the machine.

It varies with the load, and can be found only by experiment.

It is easy to see that $\dfrac{\text{mechanical advantage}}{\text{velocity ratio}}$ gives the efficiency of the machine for any particular load.

The resisting forces are often independent of the load, or nearly so, and if we assume constant resistances, the behaviour of a machine can be investigated by simple algebra.

Let the velocity ratio be k, and let the effort P move through a distance x. Then the load W moves through a distance x/k. Let the work done against resistances be Rx.

Then
$$Px = Wx/k + Rx$$
$$\therefore\ P = W/k + R$$

The graph of P against W will be a straight line. If this is verified in practice the values of k and R can be found from the graph.

The mechanical advantage

$$= \frac{W}{P} = \frac{k(P-R)}{P} = k\left(1-\frac{R}{P}\right),$$

so that as load and effort increase, the mechanical advantage tends towards equality with the velocity ratio, and the efficiency, which is $1/k$ times the mechanical advantage, approaches unity or 100 per cent. (The strength of the machine may impose a limit on this tendency.)

If the effort is removed before the load, it is important that the load should not move under its own weight. This can be ensured by various devices, but they are not necessary if the efficiency of the machine is less than 50 per cent. When the machine is operating normally,

work done by effort
$$= \text{work done on load} + \text{work done against resistances}$$

and if the efficiency is less than 50 per cent, the work done on the load is less than that done against resistances. Consequently, the machine cannot operate in reverse under the influence of the load only, for the resistances are reversed in direction, and the load alone is not sufficient to overcome them; a reversed effort would be required in addition.

Example 5.—The diagram shows a lifting tackle, W and P indicating load and effort respectively. The pulleys have been drawn of different sizes to show the run of the cord; they are in practice equal. The lengths of cord between the pulleys may be taken as vertical, and the weight of the lower block is neglected.

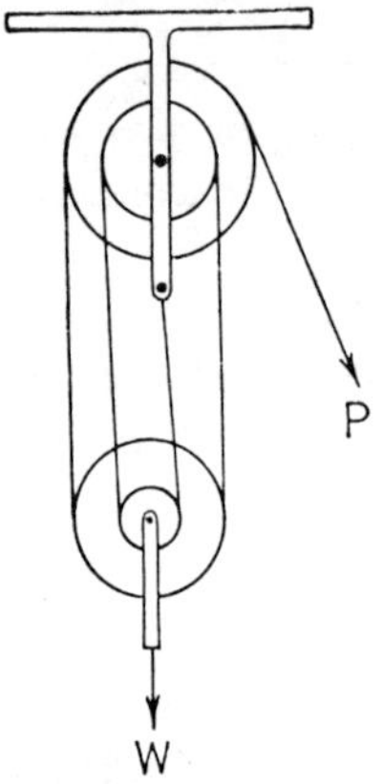

(i) Find the velocity ratio.

(ii) A load of 100 N requires an effort of 40 N. Find the efficiency.

(iii) If the resisting forces are constant, what effort will be required for a load of 160 N, and what will be the efficiency for this load?

(i) If the effort P moves 1 m, the total length of cord between the pulleys is reduced by 1 m.

$\therefore$ Each of the straight lengths is reduced by $\frac{1}{4}$ m.

$\therefore$ Velocity ratio $= 4$.

(ii) Mechanical advantage $= \frac{100}{40} = 2\frac{1}{2}$.

$$\text{Efficiency} = \frac{\text{M.A.}}{\text{V.R.}} = \frac{5}{8} = 62\tfrac{1}{2} \text{ per cent.}$$

(iii) When the effort moves x m, let the work done against resistances be Rx J.

Work done by effort = work done on load + work done against resistances.

With data of (ii), $40x = 100x/4 + Rx$

$$\therefore R = 15$$

so that 15 N of the effort is used in overcoming resistances.

If P_1 N is the effort for a load of 160 N,

$$P_1x = 160x/4 + 15x$$

$$\therefore P_1 = 55$$

and efficiency $= \frac{160}{55} \times \frac{1}{4} = \frac{8}{11} \simeq 73$ per cent.

EXERCISE 21

1. A barrel is rolled up a pair of rails inclined at 45° to the horizontal by a rope fixed at the top of the rails, passing under and round the barrel, and pulled in a direction parallel to the rails. Find the velocity ratio of the arrangement as a means of lifting a weight. If a pull of $40g$ N will move a barrel of mass 100 kg, what is the efficiency?

2. A winch has a cylinder of 20 cm diameter on which a rope is wound. A gear wheel with 150 teeth, fixed to the axis of the cylinder, is driven by a wheel with 8 teeth on a shaft turned by a handle 60 cm long. An effort of 10 N is applied at the end of the handle and perpendicular to it. If the efficiency is 40 per cent find the load that can be moved.

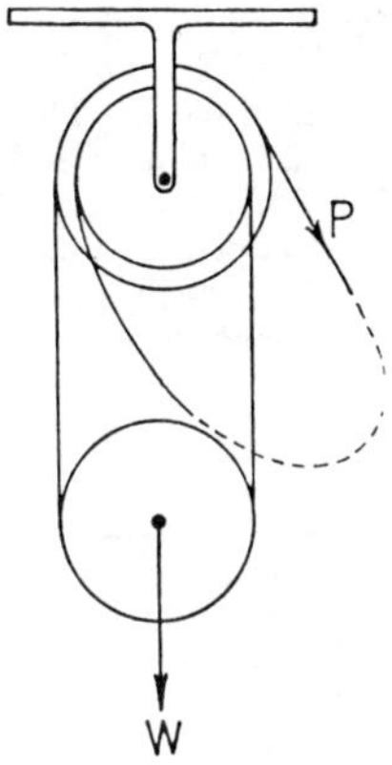

3. The diagram shows a 'differential' block and tackle. The pulleys in the upper block have radii a and b ($a > b$) and are fixed together. An endless chain moves the upper pulleys without slip. Consider what happens to the loop in which the lower loose pulley hangs when the upper pulleys make one revolution, and so prove that the velocity ratio is $2a/(a - b)$.

4. In a car jack, a screw of 5 mm pitch is turned once (i.e. the screw advances 5 mm) for every rotation of a shaft which is turned by two equal and opposite forces P applied to a handle, at distances of 18 cm from the shaft, on opposite sides of it. If the efficiency is 40 per cent, what forces P would be required to lift 500 kg?

5. A system like that in example 5 has 3 pulleys in each block. Experiment gives the following results:

Load in N	10	20	30
Effort in N	2·10	3·77	5·43

Find the mechanical advantage and efficiency in each case (to 2 significant figures). Show that the results are consistent with the hypothesis of a constant resistance, and find an expression for the effort P in terms of the load W.

6. A mass of 40 kg is pushed up a plane inclined at 30° to the horizontal. Regarding the plane as a machine, find its efficiency (i) if the effort is parallel to the plane, (ii) if the effort is horizontal. The coefficient of friction between the mass and the plane is 0·4.

7.

Load W N	20	40	60	80	100
Effort P N	4·8	6·2	7·5	8·8	10·2

The table gives the results of experiments on a machine. Draw a graph and find an equation which approximately gives P in terms of W. Estimate the velocity ratio, and find the efficiency when the load is (i) 20 N, (ii) 100 N. (iii) What is the greatest load that can be left on the machine in the absence of effort, without running backwards?

8. A car wheel is attached to the hub by five bolts with thread of 1 mm pitch. The nuts are tightened by force applied to a wrench, at a distance of 20 cm from the axis of the bolt. Regarding the wrench and nut as a machine, find its velocity ratio, and if the efficiency is 30 per cent find the total force holding the wheel against the hub when all the nuts are finally tightened by a force of 80 N applied to the wrench.

9. A sluice gate is raised and lowered by a hand wheel 0·72 m in diameter. A screwed rod with a thread of 1 cm pitch goes through a nut in the centre of the hand wheel, and the sluice gate is attached to the lower end of the rod. The efficiency of the screw and nut is 30 per cent. The weight of the sluice gate and rod is 800 N, and the pressure of water against it causes friction with its guides amounting to 960 N. What force applied tangentially to the rim of the hand wheel will (i) raise, (ii) lower, the gate?

10. A man working on the face of a building stands on a movable platform suspended from the roof by tackle like that in example 5, but with 3 pulleys in each block. The free end of the rope is coiled on the platform, and the man raises himself and the platform, a total **load of 1200 N** by pulling on the rope. When he has increased the length coiled on the platform by **1 m,** how far has the platform moved upwards? If the efficiency of the tackle is 80 per cent, how much work has he done? How far down has the point of application of the effort moved? What (uniform) force has the man been exerting?

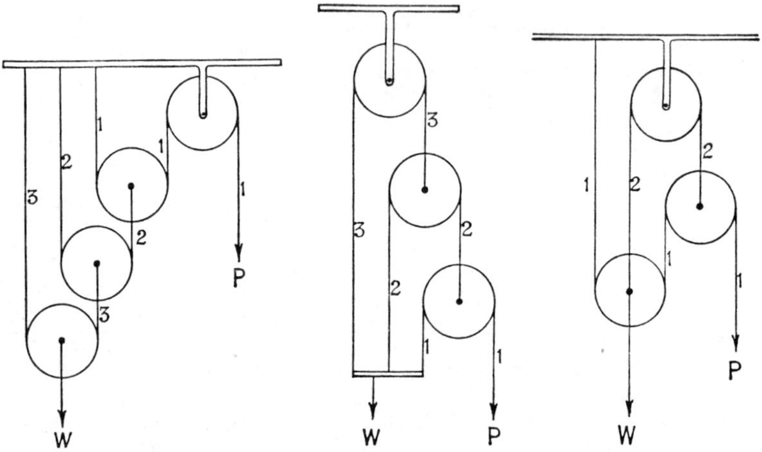

11. In the systems of pulleys shown, find the velocity ratios. Assuming that there is no friction, find in each case the tensions in the separate strings, numbered 1, 2, 3.

12. A is a gear wheel, engaging with three gear wheels B, equal in size to A, and freely mounted in a ring R. C is an internally toothed wheel, engaging with the wheels B.

(i) What is the ratio of the numbers of teeth in wheels A and C?

(ii) If R is fixed, and A makes one revolution anticlockwise, how far does C turn, and in which direction?

(iii) If A is fixed, and R makes one revolution clockwise, how far does C turn, and in which direction? [Consider first the motion relative to R.]

(iv) If A is fixed, and C makes one revolution clockwise, how far does R turn, and in which direction?

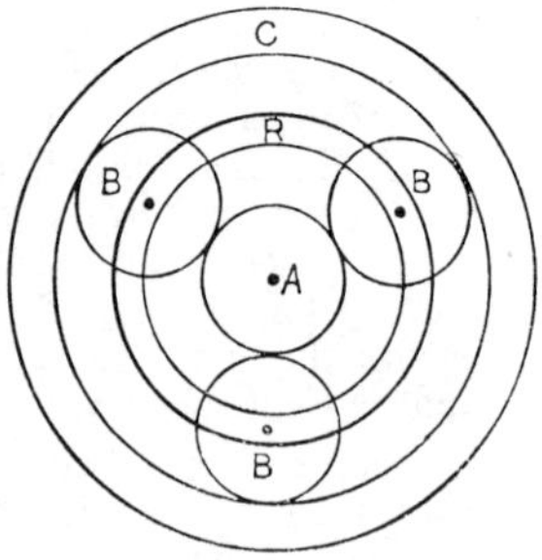

11

Impact

1. Impulse

If a constant force P acts for a time t, the product Pt is called the *impulse* of the force. If m is the mass of the body on which the force acts, it follows that

$$Pt = mft = mv - mu$$

where u and v are the initial and final velocities. Thus

$$\text{impulse} = \text{change of momentum}$$

It is to be particularly noted that impulse is a vector, having direction as well as magnitude. Therein it differs from work, which is a scalar quantity, having magnitude only.

When the force P is variable, the impulse is given by $\int P\,dt$ between proper limits, and

$$\int P\,dt = \int m\frac{dv}{dt}\,dt = \int m\,dv$$

giving the change of momentum as before. If the force varies in direction it will be necessary to consider separately its components in fixed directions, but this case will not often arise.

We are usually concerned with impulse when a very large force acts for a very short time, as when two bodies collide. It is not easy in such cases to estimate either the magnitude of the force or the duration of the impact, and the effect is measured by the change of momentum produced.

There is no special name for units of impulse. They may be described as newton seconds. The equations above show that when mass and velocity are measured in kilogrammes and metres per second the momentum will be given in newton seconds.

2. Conservation of Momentum

When two bodies collide they exert equal and opposite forces on each other, for the same time. The impulses on the two bodies are therefore equal and opposite, and so are the changes in momentum. Thus the total momentum of the two bodies together is unchanged by the impact.

This is an example of the *Principle of Conservation of Momentum*, which states that the total momentum of a system is constant in any direction in which no *external* force acts.

The principle is usually employed to compare velocities before and after an impact, and the time elapsing is so short that only the very large ' impulsive ' forces involved in the impact produce any measurable effect. Forces of ordinary size, such as the weights of bodies, may be neglected in this application of the principle, but the possibility of external impulsive forces must be remembered.

Impulsive forces act not only when bodies collide, but also, for example, when an inextensible string suddenly tightens, when a gun is fired, and when a moving body is suddenly stopped by an obstacle. In connection with collisions it should be observed that we consider at present only inelastic impacts, that is, impacts after which there is no rebound.

3. Energy

The effect of an impact is not only to change the velocity of a body. The particles composing the body are set in motion among themselves, and the result may be a change of shape, or heating of the body, or both. These changes absorb mechanical energy, and it will be shown that there is in practical cases always a loss of kinetic energy in a system in which impacts occur. The principle of energy must therefore not be used so as to include any time in which an impact takes place, though it may be used up to the moment of impact, and on from that moment.

Example 1.—A pile of mass $\frac{1}{2}$ t is driven into the ground by a mass of 2 t which is dropped on to it from a height of 2·5 m. Find the magnitude of the impulse, in N s, and the loss of kinetic energy, in kJ. If the blow drives the pile 10 cm into the ground, find the resisting force of the ground in kN.

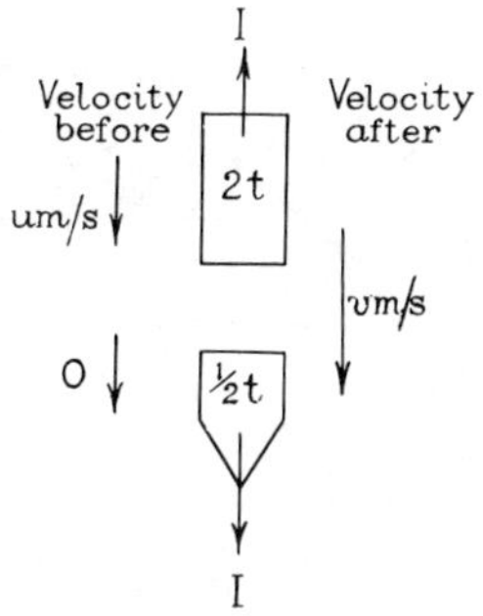

If I N s is the impulse, then

for the mass, $\quad -I = 2000(v - u)$

for the pile, $\quad I = 500v$

$$\therefore 0 = 2500v - 2000u$$

$$\therefore 2\tfrac{1}{2}v = 2u$$

Note that if the impulse had not to be found, we could write down this equation at once, using the principle of Conservation of Momentum.

The velocity u is given by $\quad u^2 = 2 \times 9{\cdot}8 \times 2{\cdot}5$

$$\therefore u = 7$$

$$\therefore v = \tfrac{4}{5}u = 5{\cdot}6$$

and $\quad I = 1000 \times 5{\cdot}6 = 5600$ N s

(given to 2 figures, since g has been taken as 9·8)

KE before impact

$$= \tfrac{1}{2} \times 2000 \times 49 \text{ J} = 49 \text{ kJ}.$$

KE after impact

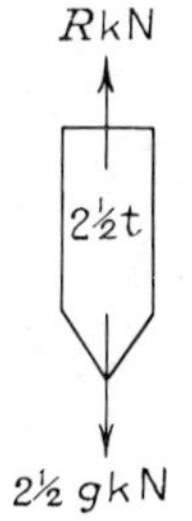

$$= \tfrac{1}{2} \times 2500 \times (5{\cdot}6)^2 \text{ J} = 39{\cdot}2 \text{ kJ}$$

$$\therefore \text{Loss of KE} = 9{\cdot}8 \text{ kJ}$$

Let R kN be the resistance of the ground. Since the pile and weight are brought to rest in 0·1 m, we have, using the Principle of Energy, and working in kN and kJ,

$$-39{\cdot}2 = -(R - 2\tfrac{1}{2}g) \times 0{\cdot}1$$

$$\therefore R - 24{\cdot}5 = 392$$

$$\therefore R \triangleq 420 \text{ kN}$$

Example 2.—A gun of mass M kg and free to move horizontally discharges a shell of mass m kg when the barrel is at an angle of elevation α. Find the direction in which the shell begins to move.

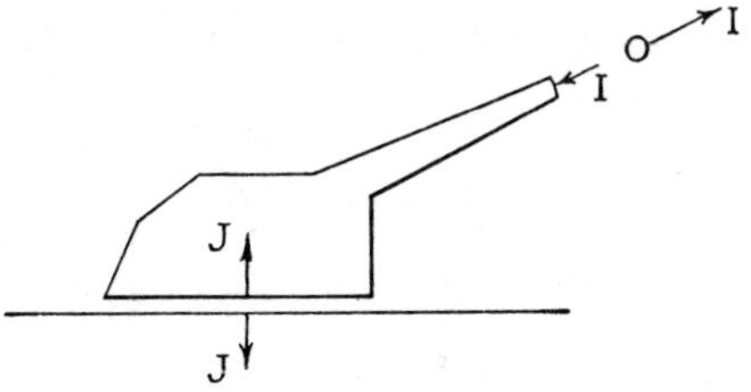

The shell is made to leave the gun by an impulsive action along the line of the barrel, but because the gun does not move vertically, the vertical component of the reaction on the gun must be balanced by a vertical impulsive action between the gun and the ground.

Let the velocity of the shell relative to the gun be u, at elevation α, and let the horizontal velocity with which the gun begins to recoil be v.

The initial velocity of the shell relative to the ground is compounded of u and v, as in the figure (p.162).

The system of gun and shell is acted on by an external vertical impulse, so that we may apply the principle of Conservation of Momentum in the horizontal direction only. The horizontal velocity of the shell relative to the ground is $u \cos \alpha - v$, and the principle gives

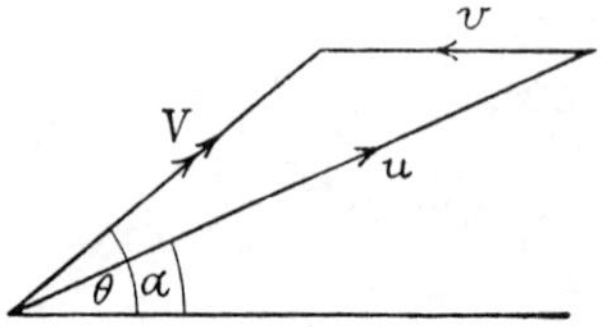

$$m(u \cos \alpha - v) - Mv = 0$$

$$\therefore\ mu \cos \alpha = (M + m)v$$

Now if the initial velocity of the shell relative to the ground is V, at elevation θ,

$$V \sin \theta = u \sin \alpha$$

and

$$V \cos \theta = u \cos \alpha - v$$
$$= u \cos \alpha - \frac{m}{M + m} u \cos \alpha$$
$$= \frac{M}{M + m} u \cos \alpha$$
$$\therefore\ \tan \theta = \frac{M + m}{M} \tan \alpha$$

Example 3.—Masses A, B, of 1 kg and 6 kg respectively, are connected by an inextensible string 6 m long which passes over a small smooth pulley 3·5 m above the floor. When B is resting on the floor, A is dropped from the level of the pulley. Find the impulsive tension in the string, the height to which B rises, and the time which elapses before B again hits the floor. Describe the subsequent motion of the system.

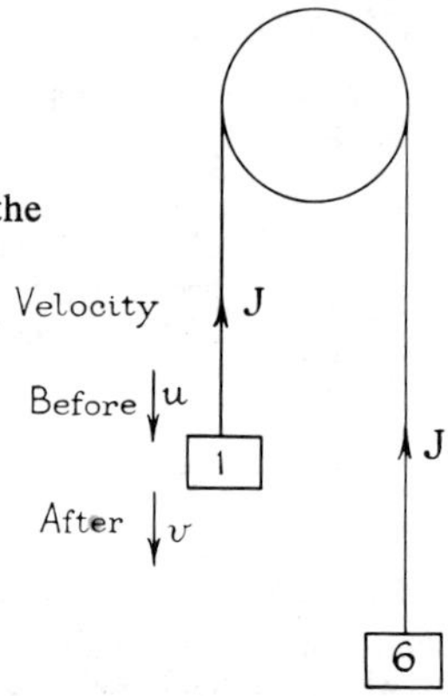

Let J N s be the jerk, or impulsive tension, in the string.

For A, $\quad v - u = -J$

For B, $\quad 6v = J$

$$\therefore\ 7v - u = 0 \quad . \quad . \quad . \quad . \quad . \quad . \quad . \quad . \quad \text{(i)}$$

Now $\quad u = \sqrt{(2g \times 2{\cdot}5)} = 7$

$$\therefore\ v = 1$$

and $\quad J = 6$ N s

It is to be noted that if the magnitude of the jerk had not been required, equation (i) could have been derived by considering the equivalent system given by regarding the string as straight (see Chapter 8, example 2, where acceleration is found in this way).

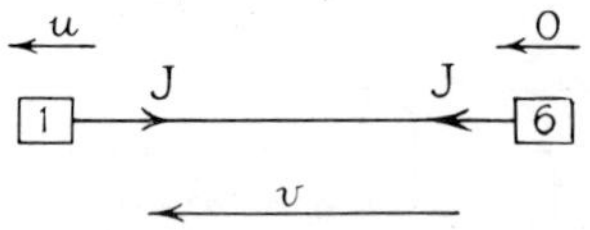

The impulses J are now internal to the system, and by Conservation of Momentum, $7v = u$.

The system now has an acceleration, which may be found from the equivalent system shown below.

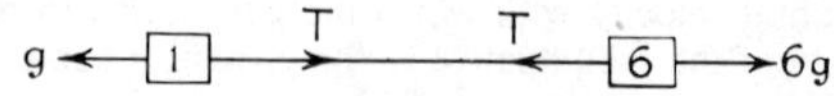

The acceleration is seen to be $\frac{5}{7}g \simeq 7$ m/s², in the opposite direction to v.

B will then rise to a height given by the use of the standard formula $v^2 - u^2 = 2fs$.

We have $$0 - 1 = -2 \times 7h$$

and $$h = \tfrac{1}{7}$$

The time to reach this height is given by $v = u + ft$.

$$0 = 1 - 7t$$

$$t = \tfrac{1}{7}$$

and B returns to the floor after twice this time, i.e. after $\frac{2}{7}$ sec.

The string now slackens, and A moves under gravity, with the initial upward velocity of 1 m/s, for the velocity of the system is reversed at the end of B's excursion. A will return to the position where the string slackened after a time given by $0 = t - \frac{1}{2}gt^2$, i.e. after approx. 0·452 sec, and will then have a downward velocity of 1 m/s.

The string now tightens again, and the whole cycle is repeated, but since the initial velocity is now 1 m/s instead of 7 m/s, and the masses and accelerations are unaltered, all velocities and times will be divided by 7.

The time from the first to the second jerk is $0{\cdot}143 + 0{\cdot}452 \simeq 0{\cdot}6$ sec. There will then be jerks at intervals given by the terms of the series $0{\cdot}6[1 + \frac{1}{7} + \frac{1}{49} + \ldots]$ sec.

The limiting sum of this series is $0{\cdot}6 \times \dfrac{1}{1-\frac{1}{7}} = 0{\cdot}7$ sec, and after this time (measured from the first jerk) all motion will have ceased, and B will remain on the floor.

Example 4.—Water is discharged horizontally from a nozzle of area A m² at a speed of v m/s. It strikes a vertical wall, and its horizontal velocity is thereby destroyed. Find the force exerted on the wall. The density of water is to be taken as k kg/m³.

In one second Av m³ of water strike the wall. This water has mass kAv kg, and velocity v m/s horizontally, so that horizontal momentum is destroyed at the rate of kAv^2 units per sec.

This change of momentum is the force exerted by the wall on the water, and the equal and opposite reaction to it is the force exerted by the water on the wall.

The unit of momentum being the newton second, the required force is kAv^2 N.

Example 5.—An inextensible string AB has particles of masses m, M at A, B respectively. An impulse of magnitude I is applied to B, at an acute angle α with AB. Find the initial velocities of A and B, and the impulsive tension in the string.

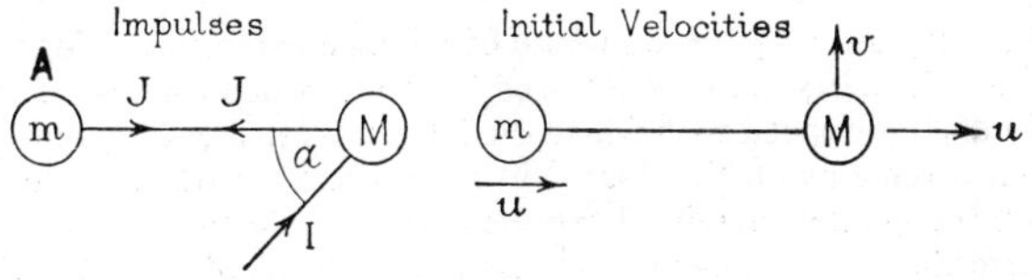

The particle at A can move only in the direction AB. Let its initial velocity be u.

Since AB is inextensible, the velocity of B in the direction AB must be equal to that of A. Let its velocity perpendicular to AB be v, and let the impulsive tension in the string be J.

Using the principle that impulse = change of momentum for B, perpendicular to AB,

$$I \sin \alpha = Mv \quad \text{(i)}$$

for B, parallel to AB,

$$I \cos \alpha - J = Mu \quad \text{(ii)}$$

for A, parallel to AB,

$$J = mu \quad \text{(iii)}$$

These equations easily give

$$u = \frac{I \cos \alpha}{M + m}$$

$$v = \frac{I \sin \alpha}{M}$$

and

$$J = \frac{Im \cos \alpha}{M + m}$$

Note that if J is not to be found the values of u and v can be found by considering the external impulse and the momentum of the whole system, perpendicular and parallel to AB, giving

$$I \sin \alpha = Mv$$

$$I \cos \alpha = (M + m)u$$

EXERCISE 22A

1. A railway truck of mass 10 t and travelling at 10 m/s overtakes another truck of mass 5 t, moving at 4 m/s, on the same line, and the two travel on together. Find (i) their common speed, (ii) the impulse in N s, (iii) the loss of kinetic energy in J.

2. A truck of mass 10 t and moving at 8 m/s on the level runs into a stationary truck of mass 6 t on the same line. The wheels of the second truck are locked, and the coefficient of friction between them and the rails is $\frac{1}{4}$. How far will the trucks move after the impact? The friction on the first truck is negligible.

3. A gun of mass 2 t discharges a 10 kg shell at 800 m/s. If the gun is free to recoil in the direction of the axis of the barrel, what constant force will bring it to rest after it has moved 1 m?

4. If the recoil of the gun in question 3 is opposed by a spring which is compressed 1 cm by a force of 80 N, what distance will the gun recoil?

5. A gun of mass 2 t is mounted on a slope of 30° to the horizontal and is free to move up the slope. It discharges a shell horizontally, in a vertical plane through a line of greatest slope. The shell has mass 10 kg and its initial velocity is 700 m/s. How far up the slope will the recoil take the gun?

6. Masses of 3 kg and 5 kg are connected by a long inextensible string which passes over a smooth pulley, and are released from rest when each is 2·5 m above the floor. Find (i) the speed with which the 5-kg mass strikes the floor, (ii) the time for which it remains on the floor, (iii) the speed with which it then leaves the floor, (iv) the greatest height above the floor reached by the 3-kg mass.

7. A uniform chain 20 m long and of mass 40 kg is lowered on to a horizontal floor at a uniform speed of 8 m/s. Find the force on the floor when half of the chain is on the floor.

8. An aircraft fires straight ahead eight machine guns. Each gun fires 600 rounds per minute. The mass of a bullet is 10 g, and its initial velocity 800 m/s. Find the backward force on the aircraft produced by firing the guns.

9. Particles of masses 4 kg and 3 kg hang in contact at the ends of strings of length 1·25 m. The 4-kg mass is drawn aside so that the string moves through 60°, and released. When the particles collide they stick together. Find (i) the kinetic energies of the system just before and just after the collision, (ii) the angle through which the strings move after the impact before coming instantaneously to rest.

10. A particle of mass M hangs in equilibrium at the end of a light inextensible string of length l. A particle of mass m moving horizontally with velocity v collides with and coalesces with the first particle, causing it to swing from the equilibrium position through an angle whose chord is of length c. Show that

$$v = \frac{c(M + m)}{m}\sqrt{\frac{g}{l}}$$

11. A string of length $4a$ carries at one end a particle of mass M and at the other end a particle of mass m ($<M$). If the string is slung over a small smooth pulley at a height $3a$ above a table and the system is released from rest when each particle is at a height a above the table, find the velocity of the mass M when it strikes the table. If on striking the table the mass M is instantly reduced to rest, prove that the mass m never reaches the pulley and that the mass M is subsequently jerked into motion with velocity $\dfrac{m}{M + m}\sqrt{\left\{\dfrac{2(M - m)ga}{M + m}\right\}}$

12. Three equal particles A, B, C rest on a smooth horizontal plane at the corners of an equilateral triangle. Equal light taut strings of length l connect A to B, and B to C. The particle A is given a speed u in a direction parallel to CB. Show that C will eventually begin to move with a speed $\frac{1}{15}u$, and that CB will begin to turn with an angular velocity $\frac{2}{15}\sqrt{3}(u/l)$.

EXERCISE 22B

1. A railway truck of mass 20 t collides with another of mass 5 t moving in the opposite direction on the same level track at 3 m/s. After the collision the trucks move together, and the velocity of the 5-t truck has been exactly reversed. Find (i) the velocity of the 20-t truck before the impact, (ii) the magnitude of the impulse in N s, (iii) the percentage of kinetic energy lost in the impact.

2. A hammer head of mass 0·5 kg moving horizontally at 3 m/s strikes a nail of mass 10 g and drives it 1 cm horizontally into a block of wood. Find (i) the average resistance exerted by the block, (ii) the percentage of KE lost in the impact.

3. Prove that when a mass M strikes a mass m which is at rest, and moves on with it, the fraction of its kinetic energy lost is $m/(M + m)$.

4. A pile of mass 0·4 t is driven into the ground by a mass of 1 t dropped on it from a height of 2·5 m. If the resistance of the ground to penetration by the pile is 140 kN, find the depth of penetration due to one blow of the mass.

5. Under the conditions of question 4, find an expression for the depth of penetration when W t is substituted for 1 t as the dropping mass ($W < 13$).

6. Masses A, B, of 990 g and 970 g respectively, are attached to the ends of an inextensible light string which passes over a smooth pulley. A bar C of mass 40 g rests on a ring through which B can pass. When B moves upwards through the ring it picks up C; when B moves downwards through the ring it deposits C on the ring. The system is started so that B approaches the ring from below at 100 cm/sec. Find the time which elapses (i) before C is again deposited on the ring, (ii) before C is picked up for a second time. Take g as 980 cm/s^2. Describe the subsequent behaviour of the system, and show that all motion will have ceased after 33 min.

7. A bullet of mass 10 g is fired horizontally into a block of wood of mass 2 kg, which hangs by parallel strings 2 m long. The bullet is moving at 300 m/s just before it hits the block. Through what angle will the strings move before coming instantaneously to rest?

8. Under the conditions of question 7, if the strings swing through 60° before coming to rest, what is the speed of the bullet before the impact? Answer correct to 2 figures.

9. Particles of masses $2m$, $3m$ are connected by a light inextensible string passing round a light smooth pulley, and pulley and particles lie on a smooth horizontal table. A second string has one end attached to the axle of the pulley, and carries at the other end a particle of mass m. Initially the two portions of the first string are taut and perpendicular to an edge of the table, and the particle of mass m is held just off this edge, with the second string slack and with the line joining the particle to the axis of the pulley parallel to each portion of the first string. The particle is then released, and after it has fallen a distance h the particles on the table are jerked into motion. Find the impulses in the two strings and show that the loss of kinetic energy due to the jerk is $\frac{24}{29}\,mgh$.

10. Four particles, each of mass 1 kg, are in contact in a straight line on a smooth horizontal floor. Adjacent particles are connected by lengths of 1 m of inextensible light string, initially slack. The particle at one end is given a velocity of 4 m/s in a line with the other particles and away from them. Find (i) the intervals of time at which the other three particles are jerked into motion, (ii) the magnitudes of the jerks in each length of string as successive particles start, (iii) the amounts of kinetic energy lost as each particle starts.

11. A block of mass 5 kg is moving in a straight line at 6 m/s when it is struck by a particle of mass 1 kg, moving in a perpendicular line at 4 m/s, which then adheres to the block. Find the angle through which the direction of motion of the block is deflected.

12. Four particles, each of mass m, are placed at the corners of a rhombus ABCD, and connected by light inextensible strings which lie along the sides of the rhombus. The angle ABC $= 2\alpha$. Impulses of magnitude I are applied to B and D, in the line BD and outwards. Find the initial speeds of B and C.

4. Restitution

When two bodies collide, they may remain in contact after the collision. This is the kind of impact we have so far considered, and it is called *inelastic*. On the other hand, it is frequently observed that after a collision the bodies separate again. In this class of impact, consideration of momentum alone does not enable us to find the final velocities. We use, in addition to the Principle of Momentum, an experimental result due to Newton, which states that

Relative velocity of bodies along line of impulse after impact
$= -e$ (relative velocity before impact)

The law may also be stated in the form

Relative velocity of separation $= e$ (relative velocity of approach).

The number e is called the *coefficient of restitution*, and depends on the materials of the bodies colliding. It has for different materials values ranging from 0 to 1. The value $e = 0$ corresponds to the inelastic impacts already considered. The value $e = 1$ is described as 'perfect restitution', and is not realized in practice.

5. Loss of energy

In any particular case the loss of kinetic energy can be found by calculating the values before and after impact, but it is instructive to deal with the matter generally, as follows.

Let m, km be the masses of two particles, moving along the same straight line.

Their distances from their centre of gravity are at all times in the ratio $k : 1$, and by differentiation, taking account of sign, we see that their velocities relative to the centre of gravity are in the ratio $k : -1$.

If V is the velocity of the centre of gravity we may then write the velocities of the particles as $V + ku$, $V - u$ respectively.

$$\begin{aligned}\text{The total momentum} &= m(V + ku) + km(V - u)\\ &= (m + km)V\end{aligned}$$

which is equal to the momentum of the whole mass concentrated at the centre of gravity. This is an instance of a general result, true for all systems of particles, proved in Chapter 16.

If the particles collide, Newton's Experimental Law is expressed by putting the velocities relative to the centre of gravity after impact as kv, $-v$, where $v = -eu$.

The total momentum is unchanged at the value $(m + km)V$, so that the velocity of the centre of gravity is still V, and the actual velocities after impact are $V - keu$, $V + eu$ respectively.

Now the initial KE

$$\begin{aligned}&= \tfrac{1}{2}m(V+ku)^2+\tfrac{1}{2}km(V-u)^2\\ &= \tfrac{1}{2}m[V^2+2kVu+k^2u^2+k(V^2-2Vu+u^2)]\\ &= \tfrac{1}{2}(m+km)V^2+\tfrac{1}{2}m(ku)^2+\tfrac{1}{2}kmu^2 \quad . \; . \; . \; . \; . \quad (1)\end{aligned}$$

which is equal to the kinetic energy of the whole mass concentrated at the centre of gravity and moving with it, together with the kinetic energies of the separate masses calculated from their velocities relative to the centre of gravity. This again is an instance of a general rule, briefly expressed as:

$$\text{Total KE} = \text{KE of CG} + \text{KE relative to CG}$$

The general rule is proved in Chapter 16.

After impact, the first term in (1) is unchanged, and each of the others is multiplied by e^2, so that the loss of KE may be expressed as $(1-e^2)$(KE relative to CG). Note that when $e<1$ this loss must be a positive quantity.

From the values just found for the velocities of the particles relative to the centre of gravity before and after impact, we see that the impulse on either particle is $kmu(1+e)$. The short time during which the particles are in contact can be considered in more detail as consisting of two stages. At first the two bodies are compressed, and at the end of this stage both are moving with the same velocity, which is that of the centre of gravity. The impulse on either particle during this stage is *kmu*. In the second stage the bodies wholly or partially regain their original shapes, and the impulse during this stage is *kmue*.

Example 6.—Two spheres, of masses 5 kg and 3 kg are moving in opposite directions along the same straight line with velocities 8 m/s, 4 m/s respectively. The coefficient of restitution between them is $\frac{2}{3}$. Find their velocities after impact, and the loss of kinetic energy.

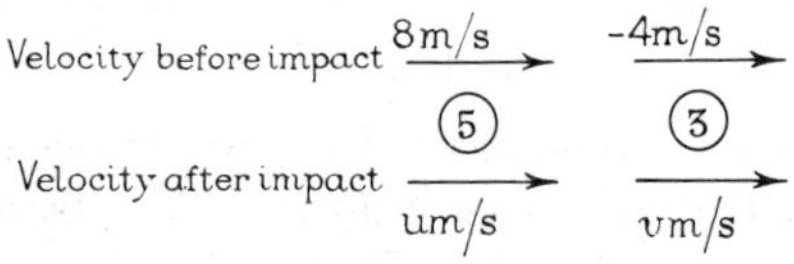

Note.—It is advisable to use one direction for all velocities, with negative signs if required, as shown.

By Conservation of Momentum,

$$5u + 3v = 40 - 12$$

By Newton's Experimental Law,

$$v - u = -\tfrac{2}{3}(-12)$$

$$\therefore\ 5u + 3v = 28$$

$$-3u + 3v = 24$$

$$\therefore\ 8u = 4$$

$$u = \tfrac{1}{2}$$

and $$v = 8\tfrac{1}{2}$$

KE before impact $= \frac{5}{2} \times 64 + \frac{3}{2} \times 16$

$= 184$ J

KE after impact $= \frac{5}{2} \times \frac{1}{4} + \frac{3}{2} \times \frac{289}{4}$

$= 109$ J

$\therefore$ Loss of KE $= 75$ J

The example may alternatively be worked as follows: Let V m/s be the velocity of the centre of gravity.

Total momentum $= 8V = 40 - 12$

$$\therefore \ V = 3\tfrac{1}{2}$$

Therefore initial velocities relative to the centre of gravity are $4\frac{1}{2}$, $-7\frac{1}{2}$ m/s respectively. Multiplying by $-e$, we have that the final velocities relative to the centre of gravity are

$$-3, \ +5 \text{ m/s}$$

giving final velocities of $\frac{1}{2}$, $8\frac{1}{2}$ m/s

Initial KE relative to centre of gravity

$$= \tfrac{5}{2} \times \tfrac{81}{4} + \tfrac{3}{2} \times \tfrac{225}{4} = 135 \text{ J}$$

Loss of KE $= (1 - e^2)$(KE relative to CG)

$= \frac{5}{9} \times 135$

$= 75$ J

EXERCISE 23A

1. A sphere moving at 20 m/s overtakes another sphere of equal mass moving in the same direction at 4 m/s. If the coefficient of restitution is $\frac{3}{4}$, find the velocity of each after the impact.
2. A sphere moving at 14 m/s collides with a sphere of equal mass moving in the opposite direction at 4 m/s. The coefficient of restitution between them is $\frac{2}{3}$. Find the velocities after impact.
3. A sphere of mass 3 kg moving at 30 m/s strikes a sphere of mass 2 kg which is at rest. The coefficient of restitution is $\frac{5}{6}$. Find the velocities after impact.
4. A sphere of mass 200 g moving at 65 cm/s strikes a sphere of mass 500 g moving in the same line but in the opposite direction at 5 cm/s. The coefficient of restitution is 0·7. Find the velocities after impact.
5. A sphere of mass 300 g moving at 76 cm/s overtakes a sphere of mass 400 g moving in the same direction at 6 cm/s. The first sphere is reduced to rest by the impact. Find the coefficient of restitution.
6. A truck of mass 10 t, moving at 8 m/s, overtakes another truck moving in the same direction at 2 m/s. After the impact the speeds are 6·8 m/s and 8 m/s respectively. Find (i) the mass of the second truck, (ii) the coefficient of restitution between the trucks, (iii) the loss of kinetic energy during the impact, in kJ.

7. Three particles, of mass $3m$, $2m$, and m lie at rest in this order at three points in a straight line on a smooth horizontal table. The coefficient of restitution for each pair is e. If the particle of mass $3m$ is projected towards that of mass $2m$ with velocity V, find their velocities after impact. Find also the velocities of the particles of mass $2m$ and m after their impact. Show that there will be no further collisions if $e > \frac{1}{2}(3 - \sqrt{5})$.

8. Equal spheres, n in number, are placed at equal intervals in a straight line on a smooth horizontal floor. The sphere at one end is projected towards the others with velocity u, and the coefficient of restitution between each pair is e. Show that the time-intervals between successive impacts form a geometrical progression, and find its common ratio. With what velocity will the last sphere begin to move?

9. Two particles, of masses 3 kg and 2 kg, are at rest 1 m apart on a rough horizontal floor, the coefficient of friction between either particle and the floor being $\frac{1}{4}$. The heavier particle is projected towards the lighter at 2·5 m/s. Find its velocity just before the collision. If the coefficient of restitution between the particles is $\frac{2}{3}$, find the velocities after impact, and the distance between the particles when they finally come to rest.

10. Two elastic particles A and B, of masses m_1 and m_2 respectively, are situated in a smooth horizontal circular groove. At a given instant B is at rest and A is moving in the groove with constant speed v_1. If the coefficient of restitution between the particles is e, and if $m_1 > em_2$, show that between their first and second collisions A will move through an arc which subtends an angle $\dfrac{2\pi(m_1 - em_2)}{e(m_1 + m_2)}$ at the centre of the circle, and that the speed of B after the second collision will be $\dfrac{m_1(1 - e^2)v_1}{m_1 + m_2}$

EXERCISE 23B

1. Prove that if two particles interchange velocities after a direct impact their masses must be equal and restitution between them must be perfect.

2. A particle of mass 4 kg moving at 19 m/s overtakes another of mass 5 kg moving in the same line at 1 m/s. The coefficient of restitution is 0·3. Find their velocities after impact.

3. A particle of mass 8 kg moving at 30 m/s overtakes another of mass 5 kg moving in the same line at 4 m/s. The coefficient of restitution is $\frac{1}{4}$. Find the magnitude of the impulse between them, and the loss of kinetic energy.

4. Two particles moving in the same line collide. The coefficient of restitution between them is 0·2. One, of mass 7 kg, is reduced to rest. The other, of mass 4 kg, moves with speed 4·4 m/s after the impact. Find their velocities before the impact.

5. A particle of mass m_1, moving at 30 m/s, overtakes another of mass m_2, moving in the same line at 6 m/s. The coefficient of restitution being $\frac{1}{3}$, find the ratio $m_1 : m_2$ for the first particle to be reduced to rest.

6. A particle of mass 6 kg, moving at 8 m/s, collides with another of mass 3 kg moving in the opposite direction along the same line at 12 m/s. The first particle is reduced to rest. Find the coefficient of restitution between the particles.

7. A sphere A is dropped from a height, and at the same instant an equal sphere B at ground level exactly below A is projected vertically upwards. Restitution between the spheres is perfect. Show that after impact A will reach the ground at the same time as B would have returned to it had there been no collision.

8. Three smooth spheres A, B, C, of equal size and of masses $2m$, $7m$, $14m$ respectively, are at rest on a smooth horizontal table, with their centres in that order on a straight line. The coefficient of restitution between each pair is $\frac{1}{2}$. The sphere A is now projected so that it impinges directly on B. Prove that after two impacts have taken place B is at rest, and A and C are moving with equal speeds in opposite directions. Find what fraction of the original kinetic energy remains after both impacts.

9. Three smooth spheres A, B, C have equal radii and masses $9m$, m, $9m$ respectively. They are at rest on a smooth horizontal table with their centres in a straight line, and B is between A and C. The sphere B is now projected directly towards C with speed V. The coefficient of restitution at all impacts is $\frac{2}{3}$. Find the velocity of each sphere after three impacts have taken place, and show that there can be no more collisions.

10. A hammer of mass $5m$, moving horizontally with velocity V, strikes a stationary horizontal nail of mass m. If the coefficient of restitution between them is $\frac{3}{5}$, find the velocity of the nail just after the blow. Immediately after the blow the nail begins to penetrate a block of mass nm which is free to move on a smooth horizontal table. Penetration is resisted by a constant force R. Find the common velocity of the block and nail when the nail ceases to penetrate the block. Show that penetration ceases at time $4mnV/\{3(n+1)R\}$ after the blow. (It may be assumed that there is only one impact between the hammer and the nail.)

6. Oblique impact

The impacts so far considered have been direct; that is, the motion has been entirely perpendicular to the plane which touches both bodies at their point of contact. We now consider other impacts, which are called *oblique*, and we shall deal only with smooth bodies.

In the impact between a moving body and a fixed obstacle, both smooth, the impulse, or change of momentum, is entirely normal to the surface of contact. It follows that the component velocity of the moving body parallel to the surface of contact is unaltered. The component perpendicular to the surface of contact is, in accordance with Newton's Experimental Law, multiplied by $-e$, where e is the coefficient of restitution between the body and the obstacle.

Similar considerations apply when two smooth moving bodies collide obliquely. In the very frequent case of two smooth spheres, for example, the component velocities perpendicular to the line of centres are unaltered. The components parallel to the line of centres are treated just like velocities in direct impact, using the Principle of Conservation of Momentum and Newton's Experimental Law.

Example 7.—A smooth particle strikes a fixed plane. Its velocity before impact is of magnitude u, at an angle α with the normal to the plane. Find its velocity just after the impact, and the loss of kinetic energy.

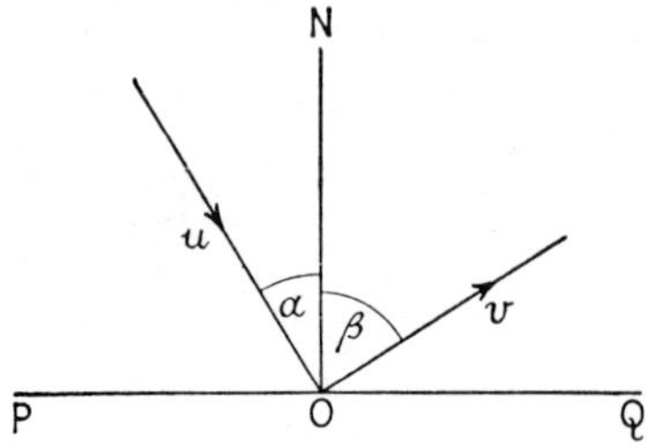

Let POQ represent the plane, and ON the normal, and let the velocity after impact be given by v, β as shown.

Velocity parallel to plane is unaltered:

$$v \sin \beta = u \sin \alpha$$

Velocity perpendicular to plane is subject to Newton's Law:

$$v \cos \beta = eu \cos \alpha$$

By division, $\cot \beta = e \cot \alpha$

By squaring and adding, $v^2 = u^2(\sin^2 \alpha + e^2 \cos^2 \alpha)$

$$\therefore \; v = u\sqrt{(\sin^2 \alpha + e^2 \cos^2 \alpha)}$$

$$\begin{aligned}\text{Loss of KE} &= \tfrac{1}{2}m(u^2 - v^2) \text{ where } m \text{ is mass of particle} \\ &= \tfrac{1}{2}mu^2(1 - \sin^2 \alpha - e^2 \cos^2 \alpha) \\ &= \tfrac{1}{2}mu^2 \cos^2 \alpha(1 - e^2)\end{aligned}$$

Example 8.—Two smooth spheres, of masses 10 kg and 8 kg, collide when their speeds are 12 m/s and 16 m/s, inclined to the line of centres at 30° and 60° as shown in the diagram. Find their velocities after impact, if the coefficient of restitution between them is $\frac{1}{2}$.

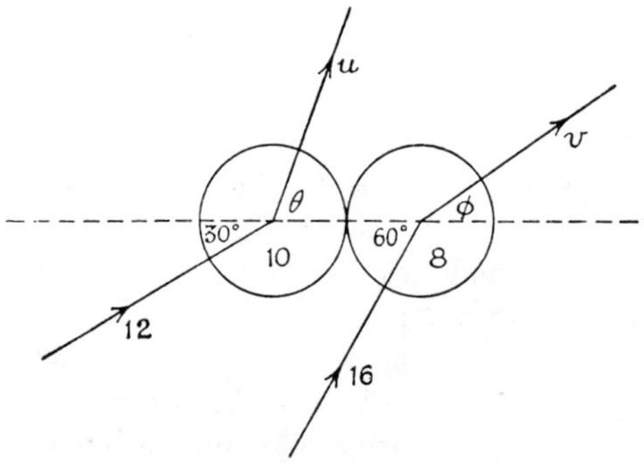

Let the velocities after impact be denoted by u, θ and v, ϕ as shown.
Since the velocities perpendicular to the line of centres are unaltered,

$$u \sin \theta = 12 \sin 30° = 6 \quad \text{(i)}$$

$$v \sin \phi = 16 \sin 60° = 8\sqrt{3} \quad \text{(ii)}$$

For the velocities parallel to the line of centres, by the Principle of Momentum,

$$\begin{aligned}10u \cos \theta + 8v \cos \phi &= 10 \times 12 \cos 30° + 8 \times 16 \cos 60° \\ &= 60\sqrt{3} + 64 \quad \text{(iii)}\end{aligned}$$

By Newton's Law,

$$\begin{aligned}v \cos \phi - u \cos \theta &= -\tfrac{1}{2}(16 \cos 60° - 12 \cos 30°) \\ &= -4 + 3\sqrt{3} \quad \text{(iv)}\end{aligned}$$

Solving equations (iii) and (iv) we find

$$u \cos \theta = 2\sqrt{3} + \tfrac{16}{3} \quad \ldots \ldots \quad \text{(v)}$$
$$v \cos \phi = 5\sqrt{3} + \tfrac{4}{3} \quad \ldots \ldots \quad \text{(vi)}$$

From (i) and (v), by division

$$\cot \theta = \tfrac{8}{9} + \tfrac{1}{3}\sqrt{3}$$
$$\simeq 1{\cdot}4662$$
$$\therefore \theta \simeq 34^\circ\, 18'$$

Squaring and adding,

$$u^2 = 36 + 12 + \frac{256}{9} + \frac{64\sqrt{3}}{3}$$
$$\simeq 113{\cdot}4$$
$$\therefore u \simeq 10{\cdot}65 \text{ m/s}$$

From (ii) and (vi) similarly,

$$\cot \phi = \frac{\sqrt{3}}{18} + \frac{5}{8}$$
$$\simeq 0{\cdot}7212$$
$$\therefore \phi \simeq 54^\circ\, 12'$$
$$v^2 = 192 + 75 + \frac{16}{9} + \frac{40\sqrt{3}}{3}$$
$$\simeq 291{\cdot}9$$
$$\therefore v \simeq 17{\cdot}09 \text{ m/s}$$

Example 9.—A smooth sphere collides with an equal sphere which is at rest. The velocity of the moving sphere before impact makes an angle α with the line of centres at impact. The coefficient of restitution between the spheres being $\frac{1}{3}$, find the value of α which makes the angle through which the moving sphere is deviated a maximum.

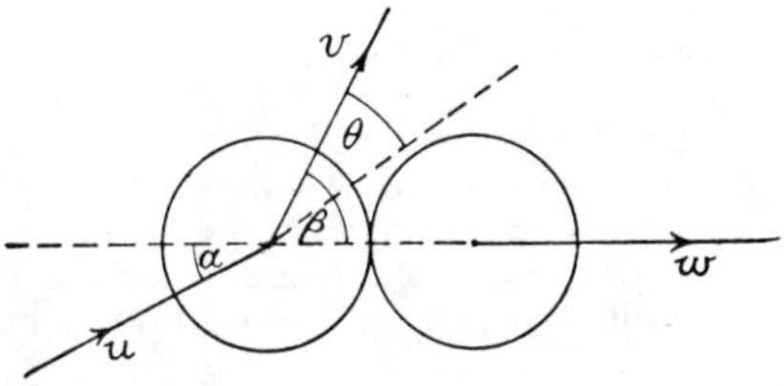

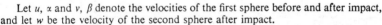

Let u, α and v, β denote the velocities of the first sphere before and after impact, and let w be the velocity of the second sphere after impact.

It is to be noted that w is in the line of centres, since the impulse between the spheres is in this line, and the second sphere is initially at rest.

Since velocity perpendicular to the line of centres is unaltered,

$$v \sin \beta = u \sin \alpha \quad \ldots \ldots \quad \text{(i)}$$

By Conservation of Momentum along line of centres, taking the mass of each sphere as unity,

$$v \cos \beta + w = u \cos \alpha \quad \ldots \ldots \quad \text{(ii)}$$

By Newton's Law,

$$w - v \cos \beta = \tfrac{1}{3}u \cos \alpha \quad \ldots \ldots \quad \text{(iii)}$$

Subtracting (iii) from (ii) to eliminate w, we find

$$v \cos \beta = \tfrac{1}{3}u \cos \alpha$$

From (i), by division

$$\tan \beta = 3 \tan \alpha$$

Now if θ is the angle through which the first sphere is deviated,

$$\theta = \beta - \alpha$$

$$\therefore \tan\theta = \frac{\tan\beta - \tan\alpha}{1 + \tan\beta\tan\alpha}$$

$$= \frac{2\tan\alpha}{1 + 3\tan^2\alpha}$$

Let $t \equiv \tan\theta$, $a \equiv \tan\alpha$

Then
$$t = \frac{2a}{1 + 3a^2}$$

$$\frac{dt}{da} = \frac{2(1 + 3a^2) - 2a(6a)}{(1 + 3a^2)^2}$$

$$= \frac{2 - 6a^2}{(1 + 3a^2)^2}$$

$\therefore$ t has a stationary value when $a^2 = \frac{1}{3}$

$$\tan\alpha = \pm\frac{1}{\sqrt{3}}$$

$$\alpha = \pm 30^\circ$$

To show that this stationary value is a maximum, we write the value of $\dfrac{dt}{da}$ as $\dfrac{p}{q^2}$

Then
$$\frac{d^2t}{da^2} = \frac{\dfrac{dp}{da}q^2 - p\dfrac{dq^2}{da}}{q^4}$$

so that when $p = 0$ the sign of $\dfrac{d^2t}{da^2}$ is that of $\dfrac{dp}{da}$, which is $-12a$, and negative for all positive values of a. Hence, when $\alpha = 30^\circ$, *the deviation is a maximum.*

Note.—The value $a = -1/\sqrt{3}$ also gives t a stationary value, when $\alpha = -30^\circ$. In this case β and θ are both negative, and θ has a minimum value. But this case is simply the result of reflecting the whole figure in the line of centres, and the deviation, though algebraically a minimum, is numerically a maximum as before.

EXERCISE 24A

1. A ball dropped on to a horizontal floor bounces to one-quarter of the height from which it was dropped. Find the coefficient of restitution between ball and floor. If the ball is dropped from a height of 2·5 m and continues bouncing, find what total time elapses before it comes to rest, and the total distance travelled by the ball.

2. A smooth inclined plane whose angle with the horizontal is $\tan^{-1}(\frac{1}{4})$ rests on a horizontal floor. A particle dropped from a height of 2·5 m strikes the plane very near its lower edge. The coefficient of restitution is $\frac{2}{3}$. Find the angle with the horizontal at which the particle rebounds, and the distance from the foot of the plane to the point where it first reaches the floor.

3. A smooth rectangular table has a smooth upstanding rim. A sphere is projected so as to strike two adjacent sides successively. The coefficients of restitution between the sphere and the two sides are equal. Prove that after the two impacts the particle is moving parallel to its original direction, and find the ratio of final to initial speed.

4. **A sphere of mass 2 kg collides with a sphere of mass 3 kg which is at rest. The** coefficient of restitution between them is $\frac{1}{2}$. Before impact the velocity of the first sphere is at 45° to the line of centres of the spheres. Find the direction in which it moves after impact.

5. A sphere of mass m_1 collides with a sphere of mass m_2 which is at rest. Prove that whatever the initial direction of the first sphere, it will move off at right angles to the line of centres if the coefficient of restitution is m_1/m_2.

6. **A sphere of mass 2 kg, with centre A, collides with a sphere of mass 1 kg with centre B. Just before the impact the velocity of A is 20 m/s in a direction making an angle $\tan^{-1}(\frac{3}{4})$ with BA produced, and the velocity of B is 10 m/s, at $\tan^{-1}(\frac{4}{3})$** with AB produced. Find the velocities of A and B after impact. The coefficient of restitution between the spheres is $\frac{1}{2}$.

7. Two equal smooth spheres of radius a are moving with equal speeds u in opposite directions along parallel lines whose distance apart is a. The coefficient of restitution between them is $\frac{1}{3}$. Find their velocities after impact.

8. Two equal smooth spheres A, B are free to move on a smooth horizontal table. Initially B is at rest and A is moving towards B with velocity u so that on impact the line of centres of the spheres is inclined at an acute angle α to the direction of motion of A. The coefficient of restitution being e, find the components of the velocities of the spheres after impact, along and perpendicular to the line of centres. Show that the angle ϕ through which the direction of motion of A is turned by the impact is given by $\tan\phi = \dfrac{(1+e)\tan\alpha}{1-e+2\tan^2\alpha}$. Show also that the greatest value of ϕ is $\tan^{-1}\dfrac{1+e}{2\sqrt{(2-2e)}}$

9. A smooth circular horizontal table is surrounded by a smooth rim whose interior surface is vertical. Two equal particles are projected simultaneously with speed V along the table from a point A of the rim in different directions each making 30° with the diameter AB of the rim. If the coefficient of restitution e of each particle with the rim of the table is greater than $\frac{1}{3}$, prove that, after one impact at the rim for each particle, they meet on AB. If then they coalesce, prove that their common velocity subsequently is $\frac{1}{4}\sqrt{3}(1-e)V$.

10. Two equal spheres of mass m are at rest in contact on a smooth horizontal table. A third sphere of equal size but of mass $3m$ is projected with speed u along the common tangent to the other two at their point of contact, so as to strike them simultaneously. The coefficient of restitution between any two spheres is e. Find the speeds of the three spheres after impact. (Note that care is required in applying the Principle of Conservation of Momentum.)

EXERCISE 24B

1. **A particle strikes a smooth horizontal floor with a speed of 7 m/s, its direction** of motion being at 30° to the horizontal. The coefficient of restitution is 0·8. Find the total time during which bouncing continues, and the distance on the floor between the first point of impact and the point where bouncing stops.

2. A particle moving on a smooth horizontal floor encounters a smooth plane inclined at 45° to the floor, and the coefficient of restitution between particle and plane is $\frac{1}{2}$. Prove that after rebounding the particle strikes the plane at right angles.

3. A staircase has steps 36 cm wide, the rise at each step being 18 cm. A ball, whose coefficient of restitution with each step is $\frac{2}{3}$, bounces down the staircase, striking each step in turn at the same distance from its edge. Find the horizontal and vertical velocities of the ball just before the first impact.

4. A sphere of mass 1 kg collides with a sphere of mass 4 kg which is at rest on a smooth horizontal table. The coefficient of restitution is $\frac{1}{2}$. Just before impact the first sphere is moving at 45° to the line of centres. Find its direction of motion after impact.

5. A sphere A of mass 3 kg moving at 10 m/s collides with a sphere B of mass 1 kg moving at 13 m/s on a smooth horizontal floor. Just before impact the velocity of A is at $\tan^{-1}(\frac{4}{3})$ with BA produced, and the velocity of B is at $\tan^{-1}(\frac{5}{12})$ with AB produced. Find the velocities after impact, if the coefficient of restitution is $\frac{2}{3}$.

6. Two small smooth spheres, of equal size and masses m, $2m$ are moving initially with equal speeds on a smooth horizontal floor. They collide when their centres are on a line AB which is parallel to and distant a from a smooth vertical wall towards which they are moving. Before collision the directions of motion are at right angles to each other, each at 45° to AB. The coefficient of restitution at all impacts is e. Find the distance apart of the points at which the spheres strike the wall. Show that after rebounding from the wall they cross the line AB simultaneously, and that their distance apart is then $2a(1 + e)$.

7. A smooth horizontal circular tray, centre O, has a smooth vertical rim round its edge. A particle is projected horizontally with velocity u from a point A of the rim, and after two impacts with the rim returns to A. If the original velocity of projection makes an angle α with AO, and V is the velocity with which the particle again reaches A, show that (i) $(1 + e + e^2)\tan^2\alpha = e^3$, (ii) $V^2 = e^3u^2$, where e is the coefficient of restitution between particle and rim.

8. Two equal particles A, B are free to move in a fixed smooth horizontal circular groove. They are projected simultaneously from the same point of the groove in the same direction with speeds u_o, v_o respectively, where $u_o > v_o$. If their coefficient of restitution is e, find expressions for their speeds u_n, v_n after the nth impact, and show that the angle swept out by the radius to the slower particle between the nth and $(n + 1)$th impacts is $\pi\left(\frac{1}{e^n} \cdot \frac{u_o + v_o}{u_o - v_o} - 1\right)$

9. The centres of three smooth spheres of the same radius are A, B, C. Spheres B, C are at rest in contact on a smooth horizontal table. Sphere A, of mass m, collides with B, of mass M, the coefficient of restitution being e. At the instant before the collision AB is perpendicular to BC, and the components of A's velocity in the directions of AB, BC are u, v respectively. Find the direction of A's velocity after the collision, and prove that A will collide with C if $m > eM$.

10. A sphere falling vertically collides with an equal sphere hanging by an inextensible string. At the moment of impact the speed of the falling sphere is u, and the line of centres of the spheres is at 45° to the vertical. The coefficient of restitution being $\frac{1}{2}$, find the velocities of the spheres after impact.

12

Circular Motion

1. Circular motion

It was shown in Chapter 3 that when a point is moving in a circle it has an acceleration towards the centre. If r is the radius of the circle, v the speed of the particle, and ω $(=v/r)$ its angular velocity about the centre, the magnitude of this central acceleration is $r\omega^2$, or v^2/r. If the speed v is uniform, this is the only acceleration.

It follows that when a particle is moving in a circle it must be acted on by a force towards the centre which provides the necessary mass-acceleration. In working examples the student is urged to adopt the practice of drawing separate diagrams showing forces and mass-accelerations, and deriving his equations from the equivalence of the two systems. Much confusion is thereby avoided.

We shall first take examples of uniform circular motion.

Example 1.—A particle of mass m is attached to a fixed point by a light inextensible string of length a, and moves uniformly in a horizontal circle with the string inclined at θ to the downward vertical. Find the time of revolution of the particle.

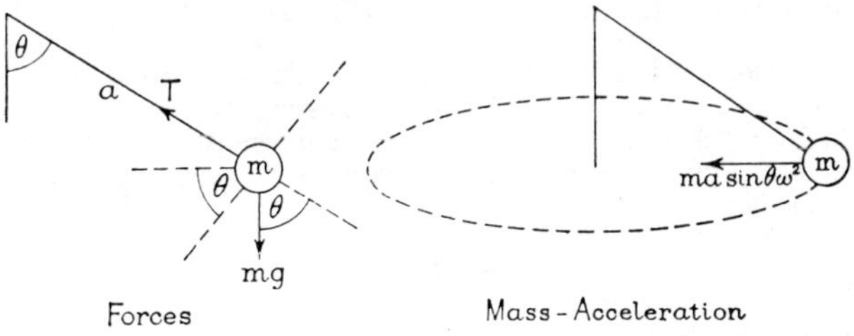

Let the vertical plane through the string turn with angular velocity ω. Then since the radius of the circle in which the particle moves is $a \sin \theta$, the acceleration towards the centre is $a \sin \theta \,.\, \omega^2$.

Since the two systems shown above are equivalent, resolving along a line in the vertical plane of the string and at right-angles to it, so as not to involve T, we have

$$mg \sin \theta = m\omega^2 a \sin \theta \,.\, \cos \theta$$

Rejecting $\sin\theta = 0$, $$\omega^2 = \frac{g}{a\cos\theta}$$

$\therefore$ Time of revolution $$= \frac{2\pi}{\omega} = 2\pi\sqrt{\frac{a\cos\theta}{g}}$$

Note that $a\cos\theta$ is the depth of the particle below the fixed point of the string.

Example 2.—A car is moving on level ground with uniform speed v; the coefficient of friction between tyres and ground is μ. Its centre of gravity, which is at a height h above the ground and midway between the wheel tracks, is describing a circle of radius r. The distance between the wheel tracks is $2a$. Find the speeds at which the car becomes liable (i) to skid, (ii) to overturn.

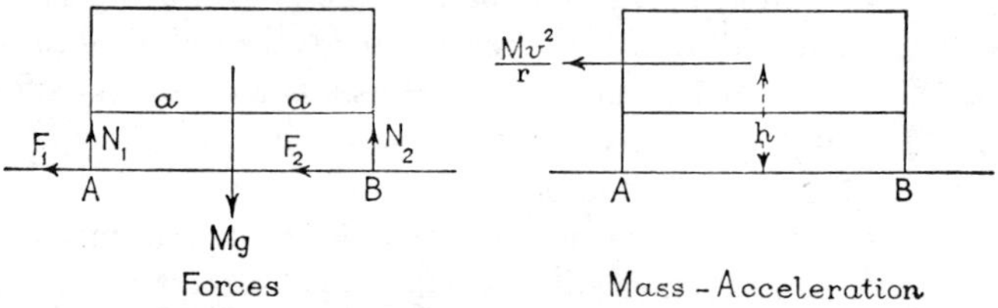

Let M be the mass of the car, and N_1, F_1, N_2, F_2 the normal and frictional forces due to the reaction of the ground, combined into single forces on each side of the car.

Resolving vertically, $$N_1 + N_2 = Mg$$

Resolving horizontally, $$F_1 + F_2 = M\frac{v^2}{r}$$

But $F_1 \leqslant \mu N_1$, and $F_2 \leqslant \mu N_2$

$$\therefore \frac{v^2}{r} \leqslant \mu g$$

$\therefore$ For the car not to skid, $$v \leqslant \sqrt{(\mu g r)}$$

It has been assumed in the above work that there is contact with the ground at A, B. We can find whether this is so by investigating the signs of N_1, N_2. A negative value for either of these would mean that a downward force was necessary to preserve contact; that is to say that in actual fact contact would cease.

Taking moments about A,

$$N_2 \,.\, 2a - Mga = M\frac{v^2}{r} \,.\, h$$

Thus N_2 is positive for all values of v.

Taking moments about B,

$$-N_1 \,.\, 2a + Mga = M\frac{v^2}{r} \,.\, h$$

$$\therefore N_1 = \frac{Mg}{2} - \frac{Mv^2h}{2ra}$$

Thus the critical speed, at which the inner wheels leave the ground, is given by

$$v = \sqrt{\frac{gra}{h}}$$

Overturning is therefore more likely with a narrow track, a high centre of gravity, or in a sharp turn.

EXERCISE 25A

1. A particle attached to a fixed point by a light inextensible string describes a horizontal circle with uniform speed once per second. The string is inclined at 60° to the vertical. Find its length, in metres, to 2 significant figures.
2. The mass of the bob of a conical pendulum (i.e. the system described in question 1) is 2 kg, and the length of the string is 1 m. If the breaking tension in the string is 200 N, find the maximum angular velocity of the bob.
3. The ' flying boats ' at a fairground are small cars attached by chains 3 m long, whose weight may be neglected, to the ends of horizontal arms 2 m long radiating from a central pillar about which the apparatus turns. When the chains are at 30° to the vertical, what is the speed of the cars, in metres per second, to 2 significant figures?
4. A cyclist travelling in a circle of radius 20 m on level ground is leaning inwards at 30° to the vertical. Assuming that the centre of gravity of cycle and rider is in the plane of the wheels, find his speed in metres per second, to 2 significant figures. Also find the least coefficient of friction between tyres and ground for there to be no side-slip.
5. At what angle to the horizontal must a railway track, in the form of part of a circle of radius 100 m, be banked for no lateral pressure to be required between the wheel flanges and the rails at 72 km/h? Find the lateral pressure on a truck travelling on this track at 36 km/h as a fraction of the weight of the truck, to 2 decimal places.
6. A smooth particle describes horizontal circles with constant speed on the inside of a hollow cone of semi-vertical angle 30°, which has its axis vertical and vertex downwards. If the speed of the particle is v, find its distance from the vertex of the cone.
7. A particle of mass m is connected by light inextensible strings, each 1·25 m long, to two points 1·5 m apart in a vertical line. The system rotates about this line with constant angular velocity ω, and both strings are taut. Find the tensions in the strings. What is the least angular velocity for which both strings will be taut?
8. Two particles are connected by a fine inextensible string which passes through a small smooth hole in a smooth horizontal table. The particle on the table has mass m_1 and describes a circle of radius r with speed v; the other particle has mass m_2 and hangs at rest. Find an expression for v in terms of m_1, m_2, r, and g.

   A downward vertical impulse I is now applied to the suspended particle. Find the impulse in the string and the direction of motion of the particle on the table just after the impulse is applied.
9. A particle of mass m rests on the rough interior of a spherical shell of internal radius a, the coefficient of friction being μ. The shell is rotating with uniform angular velocity ω about its vertical diameter, and the particle is on the point of slipping upwards. If the radius to the particle makes an acute angle θ with the downward vertical, show that the normal component of the reaction between it and the sphere is $mg/(\cos\theta - \mu\sin\theta)$, and that

$$\omega^2 a\sin\theta(1 - \mu\tan\theta) = g(\mu + \tan\theta)$$

10. The end A of a light inextensible string ABC is fixed. A particle of mass m is attached to the string at B, and a particle of mass $2m$ is attached at C. The system rotates with uniform angular velocity ω about the vertical through A, and AB, BC are inclined to the vertical at angles α, β respectively. The radii of the horizontal circles in which B, C move are in the ratio 1 : 2. Prove that A, B, C are always in a vertical plane, and that $\tan \beta : \tan \alpha = 6 : 5$.

EXERCISE 25B

1. A particle moves in a horizontal circle with uniform speed on the inside of a smooth spherical shell of radius a. If its angular distance from the lower point of the shell is θ, find an expression for its speed. If the speed is $\sqrt{(3ga/2)}$, find the value of θ.

2. An aircraft is flying at a uniform speed of 540 km/h in a horizontal circle of radius 1 km. The resultant air pressure on it may be assumed to be perpendicular to the line joining the wing tips. What must be the angle between this line and the horizontal if there is to be no side-slip?

3. Part of a cycle track, in the form of an arc of a circle of radius 40 m, is banked at 30° to the horizontal. (i) At what speed is a rider moving if the plane containing the wheels and centre of gravity is perpendicular to the track? (Answer to the nearest km/h.) (ii) If a rider travels on this track at 72 km/h, what coefficient of friction between tyres and track is needed to prevent side-slip? (Answer to 2 decimal places.)

4. An elastic string of unstretched length 1 m is fixed at one end and carries a particle of mas 1 kg at the other end. The particle is made to describe a horizontal circle at 4 radians per sec, and the length of the string is then 1·25 m. If the rotation is stopped, and the particle allowed to hang freely, what will be the length of the string?

5. A particle of mass m is connected by a light rod, freely hinged at both ends, to the top of a vertical shaft. An equal rod, freely hinged, connects the particle to a ring of mass M which can slide freely on the shaft. The system rotates about the shaft with constant angular velocity ω. Find an expression for the distance of the ring from the top of the shaft.

6. A particle A, of mass m, moves on the smooth interior surface of a circular cone of semi-angle α, fixed with its axis vertical and its vertex C downwards. An inextensible string attached to A passes through a small smooth hole at C and then vertically downwards to a particle B, of mass M, hanging freely. Show that A can revolve with speed v in a horizontal circle at a height h above C, while B remains at rest, provided that $v^2 = gh\{1 + M/(m \cos \alpha)\}$.

7. A hollow right circular cone of semi-angle α is placed with its axis vertical and vertex O downwards. Show that, if the inner conical surface is smooth, a particle can move on it in a horizontal circle with uniform angular velocity ω provided that the circle is at a height $g/(\omega \tan \alpha)^2$ above the horizontal plane through O.

If the surface is rough, with coefficient of friction $\mu(\mu < \tan \alpha < 1/\mu)$ and the cone is rotating about its axis with uniform angular velocity ω, show that in order that the particle may be carried round in a horizontal circle without moving relative to the cone, the height of the circle above the horizontal plane through O must lie between the values $\dfrac{g(\cot \alpha - \mu)}{\omega^2(\tan \alpha + \mu)}$ and $\dfrac{g(\cot \alpha + \mu)}{\omega^2(\tan \alpha - \mu)}$

8. Two particles of unequal masses are attached to the ends of a light inextensible string which passes through a small smooth ring capable of turning freely about its vertical diameter. The system rotates uniformly, with the particles and ring always in a vertical plane, and the particles are on opposite sides of the ring. Show that the particles are at the same vertical distance from the ring, and that their distances measured directly from the ring are in inverse proportion to their masses.

9. A car is moving in a horizontal circle of radius 100 m on a track banked at an angle $\tan^{-1}(\frac{5}{12})$ to the horizontal. The coefficient of friction between tyres and track is $\frac{1}{2}$. Show that, however low the speed, the car will not skid down the bank, and find the speed, in metres per second, at which it will begin to skid up the bank. Assume that the car does not overturn.

10. A light inextensible string ABC is such that AB = 2 m and BC = 1·5 m. A particle of mass m is attached to the string at C, and one of mass $7m$ is attached at B. The end A is tied to a fixed point and the whole rotates steadily about the vertical through A in such a way that B and C describe horizontal circles of radii 1·2 m and 2·4 m respectively. Show that the tension in BC is $5mg/3$. Also find the tension in AB and the speed of B.

2. Variable speed

If a particle moving in a circle is affected by any force having a component along the tangent to the circle, the circular motion will not be uniform. The acceleration along the tangent has been shown in Chapter 3 to be $\dot{v}$, or $r\dot{\omega}$, in the usual notation, and its value can be found when the tangential force is known.

This acceleration will, however, usually vary in magnitude as well as in direction, and changes in the velocity are more easily found by using the Principle of Energy, as given in Chapter 10.

Example 3.—A particle hanging at the end of a light inextensible string of length a is given a horizontal velocity u. Find the tension in the string when it has turned through an angle θ. In what position, if any, will the string become slack?

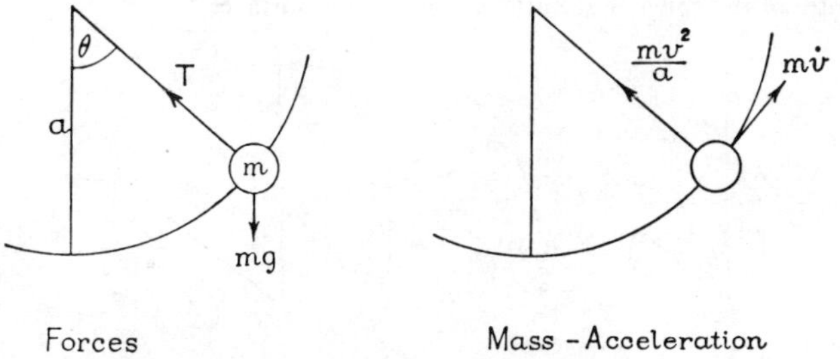

The force T does no work, since the displacement is always perpendicular to its line of action. The work done by the weight is negative.

$$\text{Final KE} - \text{initial KE} = \text{work done by gravity}$$

$$\therefore \tfrac{1}{2}mv^2 - \tfrac{1}{2}mu^2 = -mga(1 - \cos\theta)$$

$$\therefore v^2 = u^2 - 2ga(1 - \cos\theta) \quad . \quad . \quad . \quad \text{(i)}$$

Resolving along the radius,

$$T - mg\cos\theta = m\frac{v^2}{a} \quad \text{(ii)}$$

Using (i),

$$T = mg\cos\theta + \frac{mu^2}{a} - 2mg(1 - \cos\theta)$$

$$= \frac{mu^2}{a} + mg(3\cos\theta - 2) \quad \text{(iii)}$$

This equation has been established on the assumption of circular motion. But the tension in a string cannot be negative, so that any value of θ requiring T to be negative is impossible. The string will in fact become slack in the position which makes $T = 0$. From (iii) this is given by

$$u^2 + ag(3\cos\theta - 2) = 0$$

$$3\cos\theta - 2 = -\frac{u^2}{ag}$$

$$\cos\theta = \tfrac{1}{3}\left(2 - \frac{u^2}{ag}\right) \quad \text{(iv)}$$

In connection with this result, two points are to be noted.

If $u^2 > 5\,ag$, (iv) gives a value of $\cos\theta$ which is negative, and numerically greater than 1. This is impossible, and shows that if $u^2 > 5ag$ the particle will make complete revolutions.

If, on the other hand, $u^2 < 2ag$, equation (i) shows that $v = 0$ for some positive value of $\cos\theta$, i.e. for some acute angle θ. The string does not rise above the horizontal. But from (ii) T cannot be zero for an acute angle θ, as in that case v^2 would have to be negative.

This is not evident from equation (iv), because the elimination of v^2 between (i) and (ii) takes no account of the impossibility of negative values of v^2.

Thus the string will not slacken unless $2ag \leqslant u^2 \leqslant 5\,ag$.

Example 4.—A particle is slightly disturbed from rest at the highest point of a fixed smooth sphere of radius 30 cm. Find its horizontal and vertical distances from the centre of the sphere t sec after it has left the surface.

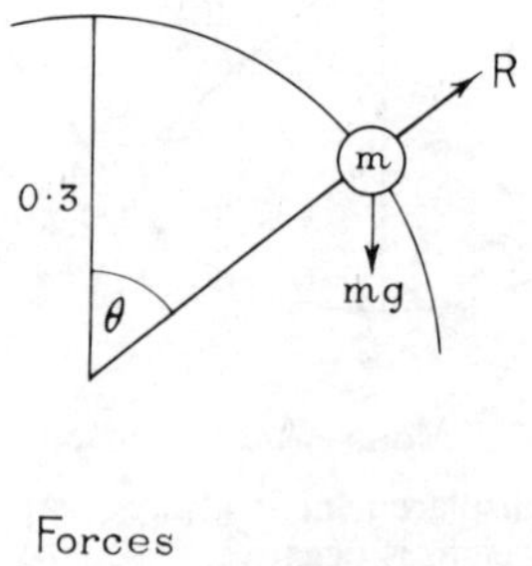

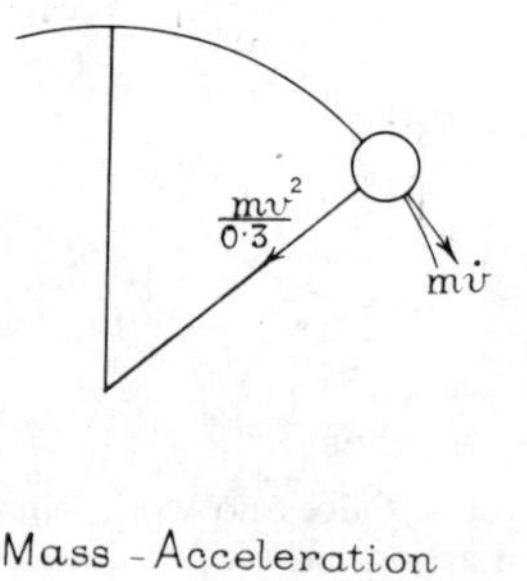

By the Principle of Energy,

$$\tfrac{1}{2}mv^2 = mg \times 0{\cdot}3(1 - \cos\theta) \quad \text{(i)}$$

Resolving along radius,

$$mg\cos\theta - R = \tfrac{10}{3}mv^2$$
$$= 2mg(1 - \cos\theta), \text{ from (i)}$$
$$\therefore\ R = mg(3\cos\theta - 2)$$

The particle will leave the surface when $R = 0$
i.e. when $\cos\theta = \frac{2}{3}$

Taking horizontal and vertical axes through the centre of the sphere, we have at this point

$$x = 0{\cdot}3\sin\theta \simeq 0{\cdot}22$$
$$y = 0{\cdot}3\cos\theta = 0{\cdot}2$$

From (i), when $\cos\theta = \frac{2}{3}$, $v^2 = 0{\cdot}2g$, and $v = 1{\cdot}4$. This velocity is at an angle θ below the horizontal, so that

$$\dot{x} = 1{\cdot}4\cos\theta \simeq 0{\cdot}93, \text{ and } \dot{y} = -1{\cdot}4\sin\theta \simeq -1{\cdot}04$$

The usual equations for motion under gravity (neglecting air resistance) then give

$$x \simeq 0{\cdot}22 + 0{\cdot}93t$$

and

$$y \simeq 0{\cdot}2 - 1{\cdot}04t - 4{\cdot}9t^2$$

Example 5.—A particle attached to a fixed point O by a light inextensible string of length a is held so that the string is horizontal, and released. When the string is vertical it encounters a small peg at a height b above the lowest point reached by the particle. The particle then makes a complete revolution about the peg. Find the greatest possible value of $b : a$.

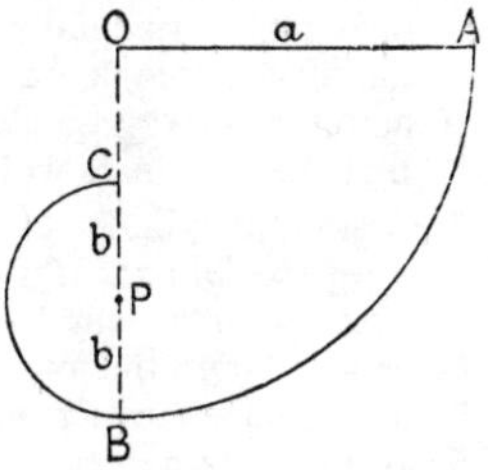

The path of the particle is shown as ABC. The peg is at P.

Since there is no sudden change in velocity at B, we may apply the Energy Equation from A to C.

If v is the speed at C, $\frac{1}{2}mv^2 = mg(a - 2b)$. The downward vertical acceleration at C is v^2/b.

Then if T is the tension in the string,

$$T + mg = mv^2/b$$
$$= 2mg(a - 2b)/b$$
$$\therefore\ T = mg\left(\frac{2a}{b} - 5\right)$$

Now T must be positive, so that for complete revolutions about P,

$$\frac{2a}{b} > 5, \text{ or } b : a < 2 : 5$$

EXERCISE 26A

1. A particle freely hinged to a fixed point by a light rod is slightly displaced from the position vertically above the fixed point. Find, in magnitude and direction, the acceleration of the particle when the rod is horizontal.

2. A particle attached to a fixed point by a light inextensible string of length a moves in a vertical circle. Its speed at the highest point is $\sqrt{(2ag)}$. Find the ratio of the tensions in the string at the highest and lowest points.

3. A particle inside and at the lowest point O of a fixed smooth spherical cavity of radius a is projected horizontally with speed $\sqrt{(4ga)}$. Prove that the particle leaves the surface of the cavity at a point which is at a height $5a/3$ above O.

4. A particle slides from rest at an angular distance α from the highest point of a fixed smooth sphere of radius r. Show that while the particle is still in contact with the sphere the reaction between them is $mg(3 \cos \theta - 2 \cos \alpha)$, where θ is the angular distance of the particle from the highest point of the sphere. If $\cos \alpha = \frac{3}{4}$, show that the particle will leave the sphere when its angular distance from the highest point is 60°. Show also that, in the ensuing motion, when the particle is at a distance $r\sqrt{3}$ from the vertical diameter of the sphere its depth below the centre of the sphere is $4r$.

5. A light inextensible string of length a is attached at one end to a fixed point O and at the other to a particle P of mass m. The particle is projected horizontally from the lowest position A with velocity v_o. Find the tension of the string when it has turned through angle θ, and show that, if the string becomes slack when $\theta = 120°$, then $v_o{}^2 = 7ga / 2$.

Show also that, with this value of v_o, the parabola in which the particle moves after the string has slackened passes through A.

6. A particle is inside a fixed smooth spherical shell of internal radius a and is projected horizontally from the lowest point O with velocity $3\sqrt{(ga)}$. Prove that it will complete a vertical circle. If this particle, on its return to the lowest point, collides and coalesces with a particle of equal mass placed at rest at O, prove that the new mass will leave the surface when it is at a height $13a/12$ above O.

7. A smooth cylinder of radius a is fixed with its axis horizontal. Two particles, of masses m and M, where $M > m$, are attached to the ends of a light inextensible string of length πa which lies over the cylinder in a vertical plane perpendicular to the axis. The particles are initially held at rest at either end of a horizontal diameter, and are then released. Prove that the lighter particle will at first remain in contact with the surface of the cylinder, but will leave it before reaching the highest point if $1 < 3m/M < \pi - 1$.

8. A particle of mass m is attached to one end of a light inextensible string of length a, the other end being fixed at a point O. A smooth peg P of negligible diameter is fixed at a distance $2a/3$ from O, P being at a lower level than O, and the angle between OP and the horizontal is α. The particle is held with the string taut at the same height as O and in the same vertical plane as OP, and released. Show that the particle will make complete revolutions about P if $\sin \alpha > \frac{3}{4}$.

9. A particle of mass M is attached to a point O by a light inextensible string of length l. The particle is held at a point P above the level of O and at a distance l from O, where OP is inclined at 60° to the upward vertical. If the particle is allowed to fall from rest at P, prove that at the instant when P is again at a distance l from O the string exerts an impulsive force of magnitude $m\sqrt{(gl/2)}$ on the particle, and that immediately afterwards the tension in the string is $2mg$.

10. A smooth circular cylinder is fixed with its axis horizontal, and a light inextensible string is placed over it in a plane perpendicular to its axis, with particles A, B, of masses m, $2m$ respectively, fastened to its ends. The particles are held against the cylinder on opposite sides of the vertical plane through the axis, and with the string taut. The radius to A is horizontal, and the radius to B is at $\tan^{-1}\frac{3}{4}$ to the upward vertical. The particles are then released. In the subsequent motion find the speed of the particles while they are still in contact with the cylinder and have moved through an angle θ from the initial position. Show that at the instant when B leaves the cylinder the value of θ is given by $28\cos\theta - 11\sin\theta = 16$.

11. A particle is attached to one end of a light inextensible string of length a, the other end being fixed. When hanging freely the particle is projected horizontally with a velocity such that the string becomes slack when it has turned through an angle $5\pi/6$. Show that when the string becomes taut again it is horizontal. Show also that if the velocity of the particle when it again reaches the lowest point of its path is U, then $U^2 = \frac{1}{8}ag(16 + 9\sqrt{3})$.

12. Two particles of masses m_1, m_2 ($m_1 < m_2$) are connected by a light inextensible string of length $2a$ and are released from rest with the string taut and horizontal. When the particles have fallen a distance a the middle point O of the string is suddenly seized and held fixed. Find the speed of the particles when the strings have rotated through an angle $\theta (\leqslant \frac{1}{2}\pi)$ in the subsequent motion, and show that the time taken from the moment of seizure until the particles collide is

$$\sqrt{\frac{a}{2g}} \cdot \int_0^{\pi/2} \frac{d\theta}{\sqrt{(1 + \sin\theta)}}$$

If the particles adhere to each other on collision and subsequently just rise to the level of O, find the value of the ratio $m_1 : m_2$.

EXERCISE 26B

1. A heavy particle hangs from a fixed point O by an inelastic string of length l. When the particle is vertically below O it is projected horizontally with speed $\sqrt{(\frac{7}{2}gl)}$. Calculate the angle through which the string has turned before it becomes slack.

2. A bead of mass m is threaded on a smooth circular wire which is held in a vertical plane. The bead is released from one end of a horizontal diameter of the wire. Find the reaction of the wire on the bead when the bead is vertically below the centre of the wire.

3. A particle is projected horizontally with speed v on the smooth inner surface of a hollow right-circular cone placed with its axis vertical and vertex downwards. If the particle describes a horizontal circle, show that the height of the circle above the vertex of the cone is v^2/g.

4. A particle P of mass m is suspended by two equal light strings PA and PB, where A and B are fixed at the same level and each string is inclined at an angle α to the horizontal. Find the tension in each string.

If the string PB is suddenly cut so that P starts to move in a circular path, find the tension in the string PA when it is inclined at an angle θ to the horizontal. If the tension in PA is suddenly halved when PB is cut, find the angle α.

5. A particle, attached by a light inelastic string of length l to a fixed point, is describing a horizontal circle of radius nl with uniform speed. The particle is now suddenly reduced to rest and then let go. If in the subsequent motion its speed when the string becomes vertical is equal to half its speed in the original motion, prove that $7n = 4\sqrt{3}$.

6. A bead of mass m is threaded on a smooth fixed vertical circular wire of radius a. The bead is started into motion with speed u at the highest point of the wire. Find (i) the reaction of the wire on the bead when the radius to it has turned through an angle $\theta°$, (ii) the greatest value of u in order that at some stage of the motion the reaction may be twice the weight of the bead.

7. A heavy particle hangs from a fixed point by a string of length l. When the particle is vertically below the point of suspension the particle is given a horizontal speed $\sqrt{(2gh)}$. Prove that if $l < h < \frac{5}{2}l$ the string becomes slack when the particle is at a vertical height above the centre of $\frac{2}{3}(h - l)$.

8. A circular cone of semi-vertical angle α is fixed with its axis vertical and its vertex upwards. An inextensible string of length l is attached at one end to the vertex and at the other to a particle of mass m resting on the smooth external surface of the cone. The particle then revolves with uniform angular velocity ω in a horizontal circle in contact with the cone. Show that $\omega^2 < g/l \,.\, \sec \alpha$ and find the tension in the string.

9. Two particles A and B of masses m and $2m$ respectively are joined by a light inextensible string of length greater than $\frac{1}{2}\pi b$. The particle A is held at rest at the highest point of a fixed smooth sphere, centre O and radius b. The particle B hangs freely in equilibrium. If A is now released and OA makes an angle θ with the upward vertical at time t after the release, show that so long as A remains in contact with the sphere

$$3b\left(\frac{d\theta}{dt}\right)^2 = 2g(1 + 2\theta - \cos\theta)$$

Deduce that if A leaves the sphere when $\theta = \alpha$ then $5 \cos \alpha = 2 + 4\alpha$.

10. The inner surface of a bowl is formed by rotating a parabola about its axis which is vertical. The vertex O of the parabola is downwards. A particle is projected horizontally along the smooth inner surface with speed V from a point P at a height h above O and at a distance r from the axis of the parabola. Find the magnitude of V if the particle describes a horizontal circle. Show also that V is the speed that would be attained at O by a particle released from rest at P.

11. A smooth fine tube is bent into a circular arc of radius a and angle $2(\pi - \alpha)$, α being acute. The tube is fixed in a vertical plane with the free ends uppermost and in a horizontal line. A particle is projected within the tube from its lowest point. It travels through the tube to one open end, describes a parabolic path under gravity, and enters the tube again at the other open end. Show that the necessary speed of projection is $\{ga(2 + 2\cos\alpha + \sec\alpha)\}^{1/2}$ and that the greatest height above the centre of the arc of the tube reached by the particle is

$$\frac{a(1 + \cos\alpha)^2}{2\cos\alpha}$$

13

Simple Harmonic Motion

In Chapter 3 Simple Harmonic Motion (SHM) was defined as the projection of uniform circular motion on a straight line, and the standard formulæ were proved. We now give a somewhat wider definition, and the same standard results will be obtained in another way.

1. Definition

If x measures the displacement of a moving point from some fixed position, and $\ddot{x} = kx$, where k is a negative constant, the point is said to have *simple harmonic motion*. Note that x need not be a distance measured along a straight line; it may be an angle, or a distance measured along a curve. The only requirement is that the value of x shall give the position of the moving point.

2. Standard results

It is convenient to write the fundamental equation as

$$\ddot{x} = -\omega^2 x \quad \ldots\ldots\ldots \quad \text{(i)}$$

This shows clearly that the constant of proportionality is negative. An equation of this type, where the first derivative $\dot{x}$ does not appear, can be integrated once if both sides are multiplied by $2\dot{x}$.

Then $$2\dot{x}\ddot{x} = -2\omega^2 x\dot{x}$$

$$\therefore\ \dot{x}^2 = -\omega^2 x^2 + A$$

To fix the constant A, more information is needed. Suppose that the moving point is instantaneously at rest where $x = a$; i.e. suppose that a is the amplitude of the motion.

Then $$0 = -\omega^2 a^2 + A$$

and $$\dot{x}^2 = \omega^2(a^2 - x^2) \quad \ldots\ldots\ldots \quad \text{(ii)}$$

(H 484)

From (ii) we see that the greatest velocity is $\pm\omega a$, and occurs where $x = 0$.

Now let $t = 0$ when $x = 0$ and $\dot{x}$ is positive.

$$\dot{x} = \frac{dx}{dt} = \omega\sqrt{(a^2-x^2)}$$

$$\therefore \frac{dx}{\sqrt{(a^2-x^2)}} = \omega\, dt$$

If the integral of the left-hand side is not recognized as a standard form, it can be found by the substitution

$$x = a \sin \theta, \; dx = a \cos \theta \, d\theta$$

giving

$$\sin^{-1}\frac{x}{a} = \omega t + B$$

The initial conditions assumed give $B = 0$

$$\therefore \; x = a \sin \omega t \qquad \text{(iii)}$$

If $t = 0$ when $x = a$, another standard integral, or the substitution $x = a \cos \theta$, gives (taking either of the square roots of (ii))

$$-\cos^{-1}\frac{x}{a} = \pm\omega t + C$$

The initial conditions now give $C = 0$, and

$$x = a \cos \omega t \qquad \text{(iv)}$$

A more general solution occasionally useful is

$$x = a \sin (\omega t + \varepsilon) \qquad \text{(v)}$$

where the value of the constant ε depends on the initial conditions.

From any of the equations (iii)–(v) we see that the values of x and $\dot{x}$ are repeated when ωt is increased by any multiple of 2π, that is, at time intervals of $2\pi/\omega$. This is the *period* of the simple harmonic motion.

Note that if $\ddot{x}$ can be expressed in the form $-\omega^2(x+k)$, where k is constant, we still have simple harmonic motion. If y is written for $x+k$, $\ddot{y} = \ddot{x}$ and the equation becomes $\ddot{y} = -\omega^2 y$. This is simple harmonic motion with the period $2\pi/\omega$, but the centre is where $y = 0$, i.e. where $x = -k$.

Example 1.—A particle of mass m hangs at the end of a light spring of natural length a, and extends it to the length $(a + b)$. The particle is pulled down a further distance c and released. Show that it executes simple harmonic motion, and find the centre, period, and amplitude.

Consider the particle at a distance x below its equilibrium position. It is essential to consider a *general* position, and not merely the initial position of the particle. The initial conditions will be used at a later stage, to find the amplitude.

When the elongation of the spring is b, the tension is mg, since this is the equilibrium position.

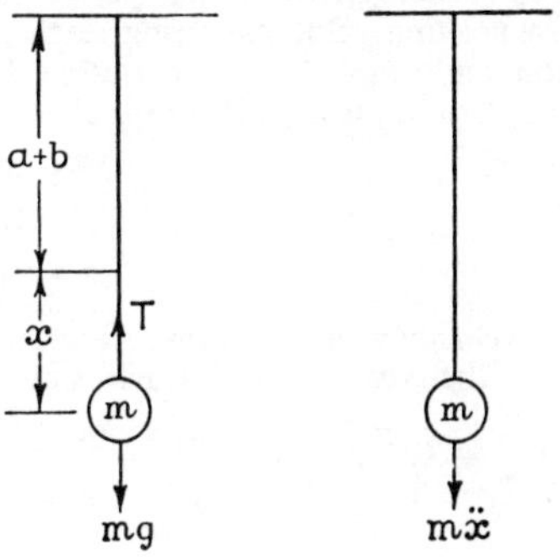

Forces Mass-Acceleration

$\therefore$ In the general position,

$$T = \frac{b + x}{b} mg$$

The equation of motion is

$$m\ddot{x} = mg - T$$
$$= mg - \left(1 + \frac{x}{b}\right) mg$$
$$\therefore \ \ddot{x} = -\frac{gx}{b}$$

This shows that the particle has simple harmonic motion. Comparing our result with the standard equation, we find $\omega^2 = g/b$, so that the period $= 2\pi\sqrt{(b/g)}$.

The centre of the motion is at the point where $x = 0$, i.e. *at the equilibrium position.*

The initial conditions tell us that $\dot{x} = 0$ when $x = c$, so that the amplitude $= c$.

If this question had referred to an elastic string instead of a spring, the work would have been the same, provided that the string did not slacken at any stage of the motion. This condition requires that $c \leqslant b$, for the minimum length of the spring is $a + b - c$. If the condition is not satisfied, the motion will be periodic, but not simple harmonic motion. The motion of the particle while the string is slack is free motion under gravity.

Example 2.—An elastic string of natural length 1 m is extended 20 cm by a particle attached to its end and hanging freely. The particle is pulled down a further distance of 40 cm and released. Find the height to which it will rise, and the time taken to reach the highest point. Take $g = 9{\cdot}8$ m/s².

Consider the particle at a distance x m below the equilibrium position.

Then while the string is taut its tension

$$T = \frac{0{\cdot}2 + x}{0{\cdot}2} mg$$
$$= mg(1 + 5x)$$

As in example 1, the equation of motion is

$$m\ddot{x} = mg - T$$
$$= -5mgx$$

Thus, while the string is taut, we have simple harmonic motion, with $\omega^2 = 5g = 49$, and $\omega = 7$.

The question of slackening must now be considered. Since the particle starts 0·4 m below the equilibrium position, which is the centre of the simple harmonic motion, it would in a complete period pass through a point 0·4 m above the equilibrium position. But the string slackens when the particle is 0·2 m above that position, and simple harmonic motion than ceases when $x = -0{\cdot}2$.

At this point, using the formula $\dot{x}^2 = \omega^2(a^2 - x^2)$ we have

$$\dot{x}^2 = 49(0{\cdot}16 - 0{\cdot}04)$$
$$= 0{\cdot}49 \times 12$$
$$\therefore \dot{x} = \pm 1{\cdot}4\sqrt{3}$$

and the velocity when the string slackens is upward.

If h is the greatest height reached above this point, the formula $v^2 = u^2 - 2gh$ gives

$$0 = 1{\cdot}96 \times 3 - 19{\cdot}6h$$
$$\therefore h = 0{\cdot}3$$

∴ Greatest height reached is 0·5 m above equilibrium position

The time to this height from the instant when the string slackens is

$$\frac{1{\cdot}4\sqrt{3}}{g} = \tfrac{1}{7}\sqrt{3} \text{ sec}$$

To find the time from the initial release of the particle to the instant when the string slackens, while simple harmonic motion continues, we express x in terms of t by $x = a \cos \omega t$. This form is chosen because it gives $x = a$ when $t = 0$. In this case $a = 0{\cdot}4$, and when the string slackens, $x = -0{\cdot}2$.

$$\therefore \cos \omega t = -\tfrac{1}{2}$$

The least positive solution is $\omega t = \tfrac{2}{3}\pi$, and since $\omega = 7$, the required time is $\tfrac{2}{21}$ sec.

∴ Total time from release of particle to reaching highest point

$$= \tfrac{2}{21}\pi + \tfrac{1}{7}\sqrt{3}$$
$$\simeq 0{\cdot}53 \text{ sec}$$

EXERCISE 27A

1. A particle of mass 500 g at the end of a vertical spring executes simple harmonic motion in a vertical line, with amplitude 20 cm, making one complete oscillation every 2 sec. Find the greatest and least forces exerted by the spring, in newtons, to 2 significant figures.
2. A particle of mass m kg rests on a platform which moves vertically with simple harmonic motion of period $\tfrac{1}{4}\pi$ sec. Find the reaction between particle and platform when the platform is x m above the centre of its motion. What is the greatest amplitude allowable if the particle is not to lose contact with the platform?
3. An elastic string carries a scale-pan of mass 100 g, which extends the string by 10 cm. A mass of 200 g is now gently placed in the pan, and released. Find the period and amplitude of the resulting simple harmonic motion.
4. An elastic string carries a scale-pan of mass 300 g, which extends the string by 60 cm. A mass of 100 g is dropped into the pan from a point 40 cm above it, and there is no rebound. Show that in the resulting motion the string does not slacken, and find the period and amplitude to 2 significant figures.

5. An elastic string of natural length 2 m and modulus λ N has a particle of mass m kg attached to its middle point. The string is stretched between two points 3 m apart on a smooth horizontal table. The particle is now displaced a distance of 3 cm in the line of the string, and released. Show that simple harmonic motion ensues, and find the period.

6. Under the conditions of question 5, the particle is displaced 1 cm in a direction perpendicular to the string, and released. It may be assumed that the tension in the string is unaltered. Show that simple harmonic motion ensues, and find its period.

7. A particle on a smooth horizontal table is connected by a spring to a fixed point on the table. The particle is held at a point A, the extension of the spring being 5 cm, and let go. It first returns to A after $\frac{1}{3}\pi$ sec. Find the speed of the particle when it has moved 2 cm from A. Find also the time taken to move through this distance, to 2 significant figures. (The unstretched length of the spring exceeds 5 cm.)

8. A heavy particle attached to a fixed point by a light elastic spring is held at rest a short distance below its equilibrium position and released. Show that the period of oscillation is $2\pi\sqrt{(d/g)}$, where d is the extension of the spring in the equilibrium position.

If the mass of the particle is m and the number of oscillations per second is n, and if a is the length of the spring when the particle hangs in equilibrium, find the natural length of the spring and show that its modulus is $m(4\pi^2n^2a-g)$.

9. A particle of mass m is fixed to one end of an elastic string of natural length a and modulus mg. The other end of the string is attached to a fixed point O. The particle is pulled vertically upwards to a point at distance $3a$ vertically above O, and released from rest. Prove that the string becomes slack after time $\sqrt{(a/g)}\cos^{-1}\frac{1}{3}$, and find the velocity of the particle when it reaches O.

10. A particle of mass 50 g on a smooth horizontal table is attached by a light elastic string to a point A on the table. The particle is initially held with the string extended by 3 cm, the tension being 0·096 N, and it is then projected directly away from A with speed 32 cm/sec. Find how far it will move before first coming to rest. Find also, to 2 significant figures, the time taken to move through this distance.

11. The ends of a light elastic string of natural length $2a$ are fixed to two points in a horizontal line, distant $2a$ apart. When a particle is attached to the mid-point of the string and hangs in equilibrium, the two parts of the string are inclined at $\pi/4$ to the horizontal. The particle is then displaced in a vertical line through a small distance x. Find the force tending to restore the particle to its original position, neglecting $(x/a)^2$ and higher powers of x/a. Show that if released the particle will execute simple harmonic motion with period

$$2\pi\sqrt{\left(\frac{2a(3-\sqrt{2})}{7g}\right)}$$

12. A light elastic string of natural length $2a$ and modulus λ is stretched between two points distant $4a$ apart on a smooth horizontal table. At the midpoint of the string a particle of mass m is attached. The particle is displaced from its equilibrium position through a distance $3a/2$ in the line of the string and released. Describe the subsequent motion and show that the time of a complete oscillation is $4\sqrt{\left(\frac{ma}{\lambda}\right)}\left(\frac{1}{\sqrt{2}}\tan^{-1}\frac{2\sqrt{2}}{3}+\tan^{-1}1\frac{3}{4}\right)$. Find also the maximum speed attained by the particle.

EXERCISE 27B

1. A particle suspended by an elastic string stretches it by 20 cm when hanging in equilibrium. The particle is projected vertically downwards from the equilibrium position at 1·05 m/s. Prove that the motion is simple harmonic motion and that the period is $\frac{2}{7}\pi$ sec. Find how far the particle will move before first coming to rest, and how long it will take to reach a point half-way between the starting point and the first position of rest.

2. A particle is moving with simple harmonic motion in a straight line, its acceleration at any point being n^2 times its distance from a point O in the line. If v is its velocity when at a distance b from O, show that the intervals of time between the instants when the velocity has the numerical value v are alternately

$$\frac{2}{n}\tan^{-1}\frac{nb}{v} \text{ and } \frac{2}{n}\cot^{-1}\frac{nb}{v}$$

3. On a certain day the depth of water in a harbour at low tide, which was at 8 a.m., was 5 m. The next high tide occurred at 2.15 p.m., and the depth was 15 m. Assuming that the surface of the water moved in simple harmonic motion, find (i) the earliest time after 8 a.m. when the depth was 8 m, (ii) the maximum rate, in m per hr, at which the water rose.

4. A particle of mass 1 kg at a point P on a smooth horizontal table is connected by an unstretched spring to a fixed point A on the table. A force of 14·4 N is needed to stretch the spring by 10 cm. The particle is projected at 3 m/s in the direction AP, and first comes to rest at Q. Find the time taken to reach Q, and the distance PQ.

5. An elastic string of natural length 2 m is such that a static load of 3·2 N stretches it by 10 cm. The string is attached at its ends to points A, B, 3 m apart on a smooth horizontal table, and a particle of mass 1 kg is attached at its mid-point. If the particle is displaced from its equilibrium position to a point in the line AB and released, show that its distance x m from the equilibrium position after t sec satisfies the equation $\ddot{x} + 64x = 0$, it being assumed that each portion of the string remains stretched during the motion. Solve the equation, subject to initial conditions $x = 0{\cdot}1, \dot{x} = 0{\cdot}6$. Find the maximum value of x and the maximum speed of the particle. Find also the time, correct to 0·01 sec, when the particle first attains its maximum displacement.

6. An elastic string of natural length a and modulus mg is fixed at one end to a point O on a smooth horizontal table, distant $2a$ from an edge. The other end of the string is attached to a particle A of mass m. An inextensible string is fixed to A, passes over the edge, and carries a particle B of mass m. The system is released from rest when A is distant a from O and from the edge, and the strings are in a vertical plane perpendicular to the edge. Prove that A will reach the edge after a time $\pi\sqrt{[a/(2g)]}$.

7. A particle on a smooth plane inclined at 30° to the horizontal is attached to a fixed point of the plane by an elastic string along a line of greatest slope. The particle is held on the plane, below the fixed point, so that the string is just taut. and released. If the particle were hanging freely it would stretch the string by a m. Prove that the particle executes simple harmonic motion, and find the period and amplitude.

8. Two particles, of masses m and $2m$, hang together at the end of an elastic string of natural length a, extending it to the length $a + 3b$. The heavier particle suddenly falls off. To what height will the other particle rise, and how long will it take to do so?

9. A particle of mass m moves in a straight line on a smooth horizontal table, starting from a point O with speed $a\omega$, and being subject to a restoring force of magnitude $m\omega^2 x$ towards O when at distance x from O. When it has covered a distance b it collides with and adheres to another particle of mass m which is at rest. Find the distance through which the combined particle will move before first coming to rest, if the restoring force remains of magnitude $m\omega^2 x$. Show also that the total time taken from the start to the first resting position will be

$$\frac{1}{\omega}\left[\sin^{-1}\frac{b}{a} + \sqrt{2}\cos^{-1}\frac{b\sqrt{2}}{\sqrt{(a^2 + b^2)}}\right]$$

10. A spring balance consists of a scale-pan of mass 100 g suspended by a light spring. The pan is empty and in equilibrium when a mass of 400 g is placed in it without impulse. The pan then descends 16 cm before beginning to rise again. Show that the time taken to descend is $\frac{1}{7}\pi\sqrt{\frac{1}{2}}$ sec and find the speed of the pan when it is 4 cm below the initial position.

11. A heavy particle at the end of a light elastic string is performing a vertical simple harmonic motion of amplitude a; the maximum speed is $\sqrt{(nga)}$, where $n > 1$. The string is cut when the particle is a height x above the centre O of the motion, and is moving upwards with the string taut. The particle then moves freely under gravity. Find the height above O reached by the particle, and show that its greatest possible value, for different values of x, is

$$\frac{a}{2}\left(n + \frac{1}{n}\right)$$

12. An elastic string of natural length a and modulus $2mg$ has one end attached to a fixed point O on a smooth horizontal table (at a distance greater than $2a$ from an edge) and the other attached to a particle P of mass m. An inextensible string PQ is attached to P, passes at right angles over the edge of the table and supports a particle Q of mass m hanging freely. The string OP is in a straight line with the part of PQ on the table, and the particles are at rest. Show that OP $= 3a/2$. The particle Q is now pulled down a distance $\frac{1}{2}a$ and released from rest. Show that in the subsequent motion, when OP $= 3a/2 + x$, the acceleration of P is given by $\ddot{x} = -\omega^2 x$, where $\omega^2 = g/a$. Hence write down in terms of the time t from the beginning of the motion and the constants a, ω, an expression for the length OP during this motion.

3. The simple pendulum

A simple pendulum is defined as a particle attached to a fixed point by a light inextensible string, and allowed to swing freely. The arrangement cannot of course be realized in practice; no string is weightless or quite inextensible, and any mass must have a finite size, and therefore must rotate as the pendulum swings. The theory is, however, important, as the motion of actual pendulums is conveniently described by comparison with the ideal simple pendulum (see Chapter 15).

Let m be the mass of the particle and l the length of the string, and suppose it drawn aside from the vertical through an angle α and released. We show a general position where the angle with the vertical is θ, and use expressions for the acceleration in terms of θ. Note that the positive direction for $\ddot{\theta}$ is in the sense of θ increasing.

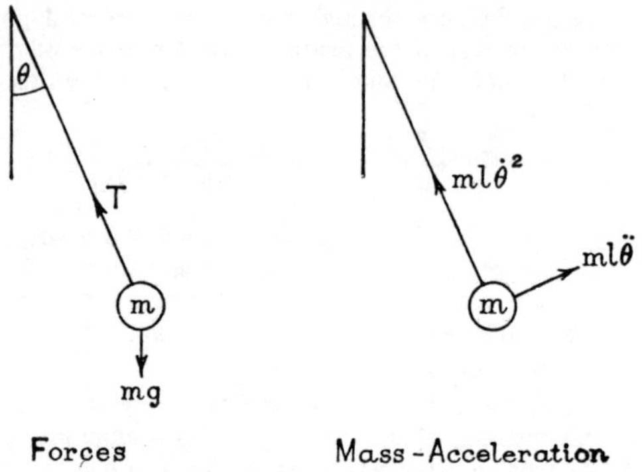

Resolving in the direction perpendicular to the string

$$ml\ddot{\theta} = -mg \sin \theta$$

This equation can be integrated once after multiplication by $2\dot{\theta}$ giving

$$ml\dot{\theta}^2 = 2mg (\cos \theta - \cos \alpha)$$

a result which can also be obtained by the Principle of Energy, but a second integration cannot be done by elementary methods. If, however, θ is restricted to be always a small angle, $\sin \theta \simeq \theta$, and the equation is approximately

$$l\ddot{\theta} = -g\theta$$

giving simple harmonic motion with $\omega^2 = g/l$, and therefore

a period of $2\pi \sqrt{(l/g)}$

The amplitude of the motion is α, but with the approximation we are using, this does not affect the period; neither has the mass of the particle any effect. In fact, it can be shown that if the amplitude is 7°, the period exceeds $2\pi \sqrt{(l/g)}$ by about 1 part in 1000; if the amplitude is 2°, the excess is about 1 part in 10,000. These figures take no account of air resistance.

4. Small changes

The length of a pendulum may change, for example on account of a change in temperature, and the value of g differs slightly from place to place. The effect of such small changes is best found by logarithmic differentiation.

With the standard notation

$$T = 2\pi\sqrt{(l/g)}$$

Taking logarithms,

$$\log T = \log 2\pi + \tfrac{1}{2}\log l - \tfrac{1}{2}\log g$$

Differentiating

$$\frac{dT}{T} = \tfrac{1}{2}\frac{dl}{l} - \tfrac{1}{2}\frac{dg}{g}$$

If then δT, δl, δg are associated small changes

$$\frac{\delta T}{T} \simeq \tfrac{1}{2}\frac{\delta l}{l} - \tfrac{1}{2}\frac{\delta g}{g}$$

To deal with the rate at which a pendulum clock gains or loses, let n be the number of seconds (or minutes) indicated by the clock in some fixed interval, such as one day. Then $n = k/T$, where k is a constant, and T is the period of oscillation of the pendulum, which need not be a ' seconds pendulum '.

Now $\quad \log n = \log k - \log T$

$$= \log k - (\log 2\pi + \tfrac{1}{2}\log l - \tfrac{1}{2}\log g)$$

$$\therefore \quad \frac{dn}{n} = -\tfrac{1}{2}\frac{dl}{l} + \tfrac{1}{2}\frac{dg}{g}$$

and

$$\frac{\delta n}{n} \simeq -\tfrac{1}{2}\frac{\delta l}{l} + \tfrac{1}{2}\frac{\delta g}{g}$$

In using these results it is necessary to be definite about signs. If actual changes are considered, as when a pendulum expands, it is natural to use the positive sign for an increase, and negative for a decrease. If, however, it is required to correct an existing error it is probably better to use the sign to give the necessary correction. Thus, if a clock is gaining, n should be decreased, and we give δn a negative value. Whichever convention is adopted should be clearly stated, and carefully kept.

5. The Energy Principle

When integrating the equation of motion for the simple pendulum we remarked that the first integral could be obtained by the Principle of Energy. In dealing with systems of particles, and later with rigid bodies, it is convenient to reverse this process, and to obtain the equation of motion by differentiating the result given by the Energy Principle. As we shall be dealing only with conservative systems of forces, we use the principle in the form

kinetic energy + potential energy = constant

(see Chapter 10).

Example 3.—A light rigid rod ABC has a particle of mass m attached at each end, and is freely pivoted at B. AB $= a - x$, BC $= a + x$. Show that if slightly disturbed from equilibrium the system executes simple harmonic motion, and find the period.

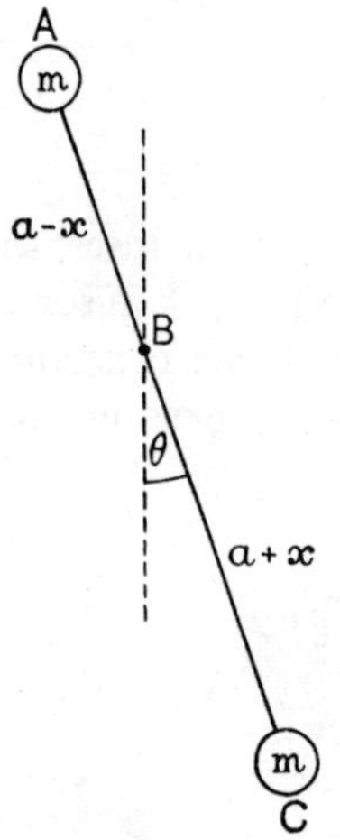

$$\text{KE} = \tfrac{1}{2}m[(a+x)^2 + (a-x)^2]\dot{\theta}^2$$
$$= m(a^2 + x^2)\dot{\theta}^2$$

The potential energy of the system is found from the height of the centre of gravity, imagining all the mass concentrated there. Since the centre of gravity is distant x from the pivot B, taking the equilibrium position as standard, we have

$$\text{PE} = 2mgx(1 - \cos\theta)$$

By the Principle of Energy, as stated above,

$$m(a^2 + x^2)\ {}^2 + 2mgx(1 - \cos\theta) = \text{constant}$$

Differentiating with respect to the time,

$$m(a^2 + x^2).\ 2\dot{\theta}\ddot{\theta} + 2mgx \sin\ \theta\ .\ \dot{\theta} = 0$$
$$\therefore\ (a^2 + x^2)\ \ddot{\theta} + gx \sin\theta = 0$$

If θ is small, this gives approximately

$$\ddot{\theta} = -\ \frac{gx\theta}{(a^2 + x^2)}$$

Hence there is simple harmonic motion, with period

$$2\pi\sqrt{\left(\frac{(a^2 + x^2)}{gx}\right)}$$

This is the period of a simple pendulum with length $(a^2 + x^2)/x$, which is called the *equivalent simple pendulum*. Note that the system may be made equivalent to a very long simple pendulum by reducing x to a small quantity.

EXERCISE 28

In this exercise all clocks are supposed to be pendulum clocks, and all pendulums are simple pendulums.

1. (i) Find the period of a pendulum of length 10 cm, taking $g = 9{\cdot}80$ m/s^2. Answer to 3 significant figures. (ii) Find the period of a pendulum of length 2 metres, taking $g = 981$ cm/s^2. Answer to 3 significant figures.

2. Find to the nearest millimetre the length of a seconds pendulum where $g = 9{\cdot}82$ m/s^2. Note that a seconds pendulum moves from one extremity of its swing to the other in 1 sec, so that its period is 2 sec.

3. A clock ticks every time the pendulum passes through its mean position. If it ticks twice per second find the length of the pendulum to the nearest millimetre. Take g as 980 cm/s^2.

4. A clock which kept correct time where $g = 9{\cdot}81$ m/s^2 was moved to a place where $g = 9{\cdot}79$ m/s^2. Find its rate of gain or loss in seconds per day.

5. A clock loses $\frac{3}{4}$ min per day. What percentage increase or decrease in the length of the pendulum is required to correct the error?

6. The length of a pendulum, which is about 15 cm, can be adjusted by turning a nut which runs on a screw having 10 threads per cm. What effect on the rate of the clock has a quarter turn of the nut, made so as to lengthen the pendulum?

7. The value of g at a place outside the earth varies inversely as the square of its distance x from the centre of the earth. Prove that $\dfrac{dg}{g} = -\dfrac{2dx}{x}$. A clock is moved from sea-level to a height of 3 km above sea-level. (Take the radius of the earth as 6400 km.) What percentage change in the length of the pendulum is needed if the rate of the clock is not to change?

8. The value of g at points below the earth's surface varies directly as the distance from the centre. A pendulum is moved from surface level to the bottom of a mine 1·6 km deep. The temperature increases from 10° C to 35° C, and the wire supporting the bob of the pendulum expands by 0·000011 of its length per degree C. Find the percentage change in the period of oscillation.

9. A pendulum hangs from the ceiling of a lift which has an upward acceleration of k m/s^2. Show that the tension in the string is the same as if the pendulum were at rest and the gravitational acceleration were changed from g to $(g + k)$. Hence find the error caused in a clock by placing it in a lift, the lift having an upward acceleration of 0·6 m/s^2. Take $g = 9{\cdot}8$ and answer in seconds per hour.

10. A pendulum is placed in a carriage of a model railway which runs in a circle of 900 m radius and makes one circuit in 5 min. Find the tension in the string when the bob is at rest relative to the carriage, as a fraction of the weight of the bob. Hence find the percentage increase or decrease in the period of oscillation of the pendulum due to its motion.

11. Prove that if the amplitude of vibration of a pendulum is small, the maximum speed of the bob is proportional to the amplitude. Find the maximum speed of a seconds pendulum swinging 6° on either side of the vertical.

12. Prove that, if the amplitude of vibration of a pendulum is small, the tension in the string when the bob is central exceeds the weight of the bob by a fraction equal to the square of the radian measure of amplitude. Find the amplitude of swing in degrees when the tension in the mean position is $\frac{1}{2}$ per cent more than the weight of the bob.

13. If in example 3 of the text x is variable, find what value of x makes the length of the equivalent simple pendulum a minimum.

14. A light rigid frame in the form of a square of side $2a$ carries a particle of mass m at each corner, and is freely pivoted at the midpoint of one side, so that it can swing in a vertical plane. Show that if slightly disturbed from the position of stable equilibrium it will perform simple harmonic motion, and find the length of the equivalent simple pendulum.

15. Repeat the work of question 14 for a regular hexagonal frame with side $2a$, carrying a particle of mass m at each vertex, and freely pivoted at a point mid-way between a vertex and the centre of the hexagon.

16. A smooth light circular cylinder of radius a can rotate freely about its axis, which is fixed and horizontal. A particle P of mass Nm is fixed to the outer surface of the cylinder, and a light inextensible string fastened to P passes over the cylinder and carries at its other end a particle of mass m hanging freely. Show that if N is large the system can be in equilibrium with the radius to P making a small angle, $1/N$ radians approximately, with the downward vertical. Show that if the cylinder is rotated through a small angle from this position and then released from rest it will perform simple harmonic motion, and find the frequency.

14

Differential Equations. The Catenary

In this chapter we shall consider a few problems which depend for their solution on simple differential equations. The differential equations will usually be ones in which the variables are separable, i.e. they will be capable of being written in the form $f(x)\,dx = \phi(y)\,dy$.

1. Motion of a particle in a resisting medium

Example 1.—A particle moves in a straight line against a resistance proportional to the speed. If the particle starts with speed u, find the speed after time t and the distance travelled in time t.

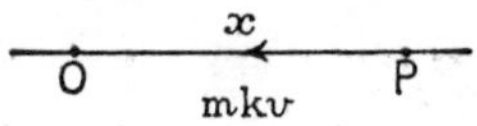

We may take the resistance in the form mkv.

Then the equation of motion is $$m\frac{dv}{dt} = -mkv$$

$$\therefore \frac{dv}{v} = -k\,dt$$

Integrating each side, $$\log v = -kt + A$$

If $v = u$ when $t = 0$, $$\log u = A$$

$$\therefore \log v - \log u = -kt$$

$$\log\frac{v}{u} = -kt$$

$$\underline{v = ue^{-kt}}$$

Now $$v = \frac{ds}{dt} = ue^{-kt}$$

$$\therefore s = -\frac{u}{k}e^{-kt} + B$$

If $s = 0$ when $t = 0$, $$0 = -\frac{u}{k} + B$$

$$\frac{u}{k} = B$$

$$\therefore s = \frac{u}{k}(1 - e^{-kt})$$

Example 2.—A particle of mass m falls from rest in a medium in which the resistance varies as the velocity. To find the velocity after t sec, the distance fallen in t sec, and the velocity v in terms of the distance travelled s.

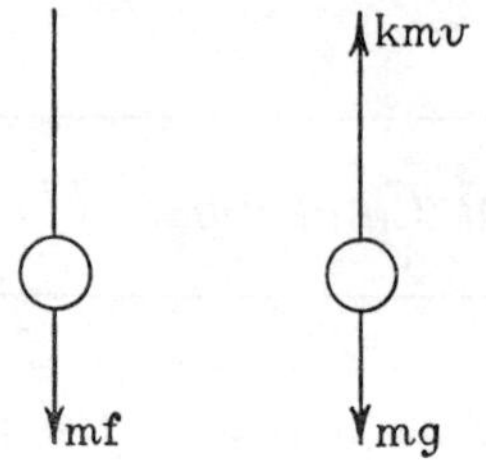

The equation of motion is $mf = mg - mkv$.

$$\therefore\ f = g - kv$$

i.e.
$$\frac{dv}{dt} = g - kv$$

Separating the variables,

$$\frac{dv}{g - kv} = dt$$

$$\therefore\ -\frac{1}{k}\log(g - kv) = t + c$$

Since $v = 0$ when $t = 0$,
$$c = -\frac{1}{k}\log g$$

$$\therefore\ \frac{1}{k}\{\log g - \log(g - kv)\} = t$$

i.e.
$$\log\frac{g}{g - kv} = kt$$

$$1 - \frac{kv}{g} = e^{-kt}$$

$$\underline{v = g/k\{1 - e^{-kt}\}}$$

(As $t \to \infty$, $e^{-kt} \to 0$, $\therefore\ v \to g/k$ as $t \to \infty$)

Now
$$v = \frac{ds}{dt} = g/k\{1 - e^{-kt}\}$$

$$\therefore\ s = \frac{g}{k}\left\{t + \frac{e^{-kt}}{k}\right\} + B$$

As $s = 0$ when $t = 0$,
$$0 = g/k^2 + B$$

$$\therefore\ s = \frac{g}{k}\left\{t + \frac{1}{k}e^{-kt}\right\} - \frac{g}{k^2}$$

$$\underline{= \frac{gt}{k} - \frac{g}{k^2}(1 - e^{-kt})}$$

If we want v in terms of s, we write $f = v\dfrac{dv}{ds}$

$$\therefore\ v\frac{dv}{ds} = g - kv$$

$$\frac{v\,dv}{g - kv} = ds$$

$$\therefore\ -\frac{kv\,dv}{g - kv} = -k\,ds$$

$$\left[1 - \frac{g}{g - kv}\right]dv = -k\,ds$$

$$v + \frac{g}{k}\log(g - kv) = -ks + A$$

When $v = 0$, $s = 0$,
$$\therefore\ \frac{g}{k}\log g = A$$

$$\therefore\ ks = \frac{g}{k}\log g - \frac{g}{k}\log(g - kv) - v$$

$$ks = \frac{g}{k}\log\frac{g}{g - kv} - v$$

Example 3.—A particle falls from rest in a medium in which the resistance is proportional to the square of the speed. To find the velocity after time *t*.

The equation of motion is
$$\frac{dv}{dt} = g - kv^2$$

$$\frac{dv}{g - kv^2} = dt$$

Put $g/k = c^2$, then
$$\frac{dv}{c^2 - v^2} = k\,dt$$

$$\frac{1}{2c}\left(\frac{1}{c + v} + \frac{1}{c - v}\right) = k\,dt$$

$$\frac{1}{2c}\log\frac{c + v}{c - v} = kt + A$$

Since when $t = 0$, $v = 0$,
$$\therefore\ A = 0$$

$$\therefore\ \log\frac{c + v}{c - v} = 2ckt$$

$$\frac{c + v}{c - v} = e^{2ckt}$$

$$\frac{v}{c} = \frac{e^{2ckt} - 1}{e^{2ckt} + 1}$$

$$v = \sqrt{\frac{g}{k}}\cdot\frac{e^{2\sqrt{(gk)}\cdot t} - 1}{e^{2\sqrt{(gk)}\cdot t} + 1}$$

Example 4.—A particle is projected vertically upwards in the medium of the previous question with initial speed u. To find the greatest height reached.

The equation of motion is $\quad f = -g - kv^2$

$$\therefore\ v\frac{dv}{dx} = -g - kv^2$$

$$\frac{v\,dv}{g + kv^2} = -dx$$

$$\frac{2kv\,dv}{g + kv^2} = -2k\,dx$$

$$\therefore\ \log(g + kv^2) = -2kx + B$$

When $x = 0,\ v = u,$ $\quad \therefore\ \log(g + ku^2) = B$

$$\therefore\ 2kx = \log(g + ku^2) - \log(g + kv^2)$$

$$2kx = \log\frac{g + ku^2}{g + kv^2}$$

At the highest point $v = 0,$ $\quad \therefore\ 2kh = \log\dfrac{g + ku^2}{g}$

$$h = \frac{1}{2k}\log\frac{g + ku^2}{g}$$

Example 5.—A particle is attracted towards a fixed point by a force inversely proportional to the square of its distance from the point. If it starts from rest at a distance a from the fixed point, find the time taken to reach the point.

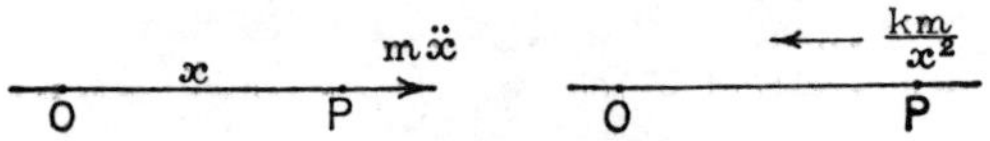

The equation of motion is $\quad m\ddot{x} = -\dfrac{km}{x^2}$

$$\therefore\ v\frac{dv}{dx} = -\frac{k}{x^2}$$

$$v\,dv = -\frac{k}{x^2}\,dx$$

$$\frac{v^2}{2} = \frac{k}{x} + A$$

When $v = 0,\ x = a,$ $\quad \therefore\ A = -\dfrac{k}{a}$

$$\therefore\ v^2 = 2k\left(\frac{1}{x} - \frac{1}{a}\right)$$

$$\frac{dx}{dt} = \pm\sqrt{(2k)}\,.\sqrt{\left(\frac{1}{x} - \frac{1}{a}\right)}$$

We must take the minus sign, for x decreases as t increases,

$$\therefore \frac{dx}{dt} \text{ is negative}$$

$$\therefore \sqrt{\left(\frac{ax}{a-x}\right)} \,.\, dx = -\sqrt{(2k)} \,.\, dt$$

Put $x = a\cos^2\theta$, $\qquad dx = -2a\cos\theta\sin\theta\, d\theta$

$$\therefore \int \sqrt{\left(\frac{a^2\cos^2\theta}{a\sin^2\theta}\right)} \,.\, (-2a\cos\theta\sin\theta\, d\theta) = -\sqrt{(2k)} \,.\, t + B$$

$$\int -2a^{3/2}\cos^2\theta\, d\theta = -a^{3/2}(\theta + \tfrac{1}{2}\sin 2\theta) = -\sqrt{(2k)} \,.\, t + B$$

When $t = 0$, $x = a$, and so $\cos\theta = 1$, $\therefore \theta = 0$, $\therefore B = 0$

$$\therefore \sqrt{(2k)} \,.\, t = a^{3/2}\{\theta + \sin\theta\cos\theta\}$$

$$= a^{3/2}\left\{\cos^{-1}\sqrt{\frac{x}{a}} + \sqrt{\left(\frac{x(a-x)}{a^2}\right)}\right\}$$

If when $t = T$, $x = 0$, $\qquad \sqrt{(2k)} \,.\, T = a^{3/2}\dfrac{\pi}{2}$

$$T = \frac{a^{3/2}\pi}{2\sqrt{(2k)}}$$

2. The compound interest law

Many cases occur in science in which the rate of increase of a quantity is proportional to the value of the quantity, i.e. if y is the value of the quantity at time t then

$$\frac{dy}{dt} = ky$$

In equations of this type the variables are separable and we get

$$\frac{dy}{y} = k\, dt$$

$$\therefore \log y = kt + \log C$$

(It is often convenient to put the constant in this form.)

$$\therefore \log y/C = kt$$

i.e. $$y = Ce^{kt}$$

1. *Newton's Law of Cooling.*

Newton's law of cooling says that 'the rate of cooling of a body is proportional to the difference between the temperature of a body and the temperature of its surroundings'.

So, if $\theta°$ is the difference between the temperature of the body and the temperature of its surroundings,

$$\frac{d\theta}{dt} = -k\theta$$

which as above gives $\theta = \theta_0 e^{-kt}$, where θ_0 is the difference of temperature when $t = 0$.

Suppose a body cools from 80° C to 60° C in 10 min, find how long it will take to cool a further 20° if the temperature of the surroundings is maintained at 20° C.

$$\theta_0 = 60°, \quad \therefore \ \theta = \theta_0 e^{-kt} = 60e^{-kt}$$

Now $\theta = 40°$ when $t = 10$, $\quad \therefore \ 40 = 60e^{-kt}$

i.e. $\quad \log_e \frac{2}{3} = -10k \quad . \ . \ . \ . \quad$ (i)

And $\theta = 20°$ when $t = T$ (say) $\quad \therefore \ 20 = 60e^{-kT}T$

i.e. $\quad \log_e \frac{1}{3} = -Tk \quad . \ . \ . \ . \quad$ (ii)

Dividing (i) by (ii)

$$\frac{\log_e \frac{2}{3}}{\log_e \frac{1}{3}} = \frac{10}{T}$$

These of course are Naperian logarithms but, since we need only their ratio, we may use logarithms to the base 10.

$$\therefore \frac{\bar{1}\cdot824}{\bar{1}\cdot523} = \frac{0\cdot176}{0\cdot477} = \frac{10}{T}$$

$$\therefore T = \frac{4\cdot77}{0\cdot176} = 27\cdot1 \text{ min}$$

2. *The speed of some chemical reactions* (*Wilhelmy's Law*).

Many chemical reactions obey this law, which states that the speed of the reaction is proportional to the concentration cf the reacting substance. Hence, if a is the original concentration and x the amount transformed at time t, then

$$\frac{dx}{dt} = K(a-x)$$

$$\frac{dx}{a-x} = K\,dt$$

$$-\log(a-x) = Kt + A$$

Since $x = 0$ when $t = 0$ $\qquad A = -\log a$

$$\therefore \log \frac{a-x}{a} = -Kt$$

$$\frac{a}{a-x} = e^{Kt}$$

$$\frac{a}{x} = \frac{e^{Kt}}{e^{Kt}-1}$$

$$x = a(1-e^{-Kt})$$

3. *The change in atmospheric pressure due to change of height above sea-level.*

Suppose p to be the pressure at height h and $p+\delta p$ the pressure at height $h+\delta h$. Consider a vertical column of cross-sectional area A.

The force at the lower end exceeds that at the upper end by the weight of the air

$$\therefore pA-(p+\delta p)A = g\rho A\delta h$$

where ρ is the density of the air.

$$\therefore -\delta p = g\rho\,\delta h$$

When $\qquad \delta h \to 0, \quad \dfrac{dp}{dh} = -g\rho$

But $\qquad \rho = \dfrac{p}{K}, \quad \therefore \dfrac{dp}{dh} = \dfrac{-gp}{K}$

$$\frac{dp}{p} = \frac{-g}{K}dh$$

$$\log p = \frac{-gh}{K}+B$$

If p_1 and p_2 are the atmospheric pressures at heights h_1 and h_2,

$$\log p_1 = \frac{-gh_1}{K}+B$$

$$\log p_2 = \frac{-gh_2}{K}+B$$

$$\therefore \log p_1 - \log p_2 = \frac{-g}{K}(h_1-h_2)$$

$$\frac{p_1}{p_2} = e^{-(h_1-h_2)g/K}$$

EXERCISE 29A

1. A particle of unit mass falls from rest in a medium in which the resistance is kv where v is the velocity at any time t. Find v in terms of t and prove that the velocity can never exceed g/k.

2. A car is travelling along a straight road and when it passes a certain point O the engine is switched off. At time t sec after passing O the speed v m/s is given by $\frac{1}{v} = a + bt$, where a and b are positive constants. Show that the retardation is proportional to the square of the speed. If when $t = 0$ the retardation is 1 m/s² and the speed is 20 m/s find a and b. If x is the distance moved from O in t sec, find x in terms of t, and v in terms of x.

3. A particle moves in a straight line so that its velocity v m/s when it has travelled a distance x m is given by $x = \frac{2v^2}{60 - v}$. Show that the acceleration is $\frac{(60 - v)^2}{2(120 - v)}$. Show also that the time taken to reach a speed of 15 m/s from rest is
$$2\{\log_e \tfrac{4}{3} + \tfrac{1}{3}\} \text{ sec}$$

4. A particle of mass 5 kg starts from rest and is acted upon by a force which increases uniformly in 5 sec from zero to 20 N. Prove that t sec after the start the acceleration is $0{\cdot}8t$ m/s². Find the distance the particle travels in the first 5 sec and show that when it has moved x m its speed is v m/s, where
$$5v^3 = 18x^2$$

5. The acceleration of a car at speed v m/s is $1{\cdot}2 - \frac{v^2}{3000}$ m/s². Find the maximum speed of the car. Prove that the car attains a speed of 50 m/s in approx. 1 min whilst travelling about 1800 m.
Take $\log_e 6 = 1{\cdot}8$ and $\log_e 11 = 2{\cdot}4$.

6. A particle of mass m is projected vertically upwards with speed V. The air resistance is kmv^2 when v is the speed. If the particle returns to the point of projection with speed V_1 prove $\frac{1}{V_1^2} = \frac{1}{V^2} + \frac{k}{g}$

7. A cyclist travelling at speed V begins to free-wheel and observes that after travelling a further distance a in time T his speed is V'. If his retardation is proportional to his speed, show that $\log\left(\frac{V}{V'}\right) = \frac{T(V - V')}{a}$

8. The displacement of a particle at time t is x measured from a fixed point and $\frac{dx}{dt} = a(c^2 - x^2)$, where a and c are constants, and both are positive. If $x = 0$ when $t = 0$, prove $x = \frac{c(e^{2act} - 1)}{e^{2act} + 1}$. If $x = 3$ when $t = 1$ and $x = \frac{75}{17}$ when $t = 2$, prove $c = 5$ and $a = \frac{1}{5} \log_e 2$.

9. A particle of unit mass travels horizontally against a resistance $k\sqrt{v}$ N. If the particle has an initial velocity u prove that the time taken to come to rest is $\frac{2\sqrt{u}}{k}$ and that the distance travelled is $\frac{2u^{3/2}}{3k}$.

10. The rate of decay at any instant of a radioactive substance is proportional to the amount of the substance remaining at that instant. If the initial amount is A and the amount remaining after time t is x prove $x = Ae^{-kt}$. If the amount remaining is reduced from $\frac{1}{4}A$ to $\frac{1}{8}A$ in 8 hours, prove that the initial amount A was reduced to $\frac{1}{2}A$ in about 13·7 hours.

11. Two liquids are boiling in a mixture. The ratio of the quantities of each passing off as vapour at any instant is proportional to the ratio of the quantities still left in the liquid state. Prove that the quantities, say x and y, are connected by an equation of the form $y = cx^k$.

12. In a certain chemical reaction the rate of decomposition at any time is proportional to the amount that remains. In any small change the percentage increase in the pressure p inside the vessel in which the reaction takes place is proportional to the percentage decrease in the amount m of the substance. Write m in terms of m_o, a and t, and p in terms of p_o, m_o, b and m, where a and b are constants. Hence find p in terms of p_o, a, b and t.

13. Water flows from a tank at a rate proportional to the amount remaining in the tank. If the tank is initially full and after 1 hour is half-full, after how many more minutes will it be one-third full?

14. The capacity of a condenser is C. It is charged so that the potential difference between the plates is V_o. The plates are then connected by a resistance R which is of negligible inductance. If at any time t the current is i and the charge Q then $Q = CV$, $i = -dQ/dt$ and $V = Ri$. Prove

$$\text{(i) } CR\frac{dV}{dt} + V = 0, \quad \text{(ii) } V = V_0e^{-t/(CR)}, \quad \text{(iii) } R = \frac{t}{C\log(V_0/V)}$$

15. In a certain chemical reaction the amount x of one substance at time t is related to the speed of the reaction dx/dt by the equation $dx/dt = k(a - x)(2a - x)$, where a and k are constant and $x = 0$ when $t = 0$. If $x = 2$ when $t = 1$ and $x = 2{\cdot}8$ when $t = 3$, prove $a = 3$ and find x when $t = 2$.

EXERCISE 29B

1. A particle of mass m falls from rest; the resistance of the air when the speed is v is kv^2 where k is a constant. Show that after time t the velocity is given by $t = \frac{V}{2g}\log_e\frac{V+v}{V-v}$ where $V^2 = mg/k$. If s is the distance fallen in time t prove

$$s = \frac{V^2}{2g}\log_e\frac{V^2}{V^2 - v^2}$$

2. A particle is projected vertically upwards with speed u. The air resistance is v^2/k^2 times the weight, where k is constant. Prove that the particle reaches a height $\frac{k^2}{2g}\log\frac{k^2 + u^2}{k^2}$ above the point of projection.

3. A particle moves from rest along a straight line under a force $(A - v/10)$ N per kg, where v is the velocity in m/s and A is constant. If the velocity at the end of 10 sec is 3 m/s, show that the displacement then is about 17·5 m.

4. A body projected vertically upwards with speed 80 m/s reaches a height of 90 m. When the speed is v m/s the retardation is kv^2 m/s². Prove that k satisfies the equation $180k = \log_e (1 + 200k)$.

5. A particle moves in a straight line against a resistance proportional to the cube of the speed. Prove that the time average of the velocity in going from A to B is equal to the velocity at the midpoint of AB and that if v_1 and v_2 are the velocities at A and B, the distance average of the velocity is $\dfrac{v_1v_2}{v_1 - v_2} \log \dfrac{v_1}{v_2}$

6. A particle of mass m is projected vertically upwards with speed u in a medium in which the resistance per unit mass is kv^2, where v is the speed. Prove that the loss of energy when it has returned to its original position is $\frac{1}{2} \dfrac{mku^4}{g + ku^2}$

7. A particle of unit mass starts from rest at a distance a from a centre which attracts it with a force k/x^3. Find in terms of a and k the time taken by the particle to reach the centre.

8. A particle of mass m moves in a straight line under the action of a force mp and a resistance mqv^4, where p and q are constant and v is the speed. Show that when the time t is small v is approximately $pt - \frac{1}{5}qp^4t^5$.

9. The rate at which liquid is flowing out of a vessel at any instant is proportional to the amount left in at that instant. If the vessel is half-emptied in 1 min, how much will flow out in 2 min?

10. The temperature of a liquid in a room of constant temperature 20° is 70°, after 5 min it is 60°. What will be its temperature after a further 30 min? After how long will its temperature be 40°?

11. The number of bacteria in a culture increases at a rate proportional to the number present. Initially there were 100 and this increased to 250 in 10 min. Find the number present after a further half-hour.

12. A radioactive element disintegrates at a rate proportional to the amount of the element present. If one-quarter of the original amount disintegrates in 10 sec, how long would it take for $\frac{3}{4}$ of the original amount present to disintegrate?

13. $L\dfrac{di}{dt} + Ri = E$, where L, R, and E are constant. Find i in terms of t if $i = 0$ when $t = 0$.

14. If the temperature of the air is constant as we ascend a mountain, the rate (per unit height ascended of the mountain) of decrease of the barometric height at any elevation varies directly as the actual barometric height there. Prove that if H is the barometric height at an elevation h and H_0 is the barometric height at sea-level, $H = H_0e^{-kh}$ where k is constant. If H is in mm and h in m, k is about 0·00013 when $H_0 = 760$. Show that the fall of the barometer is about 1 mm for an elevation of 10 m at sea-level.

15. When light passes through a pane of glass the intensity I after passing through a thickness x is given by $\dfrac{dI}{dx} = -kI$.

If a pane of glass absorbs 4 per cent of the light passing through it, how much light will get through 20 such panes? How many panes will absorb 60 per cent of the light?

3. The common catenary

The catenary is the curve in which a uniform chain or string hangs when freely suspended from two fixed points.

Denote the tension at the lowest point A by T_0; this will be horizontal. Let s be the length of chain measured from A to any point P. Let the tension at P be T and let its inclination to the horizontal be ψ. Let the weight per unit length of the chain be w.

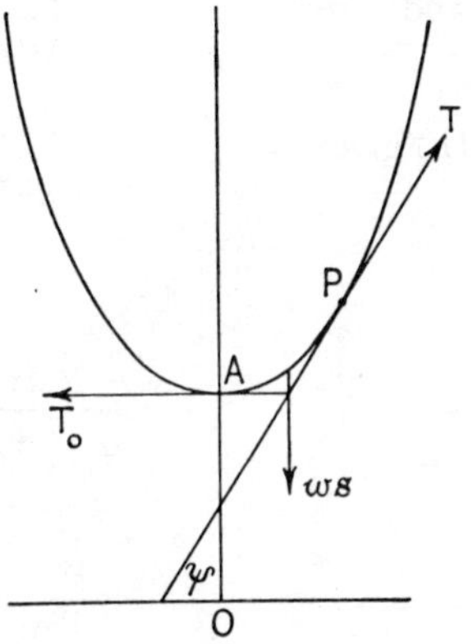

The part of the chain AP will be in equilibrium under the action of three forces: its weight ws, T_0, and T, the tensions at A and P.

Resolving vertically and horizontally,

$$T \sin \psi = ws, \; T \cos \psi = T_0$$

For convenience we introduce another constant c, which is such that $T_0 = wc$. Then

$$T \sin \psi = ws, \; T \cos \psi = wc$$

Dividing $$s = c \tan \psi \qquad \text{(i)}$$

This is the *intrinsic equation* of the curve; c is called the parameter.

To find the Cartesian equation of the curve.

$$\tan \psi = \frac{dy}{dx}, \quad \therefore \frac{dy}{dx} = \frac{s}{c}$$

Consider a small element δs of a curve joining two points P and Q on the curve. Let the coordinates of P and Q be (x, y) and $(x+\delta x, y+\delta y)$ respectively. Then

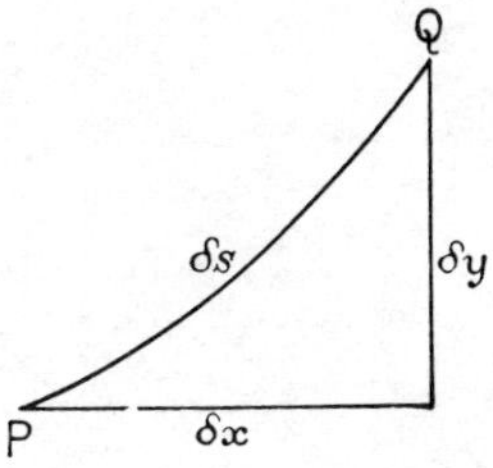

$$(\delta s)^2 \simeq (\delta y)^2 + (\delta x)^2$$

$$\therefore \left(\frac{\delta s}{\delta x}\right)^2 \simeq \left(\frac{\delta y}{\delta x}\right)^2 + 1$$

and $$\left(\frac{\delta s}{\delta y}\right)^2 \simeq 1 + \left(\frac{\delta x}{\delta y}\right)^2$$

When δs, δx and $\delta y \to 0$ these become

$$\left(\frac{ds}{dx}\right)^2 = \left(\frac{dy}{dx}\right)^2 + 1 \quad \text{(ii)}$$

and

$$\left(\frac{ds}{dy}\right)^2 = 1 + \left(\frac{dx}{dy}\right)^2 \quad \text{(iii)}$$

(ii) gives

$$\left(\frac{ds}{dx}\right)^2 = 1 + \frac{s^2}{c^2}$$

$$\therefore \frac{ds}{dx} = \frac{\sqrt{(c^2+s^2)}}{c}$$

$$\frac{c\,ds}{\sqrt{(c^2+s^2)}} = dx$$

$$\therefore \underline{x = c\sinh^{-1}\frac{s}{c}} \quad \text{(iv)}$$

or

$$\underline{s = c\sinh\frac{x}{c}} \quad \text{(v)}$$

provided $x = 0$ when $s = 0$.

(iii) gives

$$\left(\frac{ds}{dy}\right)^2 = 1 + \frac{c^2}{s^2}$$

$$\frac{ds}{dy} = \frac{\sqrt{(s^2+c^2)}}{s}$$

$$\frac{s\,ds}{\sqrt{(s^2+c^2)}} = dy$$

$$\therefore y = \sqrt{(s^2+c^2)}$$

i.e.

$$\underline{y^2 = s^2+c^2} \quad \text{(vi)}$$

provided $y = c$ when $s = 0$, i.e. when $x = 0$.

$$\therefore y^2 = c^2\left(\sinh^2\frac{x}{c}+1\right)$$

$$= c^2\cosh^2\frac{x}{c}$$

$$\therefore \underline{y = c\cosh\frac{x}{c}} \quad \text{(vii)}$$

Since $T \sin \psi = ws$ and $T \cos \psi = wc$

$$T^2 = w^2(s^2+c^2)$$

which from (vi) gives

$$T^2 = w^2y^2$$

$$\therefore \underline{T = wy}$$

Thus the tension at any point of the catenary is proportional to the height of the point above the x-axis which is usually called the *directrix*.

When c is large, from equation (vii),

$$y = c\cosh\frac{x}{c} = \frac{c}{2}(e^{x/c}+e^{-x/c})$$

$$= \frac{c}{2}\left\{1+\frac{x}{c}+\frac{x^2}{2c^2}+\ldots+\left(1-\frac{x}{c}+\frac{x^2}{2c^2}+\ldots\right)\right\}$$

$$= c+\frac{x^2}{2c}+\ldots$$

i.e. $$y-c \simeq \frac{x^2}{2c} \text{ provided } c \text{ is large.}$$

In this case the curve is approximately a parabola of latus rectum $2c$.

If k is half the span, half the length of the chain is given by

$$s = \frac{c}{2}\left\{1+\frac{k}{c}+\frac{k^2}{2c^2}+\frac{k^3}{6c^3}\ldots-\left(1-\frac{k}{c}+\frac{k^2}{2c^2}-\frac{k^3}{6c^3}\ldots\right)\right\}$$

$$= \frac{c}{2}\left\{\frac{2k}{c}+\frac{k^3}{3c^3}+\ldots\right\}$$

$$= k+\frac{k^3}{6c^2} \text{ provided } c \text{ is large}$$

$$\therefore s-k = \frac{k^3}{6c^2} \qquad \ldots\ldots\ldots \quad \text{(viii)}$$

If h is the sag, h is the difference of the ordinates when $x = k$ and when $x = 0$.

For $x = 0$, $y = c$, and for $x = k$, $y \simeq c+\dfrac{k^2}{2c}$

$$\therefore h = \frac{k^2}{2c}$$

i.e.
$$\frac{1}{c^2} = \frac{4h^2}{k^4}$$

$$\therefore s-k = \frac{k^3}{6} \cdot \frac{4h^2}{k^4} = \frac{4h^2}{6k} \qquad \text{from (viii)}$$

The span is $2k$,

$$\therefore 2(s-k) = \frac{8}{3} \cdot \frac{h^2}{2k}$$

i.e. the difference between length and span $= \dfrac{8}{3}\dfrac{(\text{sag})^2}{\text{span}}$

4. Worked examples

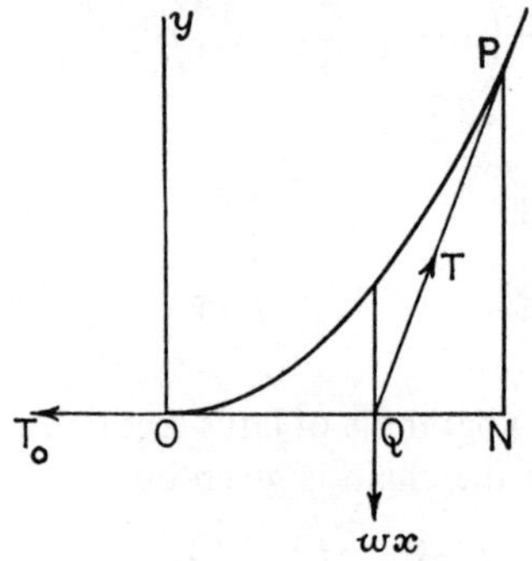

1. *The suspension bridge.*

If a chain supports a continuous load, uniformly distributed, the chain hangs in the form of a parabola. O is the lowest point of the chain and P any point of the chain whose coordinates referred to horizontal and vertical axes through O are (x, y). The weight carried by the portion OP will be proportional to ON and acts through Q the midpoint of ON. We may call it wx.

The other forces acting on the portion OP are T_0 the horizontal tension at O and the tension T at P—three of them. They must therefore meet at Q and PNQ is a triangle of forces.

$$\therefore \frac{wx}{\text{PN}} = \frac{T_0}{\text{NQ}}, \quad \therefore T_0 y = \tfrac{1}{2}wx^2$$

Hence, if we denote T_0 by wc, $y = \dfrac{x^2}{2c}$, a parabola.

The span of a suspension bridge is 96 m and the sag in the chains is 7 m. The two of them support a load of 1000 kg per horizontal metre. Find the tensions at the lowest and highest points. The load carried by OP is $24g$ kN. The triangle QPN is a triangle of forces.

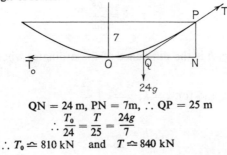

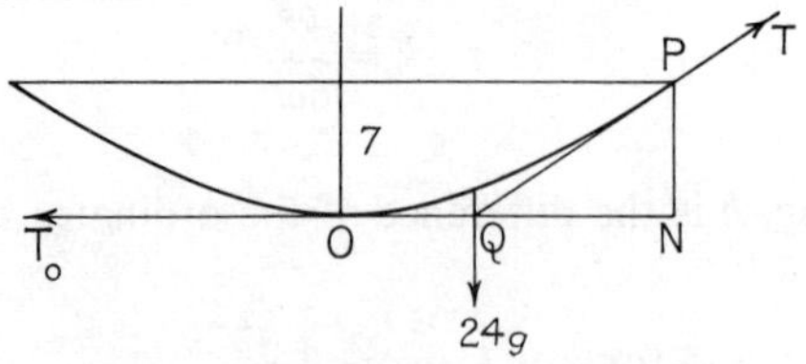

$$\text{QN} = 24 \text{ m}, \text{ PN} = 7\text{m}, \therefore \text{QP} = 25 \text{ m}$$

$$\therefore \frac{T_0}{24} = \frac{T}{25} = \frac{24g}{7}$$

$$\therefore T_0 \simeq 810 \text{ kN} \quad \text{and} \quad T \simeq 840 \text{ kN}$$

2. A uniform chain of length $2l$ and weight w per unit length is suspended between two points at the same level and has a maximum depth d. Prove that the tension at the lowest point is $\dfrac{w(l^2 - d^2)}{2d}$. If $l = 50$ m and $d = 20$ m, find the distance between the points of suspension.

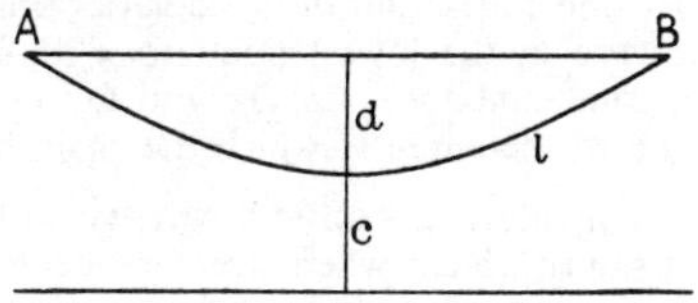

For the catenary $y^2 = c^2 + s^2$.

At B $\quad y = c + d,\ s = l$

$$\therefore (c + d)^2 = c^2 + l^2$$
$$2cd = l^2 - d^2$$
$$c = \frac{l^2 - d^2}{2d}$$

The tension at the lowest point $= wc = \dfrac{w(l^2 - d^2)}{2d}$

If $l = 50$ and $d = 20$, $\quad c = \dfrac{2500 - 400}{40} = \dfrac{105}{2}$

Now $\quad s = c \sinh \dfrac{x}{c}$

Hence if AB $= 2x$,

$$x = \frac{105}{2} \sinh^{-1} \frac{20}{21} = \frac{105}{2} \log_e \left[\frac{20}{21} + \sqrt{\left\{1 + \left(\frac{20}{21}\right)^2\right\}}\right]$$
$$= \frac{105}{2} \log_e \frac{49}{21}$$

$$\therefore \text{AB} = 105 \times 2{\cdot}303 \log_{10} \frac{49}{21} \simeq 89 \text{ m}$$

EXERCISE 30

1. A tow rope has an effective length of 20 m and has mass 5 kg per metre. One end of the rope is 4 m higher than the other. Find the maximum tension in the rope when the tangent at the lower end is horizontal.
2. A uniform chain of length $2l$ has its ends fixed at two points at the same level. The sag at the middle is h. Prove that the span is $\dfrac{l^2 - h^2}{h} \log \dfrac{l + h}{l - h}$
3. A uniform chain of length $2s$ and weight w per unit length hangs between two points at the same level. If the sag is k, find the pull on the points of support.
4. A vertical post AB of height h has a wire attached to it at B. The wire weighs w per unit length and is horizontal at the point of anchorage C. If s is the length of wire from B to C, show that the fixing couple required at A is $\frac{1}{2}w(s^2 - h^2)$.

5. A uniform wire hangs freely from two points at the same level 200 m apart. The sag is 15 m. Show that the greatest tension is approximately $348w$ and the length of wire is approximately 203 m.

6. A chain is suspended from two points at the same level and $2l$ apart. The lowest point of the chain is h below the level of the supports. In each half of the chain the load per unit horizontal distance varies uniformly with the horizontal distance from zero at the lowest point to w at the ends. Taking the origin at the lowest point and the axes Ox and Oy horizontal and vertical, show that the equation of the curve in which the chain hangs is $y = hx^3/l^3$.

7. Find approximately the greatest tension in a wire which has mass 100 g per m when it hangs with a sag of 25 cm when stretched between two points at the same level 40 m apart.

8. A uniform heavy chain of length 31 m is suspended from two points at the same level and 30 m apart. Show that the tension at the lowest point is about 1·08 times the weight of the chain.

9. A telegraph wire stretched between two points at the same level a m apart sags n m in the middle. Prove that the tension at the ends is approximately $w\left(\frac{a^2}{8n} + \frac{7}{6}n\right)$ where w is the weight per unit length.

10. A suspension chain carries a load uniformly distributed on a horizontal platform. The load is 1500 kN per metre length of the span of 200 m. The height of the point of support above the lowest point of the chain is 20 m. Neglecting the weight of the chain, find the greatest and least tensions in the chain.

11. A chain 40 m long has mass 2 kg per metre. It hangs between two points at the same level and has a sag of 5 m. Find the parameter of the catenary and the tension at a point of support.

12. A uniform string of length 6·4 m hangs in equilibrium over two smooth pegs at the same level. The lowest point of the curved part is 10 cm below the level of the pegs. Find the lengths of the portions of string which hang vertically.

15

Rigid Body rotating about a Fixed Point

1. Moment of inertia

In this and the next chapter we shall need the *moment of inertia* of various common bodies.

The moment of inertia of a body about a particular axis is defined as $m_1r_1^2+m_2r_2^2+m_3r_3^2+\ldots$, i.e. Σmr^2 where m_1, m_2, m_3, ..., are the masses of the particles of the body and r_1, r_2, r_3 ..., are their distances from the axis in question.

The finding of Σmr^2 for a continuous distribution of matter involves an integration in the same way that the finding of Σwx for centre of gravity did.

2. Finding moments of inertia

1. *For a uniform rod of mass M and length 2a about an axis perpendicular to the rod throughout its centre.*

Take the rod as x-axis and the perpendicular to the rod at G as y-axis. Let the mass per unit length of the rod be m.

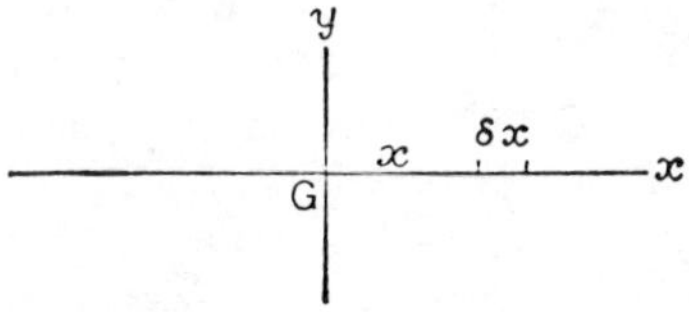

Consider a small element of the rod of length δx whose distance from G is x.

The moment of inertia of this small element about Gy is $m\delta x \, . \, x^2$

$\therefore$ The total moment of inertia is the limit of $\Sigma mx^2\delta x$ as $\delta x \to 0$

$$= \int_{-a}^{a} mx^2\,dx = [\tfrac{1}{3}mx^3]_{-a}^{a} = \tfrac{2}{3}ma^3 = \tfrac{1}{3}Ma^2$$

since $2ma = M$.

In exactly the same way we could show that the moment of inertia of the same rod about a parallel axis through one end is $\frac{4}{3}Ma^2$.

2. *For a rectangle of sides 2a and 2b and mass M about one side.*

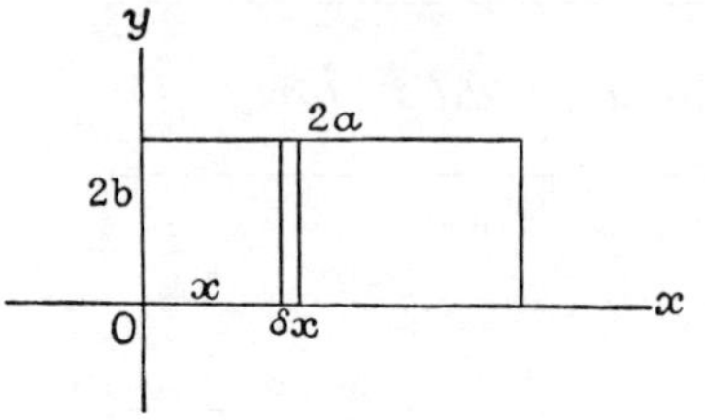

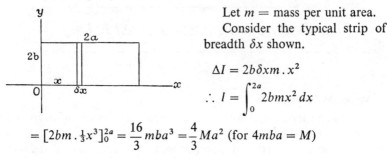

Let m = mass per unit area.

Consider the typical strip of breadth δx shown.

$$\Delta I = 2b\delta xm \, . \, x^2$$

$$\therefore I = \int_0^{2a} 2bmx^2 \, dx$$

$$= [2bm \, . \, \tfrac{1}{3}x^3]_0^{2a} = \frac{16}{3}mba^3 = \frac{4}{3}Ma^2 \text{ (for } 4mba = M)$$

Similarly about a parallel axis through the centre of gravity of the rectangle the moment of inertia is $\frac{1}{3}Ma^2$.

3. Two theorems on moment of inertia

I. *Theorem of perpendicular axes.*

For a lamina, if OX and OY are two perpendicular axes in the plane of the lamina and the moments of inertia about these axes are A and B, then the moment of inertia of the lamina about the axis OZ perpendicular to OX and OY is $A+B$.

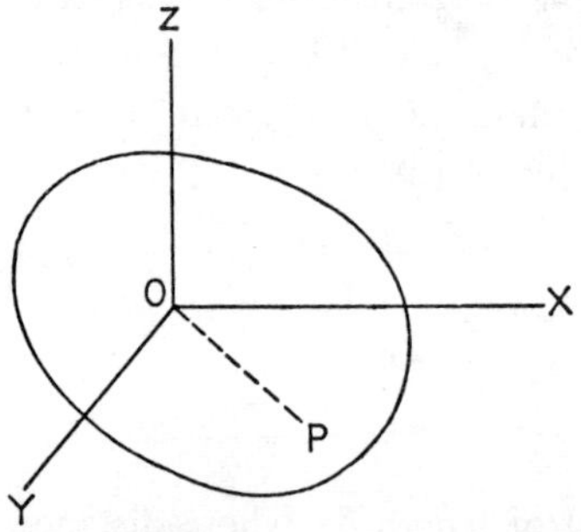

Let m be the mass of a particle P of the lamina. Let (x, y) be the coordinates of P referred to the axes OX and OY, and let OP = r. Then

$$r^2 = x^2+y^2$$

$$\therefore mr^2 = mx^2+my^2$$

and this is true for all particles of the lamina.

$$\therefore \Sigma mr^2 = \Sigma mx^2+\Sigma my^2$$

$\therefore$ moment of inertia about OZ is $A+B$.

For example, for a rectangle of sides $2a$, $2b$, the moment of inertia about an axis through its centre perpendicular to its plane will be

$$\tfrac{1}{3}Ma^2+\tfrac{1}{3}Mb^2 = \tfrac{1}{3}M(a^2+b^2)$$

For the same rectangle about an axis perpendicular to its plane through one vertex $\frac{4}{3}Ma^2+\frac{4}{3}Mb^2 = \frac{4}{3}M(a^2+b^2)$.

II. *Theorem of parallel axes.*

The moment of inertia of a body about any axis is equal to the moment of inertia about a parallel axis through its centre of gravity plus its total mass multiplied by the square of the distance between the axes,

i.e. $$I_0 = I_G+Md^2$$

Let the given axis be the z-axis. Let the coordinates of a typical particle of mass m be (x, y, z) and let the coordinates of the centre of gravity be $(\bar{x}, \bar{y}, \bar{z})$, and let the coordinates of the typical particle referred to parallel axes through G be (x', y', z'); then $x = \bar{x}+x'$, $y = \bar{y}+y'$, $z = \bar{z}+z'$.

The moment of inertia about OZ is $\Sigma m(x^2+y^2)$

$$= \Sigma m[(\bar{x}+x')^2+(\bar{y}+y')^2]$$

$$= \Sigma m \{x'^2+y'^2\} +(\bar{x}^2+\bar{y}^2)\Sigma m+2\bar{x}\Sigma mx'+2\bar{y}\Sigma my'$$

But $\Sigma mx' = \Sigma my' = 0$, since G is the centre of gravity.

$$\therefore \text{ Moment of inertia about OZ} = \Sigma m(x'^2+y'^2)+(\bar{x}^2+\bar{y}^2)\Sigma m$$

$$= I_G+Md^2$$

where I_G is the moment of inertia about a parallel axis through G and d is the distance between the axes.

4. Further moments of inertia

3. *Uniform circular ring about an axis through its centre O perpendicular to its plane.*

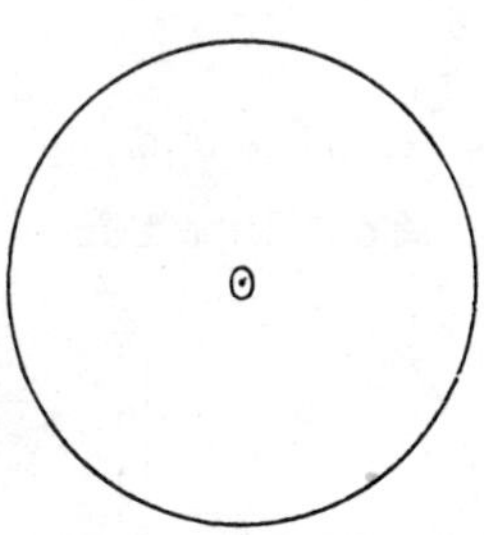

Let the radius of the ring be a. Then every particle of it is a distance a from the axis.

Hence the moment of inertia about this axis is Ma^2.

4. *Moment of inertia of a thin ring of radius a about* (*i*) *a diameter and* (*ii*) *a tangent.*

Take two perpendicular diameters. The moments of inertia of the ring about these will be the same. Call each I_D.

From the Perpendicular Axes Theorem

$$I_D + I_D = Ma^2$$

$$\therefore I_D = \tfrac{1}{2}Ma^2$$

Let the moment of inertia about a tangent be I_T, then, from the Parallel Axes Theorem,

$$I_T = I_D + Ma^2, \quad \therefore I_T = \tfrac{3}{2}Ma^2$$

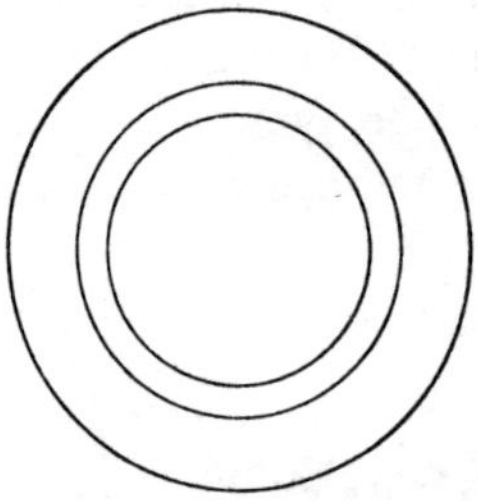

5. *Moment of inertia of a uniform thin circular disc about (i) an axis through its centre perpendicular to its plane, (ii) about a diameter, and (iii) about a tangent.*

Let the mass of the disc per unit area be m. Consider the thin ring included between two circles concentric with the disc of radii r and $r+\Delta r$.

Then $\qquad \Delta I = 2\pi r\Delta r m \,.\, r^2$

Hence $\qquad I = \int_0^a 2\pi m r^3\, dr = 2\pi m \,.\, \tfrac{1}{4}a^4$

$$= \underline{\tfrac{1}{2}Ma^2} \text{ (for } \pi m a^2 = M)$$

As for a ring $\quad I_D + I_D = \tfrac{1}{2}Ma^2, \quad \therefore I_D = \tfrac{1}{4}Ma^2$

and again $\quad I_T = I_D + Ma^2, \quad \therefore I_T = \tfrac{5}{4}Ma^2$

6. *Moment of inertia of a uniform solid sphere about a diameter.*

Consider a thin slice of the sphere made by two planes each perpendicular to the diameter. Let the planes be at distances x and $x+\Delta x$ from the centre O. Let the radius of the sphere be a and its mass per unit volume m.

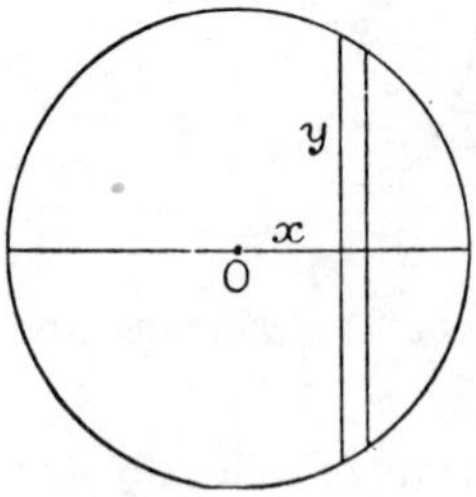

The slice of the sphere is a thin disc of radius y, and the diameter about which we are finding the moment of inertia is an axis through its centre normal to its plane,

$$\therefore \Delta I = \tfrac{1}{2}\pi y^2 \Delta x \,.\, y^2 m$$

But $$x^2+y^2 = a^2$$

$$\therefore \Delta I = \tfrac{1}{2}\pi m(a^2-x^2)^2\, \Delta x$$

$$I = \tfrac{1}{2}\pi m \int_{-a}^{a} (a^2-x^2)^2\, dx = \tfrac{1}{2}\pi m \int_{-a}^{a} (a^4-2a^2x^2+x^4)\, dx$$

$$= \tfrac{1}{2}\pi m[a^4x-\tfrac{2}{3}a^2x^3+\tfrac{1}{5}x^5]_{-a}^{a} = \tfrac{1}{2}\pi m a^5\{1-\tfrac{2}{3}+\tfrac{1}{5}-(-1+\tfrac{2}{3}-\tfrac{1}{5})\}$$

$$= \tfrac{8}{15}\pi m a^5 = \tfrac{2}{5}Ma^2$$

(for $\tfrac{4}{3}\pi m a^3 = M$).

From the theorem of parallel axes the moment of inertia of the sphere about a tangent is $\tfrac{2}{5}Ma^2+Ma^2 = \tfrac{7}{5}Ma^2$.

7. *Moment of inertia of a thin uniform spherical shell about diameter.*

Consider a band of the sphere cut off by two planes perpendicular to the diameter. Let angle POA $= \theta$ and angle QOP $= \Delta\theta$. The breadth of the band is $a\Delta\theta$ and its length is approximately $2\pi a \sin\theta$.

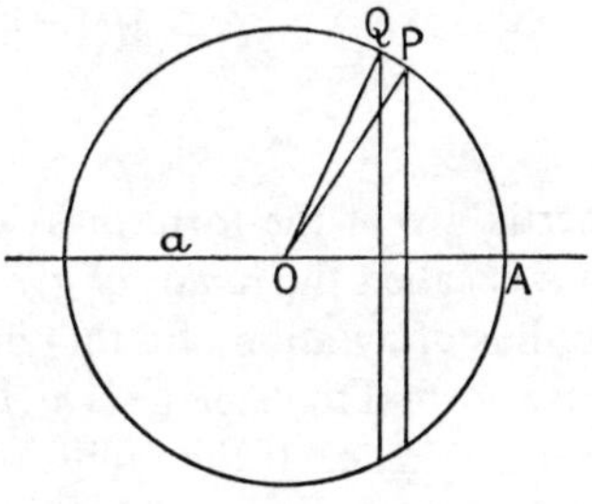

If m is the mass per unit area, then we want the moment of inertia of a thin ring about an axis through its centre normal to its plane, i.e.

$$\Delta I = 2\pi a \sin\theta \,.\, ma\,\Delta\theta\, (a \sin\theta)^2$$

$$= 2\pi m a^4 \sin^3\theta\, \Delta\theta$$

$$\therefore I = \int_0^{\pi} 2\pi m a^4 \sin^3\theta\, d\theta = 2\pi m a^4 \,.\, \tfrac{2}{3} \int_0^{\pi} \sin\theta\, d\theta$$

$$= \tfrac{8}{3}\pi m a^4 = \tfrac{2}{3}Ma^2$$

(for $4\pi m a^2 = M$).

From the Theorem of Parallel Axes the moment of inertia of the spherical shell about a tangent is $\tfrac{2}{3}Ma^2+Ma^2 = \tfrac{5}{3}Ma^2$.

8. *Moment of inertia of a uniform solid cylinder of radius r and length l about a diameter of one end.*

Take the axis of the cylinder as x-axis, then the y-axis is the diameter of an end. Take a slice of the cylinder by two planes parallel to the y-axis whose distance from it are x and $x+\Delta x$.

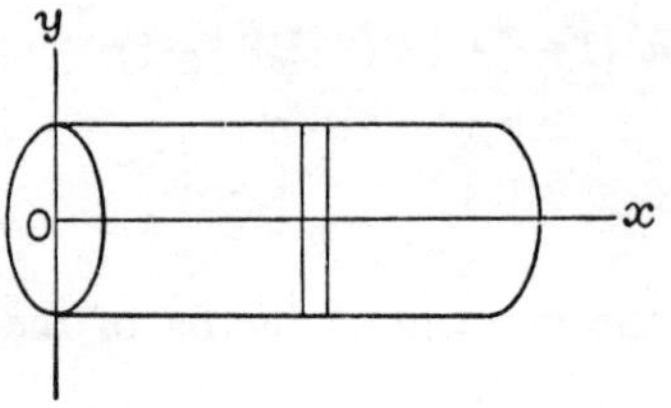

The moment of inertia of the slice, which is a thin disc, about its diameter parallel to Oy is $\pi r^2 m\Delta x \, . \, \frac{1}{4}r^2$.

By the parallel-axis theorem the moment of inertia of this slice about Oy is $\pi r^2 m\Delta x(\frac{1}{4}r^2+x^2)$. We then add the moment of inertia of all the discs about the axis Oy.

$\therefore$ moment of inertia of the whole cylinder about Oy is

$$\int_0^l \pi r^2 m(\tfrac{1}{4}r^2+x^2)\,dx = \pi r^2 m[\tfrac{1}{4}r^2 x+\tfrac{1}{3}x^3]_0^l$$

$$= \pi r^2 ml(\tfrac{1}{4}r^2+\tfrac{1}{3}l^2)$$

$$\text{But } \pi r^2 ml = M, \quad \therefore \; I = M(\tfrac{1}{4}r^2+\tfrac{1}{3}l^2)$$

5. Radius of gyration

All moments of inertia are of the form mass$\times$(length)^2, i.e. Mk^2. In this case the length k is called the *radius of gyration.*

For example, the radius of gyration of a thin disc of radius a about an axis through its centre normal to its plane is $a/\sqrt{2}$; that of a uniform sphere about a diameter is $\sqrt{\frac{2}{5}} \, . \, a$. If the radius of gyration of a body of mass M is $2a/\sqrt{3}$, then its moment of inertia is $\frac{4}{3}Ma^2$ about the same axis.

6. Routh's Rule

Routh's Rule gives a useful way of remembering the moments of inertia of a body with three mutually perpendicular axes of symmetry.

> *The square of the radius of gyration about any one of these axes is equal to the sum of the squares of the other two semi-axes divided by* 3, 4, or 5

the divisor being 3 for rectangular solids
4 for elliptical or circular solids
5 for ellipsoidal or spherical solids.

EXERCISE 31

1. Find the radius of gyration of a rod of length l about an axis perpendicular to the rod through a point of trisection of the rod.

2. Find the moment of inertia of a circular cylinder of radius a, height h, and mass M about its axis.

3. Find the moment of inertia of a rod of length $2a$ and mass M about an axis through its centre inclined at $\theta°$ to the rod.

4. Find the moment of inertia of a rectangle of sides $2a$, $2b$ about an axis through one vertex perpendicular to its plane.

5. Prove that the moment of inertia of a solid cone of mass M and radius a about the axis of the cone is $\frac{3}{10} Ma^2$.

6. Find the moment of inertia of a ring whose mass is M and whose outer and inner radii are a and b about (i) an axis through its centre perpendicular to its plane, (ii) about a diameter.

7. A wire is bent into the form of an equilateral triangle of side $2a$. Find its moment of inertia about an axis through its centre of gravity perpendicular to its plane.

8. Find the moment of inertia of a semicircular lamina about its diameter.

9. Find the moment of inertia of a thin hemispherical bowl about the radius through the centre of gravity.

10. Prove that the moment of inertia about a diameter of a hollow sphere whose external and internal radii are a and b and whose mass is M is $\frac{2}{5}M\dfrac{a^5 - b^5}{a^3 - b^3}$

11. ABC is a triangular lamina of mass M. If h is the perpendicular from A to BC, find the moment of inertia of the lamina in terms of M and h

(i) about an axis through A parallel to BC,

(ii) about an axis through its centre of gravity parallel to BC,

(iii) about BC.

12. Use the results of question 11 to find the moment of inertia of a regular hexagon ABCDEF of side a about CF.

7. Dynamics of a rigid body rotating about a fixed axis

We have already defined a rigid body in Chapter 5 as one whose shape remains unaltered by the forces acting upon it. We must now take this definition a little further and assume that a rigid body consists of heavy particles, i.e. each particle has a mass, and that all the particles are connected with rigid connections.

When such a body is acted upon by external forces we get internal forces of reaction between the particles of the body and for two adjacent particles these internal forces will be equal and opposite.

If particle A pushes particle B with a force *F*, then particle B will push particle A with an equal and opposite force *F*.

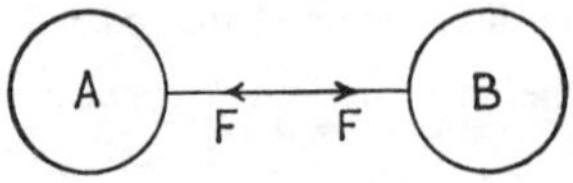

Now consider a rigid body acted upon by external forces, the body being free to rotate about a fixed axis through O.

Any particle of the body will be acted upon by external force and internal force, and these forces will give the particle an acceleration. The resultant force on the particle will be equal to the product of the mass of the particle and its acceleration, this last we shall call its *mass-acceleration.* Hence for all the particles we have a system of internal and external forces. The internal forces occur in equal and opposite pairs, and so their vector sum will be zero. Thus we are left with the external forces, which we have already shown may be reduced to a single force through any chosen point equal to the vector sum of all the external forces, together with a couple equal to the algebraic sum of the moments of the external forces about the chosen point. We shall use the point O as our chosen point.

We have also a set of mass-accelerations, another set of vectors, and these may, just like forces, be reduced to a single mass-acceleration equal to the vector sum of the constituent mass-accelerations and a mass-acceleration couple equal to the sum of the moments of all the mass-accelerations about O.

These two, the forces thus reduced and the mass-accelerations reduced in the same way, are equivalent sets, and equating them will give the equations of motion of the body.

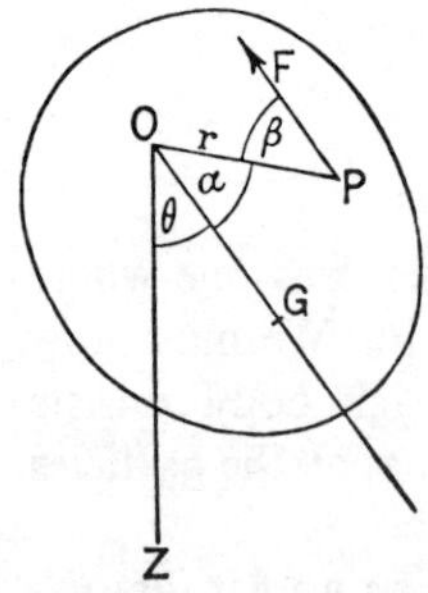

8. Equation of motion of a rigid body rotating about a fixed axis

Suppose the plane of the page cuts the fixed axis at O, and let P be any particle of the body of mass m and such that OP $= r$ in this plane. Let OZ be the perpendicular to the axis at O and let OG be a fixed line in the body in this plane. Let the angle POG $= \alpha$, ZOG $= \theta$ and ZOP $= \phi$, then $\phi = \theta + \alpha$.

Differentiating this twice and remembering α is fixed,

$$\dot{\phi} = \dot{\theta} \text{ and } \ddot{\phi} = \ddot{\theta}$$

Hence the velocity of P is $r\dot{\theta}$ perpendicular to OP, and the acceleration of P is $r\dot{\theta}^2$ along PO and $r\ddot{\theta}$ perpendicular to OP, for P moves in a circle centre O and radius r. Now let the resulting force on P be F making β with OP. Now the resultant force on P in any direction must be equal to the mass of P times the resultant acceleration in that direction. Therefore perpendicular to OP we have

$$F \sin \beta = mr\ddot{\theta}$$

$$\therefore \ Fr \sin \beta = mr^2\ddot{\theta}$$

Writing down similar equations for all particles of the body and noticing that $Fr \sin \beta$ is the moment of F about O, we have

$$\Sigma Fr \sin \beta = \Sigma mr^2\ddot{\theta} = \ddot{\theta}\Sigma mr^2$$

Now $\Sigma Fr \sin \beta$ is the sum of the moments of all the forces, both internal and external, about O; but, since the internal forces for all particles occur in equal and balancing pairs, we can neglect their moments, and so $\Sigma Fr \sin \beta$ reduces to the sum of the moments about O of all the external forces, say G.

Hence $$G = \Sigma mr^2\ddot{\theta} = I\ddot{\theta}$$

where G is the sum of the moments of all the external forces about the axis of rotation, I is the moment of inertia of the body about this axis and $\ddot{\theta}$ is the angular acceleration of the body.

9. Units

Since the equation $G = I\ddot{\theta}$ was deduced from $P = mf$ for a particle, and in the latter the force P must be in newtons, so in the former the couple G must be in newton metres (N m).

Notice that the equation $G = I\ddot{\theta}$ corresponds to $P = mf$ for a particle. In fact x, $\dot{x}$, $\ddot{x}$, m, P correspond to θ, $\dot{\theta}$, $\ddot{\theta}$, I, G and so for constant angular acceleration we have the following formulæ:

$$v = u+ft \qquad \Omega = \omega+\alpha t$$

$$s = ut+\tfrac{1}{2}ft^2 \qquad \theta = \omega t+\tfrac{1}{2}\alpha t^2$$

$$s = \tfrac{1}{2}(u+v)t \qquad \theta = \tfrac{1}{2}(\omega+\Omega)t$$

$$v^2 = u^2+2fs \qquad \Omega^2 = \omega^2+2\alpha\theta$$

10. The kinetic energy of a rigid body rotating about a fixed axis

If $\ddot{\theta}$ is the angular acceleration of the body and ω its angula velocity, then

$$\ddot{\theta} = \frac{d\omega}{dt} = \frac{d\omega}{d\theta}.\frac{d\theta}{dt} = \omega\frac{d\omega}{d\theta} = \frac{d}{d\theta}(\tfrac{1}{2}\omega^2)$$

Hence from $G = I\ddot{\theta}$, integrating with respect to θ,

$$\int G\, d\theta = I.\tfrac{1}{2}\omega^2 + C$$

If $\omega = \Omega$ when $\theta = 0$, then $0 = I.\tfrac{1}{2}\Omega^2 + C$

$$\therefore \int G\, d\theta = I.\tfrac{1}{2}\omega^2 - I.\tfrac{1}{2}\Omega^2$$

Now for the particle P in the diagram on page 218, the velocity of P is $r\dot{\theta}$ and its kinetic energy is therefore $\tfrac{1}{2}mr^2\dot{\theta}^2$.

$\therefore$ The kinetic energy of the whole body is.

$$\tfrac{1}{2}\Sigma mr^2\dot{\theta}^2 = \tfrac{1}{2}\dot{\theta}^2\Sigma mr^2 = \tfrac{1}{2}I\omega^2$$

Hence we have $\int G\, d\theta$ = the increase in kinetic energy and $\int G\, d\theta$ is the total work done by the couple rotating the body.

Hence the energy equation

work done = increase in kinetic energy

If G is a constant couple then $\int G\, d\theta = G\theta$ and we have

$$G\theta = \tfrac{1}{2}I\omega^2 - \tfrac{1}{2}I\Omega^2$$

compare this with $Ps = \tfrac{1}{2}mv^2 - \tfrac{1}{2}mu^2$ for a particle.

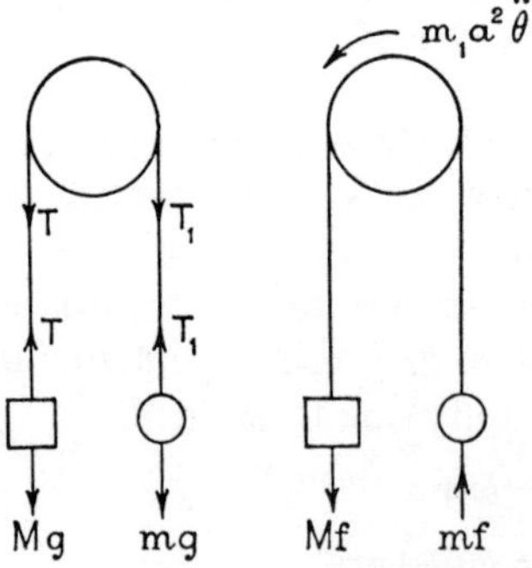

Example 1.—Two masses M and m hang over a pulley of mass m_1 and radius a, which may be regarded as a thin ring. The system is released from rest. Find the acceleration of M.

The forces and mass-accelerations are as shown.

For M, $\quad Mg - T = Mf$

For m, $\quad T_1 - mg = mf$

For the pulley, $\quad Ta - T_1a = m_1a^2\ddot{\theta}$

If a length x of string passes over the pulley while the pulley turns through θ, then $x = a\theta$. Differentiating twice

$$\ddot{x} = a\ddot{\theta} \text{ and } \ddot{x} = f$$

Adding the three equations above after dividing each side of the third by a,

$$(M - m)g = (M + m + m_1)f$$

$$\therefore f = \frac{(M - m)g}{M + m + m_1}$$

We could use the energy equation.

Suppose M falls by x, then m rises x, while the pulley turns through θ radians. Then $x = a\theta$ and $\dot{x} = a\dot{\theta}$.

$$\text{Work done} = (M - m)gx$$

$$\text{Gain in kinetic energy} = \tfrac{1}{2}M\dot{x}^2 + \tfrac{1}{2}m\dot{x}^2 + \tfrac{1}{2}ma^2\dot{\theta}^2$$

$$= \tfrac{1}{2}(M + m + m_1)\dot{x}^2$$

$$\therefore \dot{x}^2 = \frac{2(M - m)gx}{M + m + m_1}$$

Since $$2f = \frac{d}{dx}(\dot{x}^2),\quad 2f = \frac{2(M - m)g}{M + m + m_1}$$

$$f = \frac{(M - m)g}{M + m + m_1}$$

Example 2.—A uniform rod AB can rotate in a vertical plane about a horizontal axis at A. At B a particle of the same mass as the rod is attached. The rod is held at rest in the position of unstable equilibrium. Find the velocity of B when B is

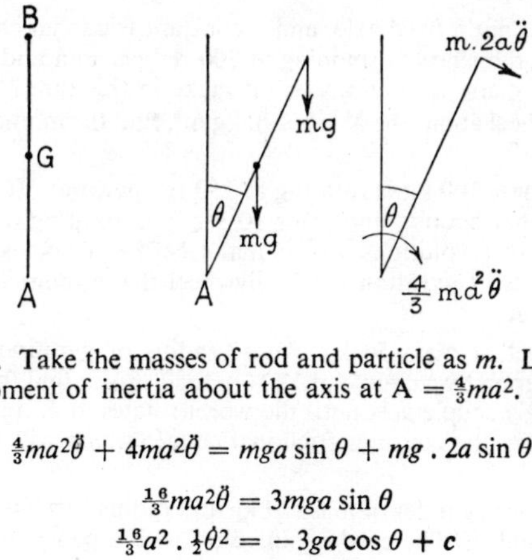

vertically below A. Take the masses of rod and particle as m. Let the length of the rod be $2a$. Its moment of inertia about the axis at A $= \frac{4}{3}ma^2$. Moments about A give

$$\tfrac{4}{3}ma^2\ddot{\theta} + 4ma^2\ddot{\theta} = mga \sin\theta + mg\,.\,2a\sin\theta$$

i.e. $$\tfrac{16}{3}ma^2\ddot{\theta} = 3mga\sin\theta$$

$$\tfrac{16}{3}a^2\,.\,\tfrac{1}{2}\dot{\theta}^2 = -3ga\cos\theta + c$$

When $\dot{\theta} = 0$, $\theta = 0$, $$\therefore c = 3ga$$

$$\therefore \tfrac{16}{3}a^2\,.\,\tfrac{1}{2}\dot{\theta}^2 = 3ga(1 - \cos\theta)$$

When $\theta = 180°$, $$\tfrac{8}{3}a^2\omega^2 = 6ga$$

$$4a^2\omega^2 = 9ga$$

$$\therefore 2a\omega = 3\sqrt{(ga)}$$

Or, by the energy equation, if ω is the angular velocity when B is vertically below A,

$$\text{kinetic energy in this position} = \tfrac{1}{2} \cdot \tfrac{4}{3}ma^2\omega^2 + \tfrac{1}{2}m \cdot (2a\omega)^2 = \tfrac{8}{3}ma^2\omega^2$$

Work done on the rod and mass $= 2mga + 4mga = 6mga$

$$\therefore \tfrac{8}{3}ma^2\omega^2 = 6mga$$
$$4a^2\omega^2 = 9ga$$
$$2a\omega = 3\sqrt{(ga)}$$

EXERCISE 32A

1. A wheel of mass 3 kg, whose radius of gyration is 50 cm, is acted upon by a couple of moment 6 N m. Find the angular acceleration produced.

2. A flywheel which may be regarded as a circular cylinder of radius 50 cm and mass 10 kg is freely mounted on its axis. Find the angular acceleration produced by a couple of moment 40 N m. Find also the speed of a point on the rim of the wheel after 5 sec, if the flywheel starts from rest.

3. A flywheel which may be regarded as a circular cylinder of radius 20 cm and mass 20 kg mounted freely on its axis is rotating at 60 radians per sec. Find in N m the frictional couple which will bring it to rest in 10 sec.

4. A wheel whose mass is 50 kg and whose radius of gyration is 50 cm is rotating at 10 rev per sec. Find the moment of the frictional couple which will bring it to rest in 1 min.

5. A wheel spins about a fixed axle, and a constant frictional couple is exerted on its bearings. If the wheel is spinning at 200 rev per min and is brought to rest in $1\frac{1}{2}$ min, how many revolutions will it make in this time? If the moment of inertia of the wheel about the axle is 270 kg m^2, find the moment of the frictional couple.

6. A flywheel of mass 100 kg is rotating at 150 rev per min. It is acted upon by a constant frictional couple, and after 10 sec it is rotating at 100 rev per min. How many more revolutions will it make before it comes to rest? Find in metres the radius of gyration of the flywheel if the moment of the frictional couple is 50 N m.

7. A uniform wheel of mass 50 kg, whose radius of gyration about its axle is 30 cm, is acted upon by a couple of 12 N m for 1 min; find the angular velocity produced. If this couple acts until the wheel rotates at 15 rev per sec and then ceases to act, find the constant frictional couple which will bring the wheel to rest in $\frac{1}{2}$ min.

8. A uniform circular cylinder of mass M kg and radius a m has a fine cord wound many times round it. The cord is pulled off with a constant pull of F N. If it takes t sec to unwind the cord, find the angular velocity of the wheel after this time.

9. A uniform cylinder of mass M can rotate freely about its horizontal axis. A light inelastic string is coiled round the cylinder and carries at its free end a particle of mass m. If the system is released from rest, prove that the particle will descend with acceleration of $\dfrac{2mg}{M + 2m}$

10. Two masses of M kg and m kg ($M > m$) hang over a smooth pulley of radius a whose moment of inertia about its axis is I. Prove that if M descends with a constant acceleration f then $(M - m)g = f(M + m + I/a^2)$.

11. The end of a thread wound round a reel is held fixed and the reel is allowed to fall. If the mass of the reel is m, its radius is a and its radius of gyration about its axis is k, find the acceleration with which the reel falls.

12. A circular hoop of radius a, whose mass may be supposed concentrated in its rim, rolls without slipping down an inclined plane of inclination α. Prove that the acceleration is $\frac{1}{2}g \sin \alpha$.

13. A flywheel which is free to turn about a horizontal axis carries a mass m suspended by a vertical string which is wrapped round an axle of radius b. If the moment of inertia of the flywheel is I, find the acceleration of the particle and the tension in the string.

14. A weight hangs from an axle of radius b and is held in equilibrium by a force P applied tangentially to the rim of a concentric wheel of radius a. If the force P were increased to Q, show that the weight will ascend with acceleration $\dfrac{(Q - P)abg}{Ig + Pab}$ where I is the moment of inertia of the wheel and axle.

15. A mass M rests on a smooth horizontal table and is connected to a mass m hanging freely by a light string which passes over a pulley of mass M_1 and radius a which may be regarded as a uniform circular disc. The string is normal to the edge of the table. Prove that the acceleration of the mass M is given by $f(2M + 2m + M_1) = 2mg$.

16. A uniform rod of length $2a$ can turn freely about one end. It is held in a horizontal position and allowed to fall; find the angular velocity with which it first reaches a vertical position and the velocity of its free end in this position.

17. A uniform rectangular plate ABCD is freely pivoted at the midpoint of AD so that it can rotate in a vertical plane about a horizontal axis there. If AB $= 2a$ and BC $= 2b$, $a > b$ and the rectangle is held with AB horizontal and allowed to fall, find the angular velocity when AB is vertical.

18. A uniform circular disc of radius a can turn in a vertical plane about a horizontal axis at a point A of its rim. At B, the point diametrically opposite A. a particle of the same mass as the disc is attached to the disc and the disc is held with AB horizontal. Find the speed of B when AB is next vertical.

19. A uniform rod AB of length $2l$ is hinged about a horizontal axis perpendicular to the rod at its centre and carries at one end A a particle whose mass is equal to that of the rod. The rod is at rest with A vertically below B, and A is given a velocity just sufficient to bring the rod to the horizontal position. Find this velocity.

20. A uniform circular disc of mass M and radius a is free to turn about a horizontal axis through its centre perpendicular to its plane. A particle of mass m is attached to the edge of the disc. If motion starts from the position in which the radius to the particle makes 60° to the upward vertical, find the angular velocity when m is in its lowest position.

21. A uniform circular disc of mass M and radius a can turn freely in its own plane about its centre. A particle of mass m is attached to a point on the rim of the disc. Show that if ω_1 and ω_2 are the angular velocities of the disc when the particle is at its highest and lowest points $a(M + 2m)(\omega_2^2 - \omega_1^2) = 8mg$.

EXERCISE 32B

1. The moment of inertia of a flywheel about its axis is 6 kg m². The flywheel is acted upon by a constant couple of moment 30 N m for 12 sec. Find the angular velocity of the flywheel after the 12 sec, and the number of revolutions made by the flywheel in this time.
2. A wheel of mass 5 kg and radius 25 cm, whose mass may be supposed to be concentrated in its rim, is rotating at 60 rad per sec when a frictional force is applied tangentially to the wheel. If the wheel stops in 10 sec, find the size of the force. Find also the number of revolutions made by the wheel before coming to rest.
3. A uniform circular disc of radius 20 cm and mass 25 kg can rotate freely about a horizontal axis through its centre, perpendicular to its plane. A light cord passes over a small rough groove in the edge of the disc. To one end of the cord is attached a block of mass 10 kg and the other end of the cord is pulled downwards with a constant tension of 143 N so that the block ascends vertically. Find the acceleration of the block and the tension in the portion of the cord attached to the block.
4. A uniform rectangular trap door ABCD of mass m kg is hinged along a fixed edge AB which is horizontal. If BC $= a$ m, find the angular velocity of the door as it reaches the vertical from rest in the horizontal position (i) if friction is negligible, and (ii) if the hinges exert a constant frictional couple L N m.
5. A uniform circular disc of radius a rolls without slipping up a line of greatest slope of a plane inclined at α to the horizontal. If the initial angular velocity is ω, show that the disc will roll a distance $\dfrac{3a^2\omega^2}{4g \sin \alpha}$ up the plane before coming instantaneously to rest.
6. A uniform rod AB of mass 1 kg and length 1 m is free to swing in a vertical plane about a fixed smooth axis at A. A mass m kg is attached to the rod at B and the rod is released from the horizontal position. Find m if the greatest velocity of B in the subsequent motion is 5 m/s.
7. The mass of a flywheel is 100 kg. It may be regarded as a uniform circular disc of radius 30 cm. The flywheel is rotating about its axis at 5 rev per sec. After 40 sec its speed has fallen to 4 rev per sec due to the action of a constant frictional couple. What constant couple should now be applied so that after a further 20 sec the angular velocity will be 8 rev per sec? Find the total angle turned through during the minute.
8. A mass M rests on a smooth horizontal table. It is connected by a light inelastic string which passes over a rough pulley at the edge of the table to a mass m hanging freely. The pulley is a uniform circular disc of mass m and radius a. Prove that the acceleration of the mass m is $\dfrac{2mg}{2M + 3m}$ and find the tension in the vertical string.
9. A mass M rests on a smooth plane inclined at 30° to the horizontal. It is connected by a light inelastic string which passes over a rough pulley at the top of the plane to a mass m hanging freely. The string is parallel to a line of greatest slope of the plane and the pulley is a uniform circular disc of mass M and radius a. Assuming the mass M is pulled up the plane find the acceleration of the system.

10. A uniform rod of mass m and length $2a$ can turn freely in a vertical plane about a horizontal axis at one end. To the other end a particle also of mass m is attached. The rod is held in a horizontal position and allowed to fall. Find the speed of the particle when the rod is vertical.

11. A pulley which may be regarded as a uniform circular disc of mass 4 kg and radius 20 cm is freely mounted on a horizontal axis. A light inextensible string passes over the pulley and carries masses of 2 kg and 3 kg at its ends. Find the acceleration with which the 3 kg mass descends assuming no slipping between the string and the pulley. If the bearings of the pulley were rough and exerted a constant retarding couple of $0{\cdot}1g$ N m, find the acceleration in this case.

12. A flywheel of mass 1 t and radius of gyration 1 m is rotating at 2π rad per sec. Find the time it takes to come to rest if it is acted upon by a couple of 50 N m.

13. A flywheel whose mass is 50 kg and whose radius of gyration is 50 cm is rotating at 120 rev per min. Find in N m the frictional couple which will bring it to rest whilst it makes 30 revolutions.

14. A bent rod made of uniform material is made of two portions AO and OB, each $2a$ long. The angle AOB $= 90°$. The rod is freely pivoted at O. When OA is horizontal and OB vertically downwards the rod is released from rest. Prove that when OA and OB are equally inclined to the vertical the angular velocity is $\left\{\dfrac{3g(\sqrt{2}-1)}{4a}\right\}^{1/2}$

15. A chain of length 12 m is hung over a pulley of mass 8 kg which may be considered a uniform circular disc. The mass per unit length of the chain is 1 kg, and initially the chain is held with the difference of level between the two ends of the chain 1 m. Calculate.
(i) the initial acceleration and
(ii) the speed of the chain when the lower end has fallen 3 m.

16. A uniform rod of length 1 m and mass m can turn freely about a horizontal axis through one end. To the other end is attached a mass m. The rod is held at 60° to the downward vertical and allowed to rotate in a vertical plane. Find the velocity of the free end of the rod when the rod is vertical.

17. A uniform circular cylinder of mass $4m$ and radius a can turn freely about its horizontal axis. Particles of mass m and $2m$ are attached to the ends of a light inextensible string which passes over the cylinder in a plane perpendicular to its axis. The system is released from rest. Assuming the cylinder is sufficiently rough to prevent the string from slipping, show that when the mass $2m$ has descended x its velocity is $\frac{2}{5}gx$. Find the acceleration of the particles and the tensions in the two vertical portions of the string.

18. A uniform circular disc of mass M and radius a can revolve freely about a horizontal axis through its centre perpendicular to its plane. A particle P of mass m is attached to the highest point of the disc and the system is slightly disturbed from rest. Show that the angular momentum at the instant when P is vertically below the centre is $\sqrt{[2m(M+2m)ga^3]}$. At this instant a constant retarding couple is applied to the disc and it comes to rest after rotating through a further 60°. Find the size of the couple.

19. A uniform circular lamina of mass m and radius a can swing about a fixed horizontal tangent. A frictional couple $\frac{1}{2}mga$ acts at the axis. The lamina starts from its lowest position with angular velocity ω. If ω is sufficiently large for a complete revolution to be made, show that the angular velocity during this revolution is least when the angle turned through is $7\pi/6$.

20. A uniform circular disc of mass m and radius a whose centre is O has a particle of mass $\frac{1}{2}m$ attached at a point P of its circumference. The disc is mounted on a fixed horizontal axis through O perpendicular to its plane, and its rotation is resisted by a constant couple. The disc is released from rest with OP horizontal and next comes to rest when P is vertically below O having turned through $\frac{1}{2}\pi$. Find the moment of the couple and show that the angular velocity of the disc is greatest when it has turned through an angle α, where $\cos \alpha = 2/\pi$.

11. The compound pendulum

The compound pendulum is a rigid body rotating about a fixed axis.

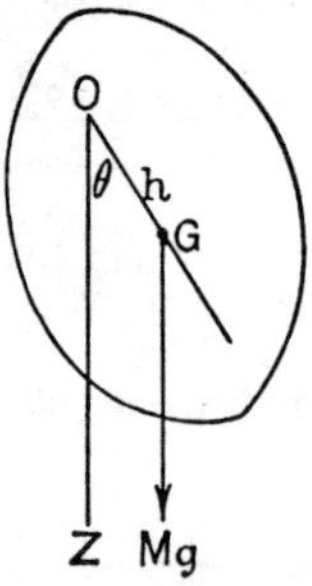

The vertical plane through the axis of rotation is taken as the plane of reference, and the plane perpendicular to the axis and through the centre of gravity of the body is taken as the fixed plane in the body. The figure shows a section of the body through the centre of gravity and perpendicular to the axis. Let the radius of gyration of the body about an axis through G parallel to the axis of rotation be k. Let angle ZOG $= \theta$ and OG $= h$. Let the mass of the body be M.

The moment of inertia about the axis is $M(k^2+h^2)$. Then from $G = I\ddot{\theta}$ we have

$$Mgh \sin \theta = -M(h^2+k^2)\ddot{\theta}$$

Notice that the moment of Mg is clockwise and $I\ddot{\theta}$ is + in the direction in which θ increases, i.e. anticlockwise.

$$\therefore \ \ddot{\theta} = -\frac{gh \sin \theta}{h^2+k^2}$$

If θ is small so that $\theta \simeq \sin \theta$,

$$\ddot{\theta} = \frac{-gh}{h^2+k^2}\theta$$

Hence the motion is simple harmonic and the period is given by $2\pi/\omega$,

$$\text{i.e. period} = 2\pi\sqrt{\left(\frac{h^2+k^2}{hg}\right)}$$

For a simple pendulum of length l the period is $2\pi\sqrt{\left(\frac{l}{g}\right)}$

Hence we may say that the length of the simple equivalent pendulum is

$$\frac{h^2+k^2}{h}$$

12. Centre of suspension and centre of oscillation

The point O is called the *centre of suspension.* If a point C is taken on OG produced, so that OC = l (where l is the length of the simple equivalent pendulum) and the body is now suspended from C, the length of the simple equivalent pendulum in this case will be l',

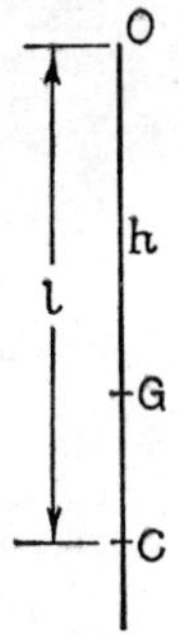

where $$l' = \frac{k^2+(l-h)^2}{l-h}$$

But $$l = \frac{h^2+k^2}{h}, \therefore k^2 = lh-h^2$$

$$\therefore l' = \frac{lh-h^2+(l-h)^2}{l-h} = \frac{l(l-h)}{l-h} = l$$

Hence the periods of oscillation about O and C are equal.

13. Minimum time of oscillation of a compound pendulum

The period T is given by $T = 2\pi\sqrt{\left(\frac{h^2+k^2}{hg}\right)}$

This is a minimum when $\frac{h^2+k^2}{h}$ is least, i.e. when $h+\frac{k^2}{h}$ is least.

If $z = h+\frac{k^2}{h}$, $\frac{dz}{dh} = 1-\frac{k^2}{h^2}$ and $\frac{d^2z}{dh^2} = \frac{2k^2}{h^3}$

When $\frac{dz}{dh} = 0$, $h = \pm k$, and for $h = +k$, $\frac{d^2z}{dh^2}$ is positive.

$\therefore$ the period is least when $h = k$, i.e. when the length of the simple equivalent pendulum $= 2k$.

EXERCISE 33

1. A uniform rod AB of length $2a$ is freely pivoted at A. Find the length of the simple equivalent pendulum.
2. A uniform disc of radius a is pivoted at a point on its circumference and makes small oscillations in a vertical plane. Find the period.
3. Calculate the period of small oscillations of a uniform rod 2 m long about a horizontal axis through one end when a mass equal to the mass of the rod is attached at its midpoint.

4. A light rod 40 cm long is freely suspended at one end A. The end B is rigidly attached to a heavy circular disc of diameter 20 cm so that the rod produced is a diameter of the disc. Find the time of a small oscillation in the plane of the disc.

5. A solid uniform hemisphere makes small oscillations about a diameter of its base as axis. Find the length of the simple equivalent pendulum if the axis is horizontal and the radius is a.

6. Two compound pendulums whose masses are M and M_1 can swing about the same horizontal axis. The distances of their centre of gravity from the axis are h and h_1, and the lengths of their simple equivalent pendulums are l and l_1. Show that if they are fastened together the length of the simple equivalent pendulum is $\dfrac{Mhl + M_1h_1l_1}{Mh + M_1h_1}$

7. A uniform rod of length l is to be suspended from some point of its length to swing freely. Find the least possible period of small oscillations about the position of stable equilibrium and the corresponding point of suspension.

8. A uniform square lamina of side $2a$ is freely pivoted at a point in one diagonal and oscillates in its own plane. Prove that when the period of small oscillations is a minimum the distance of the pivot from the centre is $\frac{1}{3}a\sqrt{6}$.

9. A uniform square plate of side $2a$ is freely hinged about one side and hangs vertically with that side horizontal. It is slightly displaced. Show that it will oscillate in simple harmonic motion and find the length of the equivalent simple pendulum. If instead of being hinged about one side the same plate is now hinged about a line parallel to the side and x from the centre of the plate, find x if the period is the same as before.

10. A particle of mass M is attached to the point of the curved surface of a hemisphere most distant from the plane face. The mass of the hemisphere is M and its radius a. If it makes small oscillations about a diameter of its plane face, find the length of the equivalent simple pendulum.

11. A uniform solid right-circular cone has mass M, radius a, and height h. It is freely pivoted at its vertex and makes small oscillations in a vertical plane. Find the length of the equivalent simple pendulum. If $a = h$ show that the length of the equivalent simple pendulum would be the same if the cone were pivoted at the centre of its base instead of at the vertex.

12. A thin uniform rod AB of mass m and length $2a$ can turn freely in a vertical plane, about a fixed horizontal axis through A. A uniform circular disc of mass $24m$ and radius $\frac{1}{3}a$ has its centre C clamped to the rod so that AC $= x$ and the plane of the disc passes through the axis of rotation. Show that the moment of inertia of the system about the axis is $2m(a^2 + 12x^2)$. The system makes small oscillations; find the period and show that the least period as x varies is $2\pi\sqrt{\left(\dfrac{a}{2g}\right)}$

13. A non-uniform rod AB of length 150 cm oscillates like a simple pendulum of length 120 cm when suspended from one end A, and like a simple pendulum of length 90 cm when suspended from B. Find the length of the simple equivalent pendulum when pivoted at the midpoint of AB. Show that there are two other points on the rod at which it may be pivoted so as to oscillate with the same period as when it is pivoted at the middle point. Find the distances of these points from A.

14. OAB is an isosceles triangular lamina in which OA = OB = $13a$ and AB = $10a$. Find the moment of inertia about an axis through O parallel to AB if the mass of the lamina is m. The lamina can rotate freely about this fixed axis which is horizontal. Two particles each of mass M are attached to the lamina at points in OA and OB at equal distances from O. Find the positions of the particles if the period of small oscillations of the loaded lamina is independent of M.

14. Motion of the centre of gravity of a rigid body

Let (x_1, y_1), (x_2, y_2), (x_3, y_3), etc., be the coordinates of particles of masses m_1, m_2, m_3, etc., which make the rigid body. Let the particles be subject to external forces X_1, X_2, X_3, etc., and internal forces X'_1, X'_2, X'_3, etc., parallel to Ox, and Y_1, Y_2, Y_3, etc., Y'_1, Y'_2, Y'_3, etc., parallel to Oy.

Hence the equations of motion of the particles are

$$m_1\ddot{x}_1 = X_1+X'_1 \text{ and } m_1\ddot{y}_1 = Y_1+Y'_1$$
$$m_2\ddot{x}_2 = X_2+X'_2 \text{ and } m_2\ddot{y}_2 = Y_2+Y'_2$$

and similarly for all other particles.

Adding these we have

$$\Sigma m\ddot{x} = \Sigma X+\Sigma X'$$

and
$$\Sigma m\ddot{y} = \Sigma Y+\Sigma Y' \quad \ldots \ldots \quad \text{(i)}$$

Since the internal forces occur in equal and balancing pairs

$$\Sigma X' = \Sigma Y' = 0$$

Hence (i) becomes $\Sigma m\ddot{x} = \Sigma X$

and
$$\Sigma m\ddot{y} = \Sigma Y \quad \ldots \ldots \quad \text{(ii)}$$

or
$$\frac{d}{dt}(\Sigma m\dot{x}) = \Sigma X \text{ and } \frac{d}{dt}(\Sigma m\dot{y}) = \Sigma Y$$

or in words, *the rate of change of the linear momentum is equal to the force, both being taken in the same direction.* Hence, if there is a direction in which the sum of the resolved parts of the external forces is zero, then the linear momentum in that direction will be constant. This is the *Principle of Linear Momentum.*

The centre of gravity of the body is given by the equation

$$M\bar{x} = \Sigma mx \text{ and } M\bar{y} = \Sigma my$$

Differentiating we have

$$M\dot{\bar{x}} = \Sigma m\dot{x} \text{ and } M\dot{\bar{y}} = \Sigma m\dot{y} \quad \ldots \quad \text{(iii)}$$

and
$$M\ddot{\bar{x}} = \Sigma m\ddot{x} \text{ and } M\ddot{\bar{y}} = \Sigma m\ddot{y} \quad \ldots \quad \text{(iv)}$$

Equation (iii) shows that the linear momentum of the body is the same as that of a particle, equal in mass to the whole body and moving with the velocity of the centre of gravity. And from (ii) and (iv) we have

$$M\ddot{\bar{x}} = \Sigma X \text{ and } M\ddot{\bar{y}} = \Sigma Y$$

showing that the motion of the centre of gravity is the same as if all the mass were collected at the centre of gravity and all the external forces were moved parallel to themselves to act at the centre of gravity.

In other words all the work done in Chapters 8 and 9 applies to the centre of gravity of a rigid body. The centre of gravity behaves just as if it were a particle of mass equal to the total mass acted upon by all the external forces, acting at the centre of gravity in their proper directions.

15. Reactions at the axis of a compound pendulum

The figures show the compound pendulum described earlier. Let X and Y be the components of the reaction at the axis (shown on the

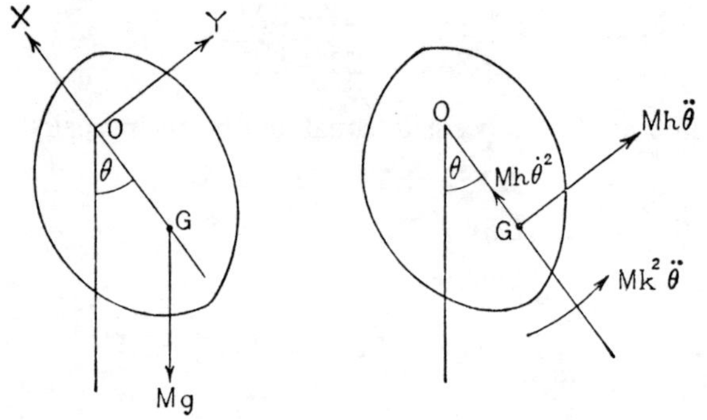

left). The components of the mass-acceleration of G are given on the right.

Moments about O give

$$Mgh \sin \theta = -M(h^2+k^2)\ddot{\theta} \quad \ldots \quad \text{(i)}$$

Resolving along the line OG,

$$X - Mg \cos \theta = Mh\dot{\theta}^2 \quad \ldots \quad \text{(ii)}$$

Resolving perpendicular to OG,

$$Y - Mg \sin \theta = Mh\ddot{\theta} \quad \ldots \quad \text{(iii)}$$

From (i) and (iii) $Y = Mg\sin\theta + Mh\ddot{\theta} = Mg\sin\theta - \dfrac{Mgh^2\sin\theta}{h^2+k^2}$

$$= \frac{Mgk^2\sin\theta}{h^2+k^2}$$

Integrating (i) with respect to θ and taking as our initial conditions that when $\theta = \alpha$, $\dot{\theta} = 0$, (i) may be written

$$Mgh\sin\theta = -M(h^2+k^2)\frac{d}{d\theta}\left(\frac{\dot{\theta}^2}{2}\right)$$

$$\therefore\ -Mgh\cos\theta = -M(h^2+k^2)\frac{\dot{\theta}^2}{2}+c$$

$$\therefore\ -Mgh\cos\alpha = c$$

$$\therefore\ M(h^2+k^2)\dot{\theta}^2 = 2Mgh(\cos\theta-\cos\alpha) \quad . \quad . \quad . \quad \text{(iv)}$$

Substituting for $\dot{\theta}^2$ in (ii)

$$X = Mg\cos\theta + \frac{2Mh^2g(\cos\theta-\cos\alpha)}{h^2+k^2}$$

$$\therefore\ X = \frac{Mg}{h^2+k^2}[(3h^2+k^2)\cos\theta - 2h^2\cos\alpha]$$

Instead of the integration to find $\dot{\theta}^2$ we might have used the energy equation.

The depth of G below O in the figure is $h\cos\theta$. Hence in passing from the position in which $\theta = \alpha$ to the position shown, the centre of gravity falls $h(\cos\theta - \cos\alpha)$. Therefore the work done is $Mgh(\cos\theta - \cos\alpha)$ and this is equal to the gain in kinetic energy.

$$\therefore\ \tfrac{1}{2}M(h^2+k^2)\dot{\theta}^2 = Mgh(\cos\theta - \cos\alpha)$$

and this is the same as (iv).

16. Angular momentum or moment of momentum

For the particle P of the rigid body in the figure on p. 218 the velocity is $r\dot{\theta}$ perpendicular to OP. Therefore the momentum of P is $mr\dot{\theta}$ and the moment of momentum of P about the axis of rotation is $mr^2\dot{\theta}$. Therefore the moment of momentum of the whole body is $\Sigma mr^2\dot{\theta}$ and this is $\dot{\theta}\Sigma mr^2 = I\dot{\theta}$ or $I\omega$.

Referring back to our fundamental equation $G = I\ddot{\theta}$ and integrating with respect to t,

$$G = I\ddot{\theta} = I\frac{d\omega}{dt}$$

$$\therefore \int G\,dt = I\omega + c$$

If $\omega = \Omega$ when $t = 0$, $c = -I\Omega$

$$\therefore \int G\,dt = I\omega - I\Omega$$

In words, *the algebraic sum of the moments of the impulses is equal to the increase in the moment of momentum, both being taken about the axis of rotation.*

If G is fixed then $\int G\,dt = Gt$

and we have $Gt = I\omega - I\Omega$

compare this with $Pt = mv - mu$ for a particle.

17. Impulsive forces

These are often large forces acting for very small times, as for example a bat striking a ball. In problems of this kind, e.g. a swinging rod picking up a particle, we may neglect the finite forces such as the weight of the rod or particle during the small time of the impact and consider only the impulsive forces—large forces acting for infinitesimal times.

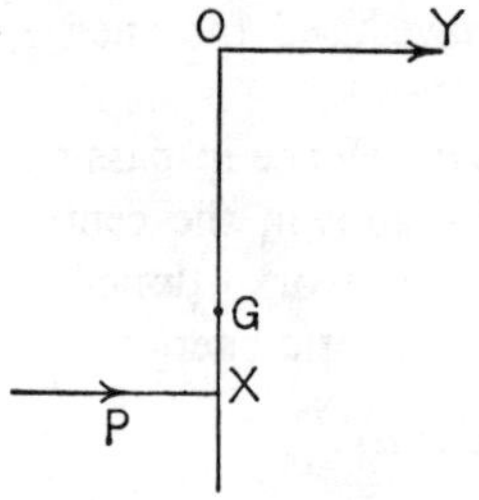

We do this by applying the idea of the previous page where we showed by integrating $G = I\ddot{\theta}$ that the moment of momentum about the fixed axis is equal to the moment of the impulses about this axis.

Consider the compound pendulum shown. The point of suspension is O, the centre of gravity G. Let OG $= h$ and let the radius of gyration about an axis through G parallel to the axis through O be k.

Suppose an impulsive force P acts on the pendulum at X where OX $= x$. Let ω be the angular velocity with which the pendulum starts to rotate, and let Y be the impulsive reaction at O. Y will be parallel to P.

The velocity of G is $h\omega$. Therefore, since the increase in the linear momentum of G equals the impulse,

$$Y + P = Mh\omega \quad \text{. . . . . (i)}$$

and since the increase in the angular momentum about O equals the moment of the impulses about O,

$$Px = M(h^2+k^2)\omega \quad . \quad . \quad . \quad . \quad . \quad . \quad . \quad . \quad \text{(ii)}$$

From (ii) $\quad \omega = \dfrac{Px}{M(h^2+k^2)}$

$\therefore$ from (i) $\quad Y = \dfrac{MhPx}{M(h^2+k^2)} - P = \dfrac{P}{h^2+k^2}[hx-h^2-k^2]$

If $Y=0$, $\quad hx = h^2+k^2, \quad \therefore\ x = \dfrac{h^2+k^2}{h}$

i.e. $x =$ the length of the equivalent simple pendulum.

In this case, when $Y = 0$, X is called the *centre of percussion.* It is the point at which the cricket ball should strike the bat for there to be no jar on the hands.

Example 3.—A lamina of mass m is free to turn in its own vertical plane about an axis through O at a distance c from its mass centre G. If the lamina just makes complete revolutions, prove that the greatest reaction at O is $mg\left(\dfrac{k^2+5c^2}{k^2+c^2}\right)$, where k is the radius of gyration about an axis through G normal to the lamina.

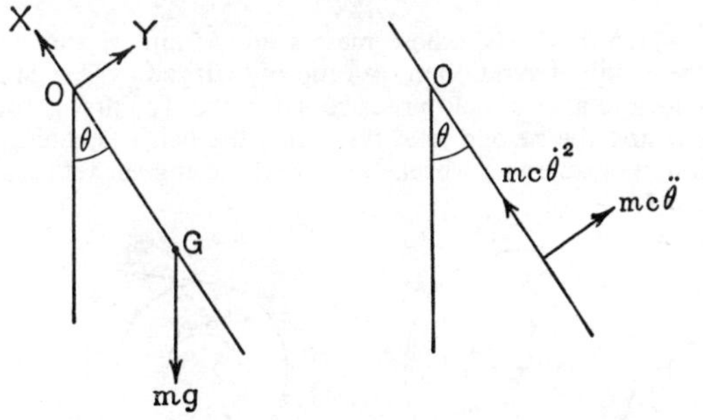

The diagrams show the line OG with the components of the reaction at O along and perpendicular to OG, namely X and Y.

Moments about O give $\quad -mgc\sin\theta = m(c^2+k^2)\ddot{\theta} \quad . \quad . \quad . \quad . \quad . \quad . \quad . \quad . \quad$ (i)

Integrating with respect to θ,

$$mgc\cos\theta = m(c^2+k^2)\,.\,\tfrac{1}{2}\dot{\theta}^2 + A$$

Since the lamina just makes complete revolutions $\dot{\theta} = 0$ when $\theta = 180°$.

$$\therefore\ A = -mgc \text{ and } mgc(1+\cos\theta) = m(c^2+k^2)\tfrac{1}{2}\dot{\theta}^2$$

Resolving along GO, $X - mg\cos\theta = mc\dot{\theta}^2$

$$\therefore\ X = mg\cos\theta + \frac{2mgc^2(1+\cos\theta)}{c^2+k^2}$$

$$= \frac{mg}{c^2+k^2}\{(3c^2+k^2)\cos\theta + 2c^2\}$$

Resolving perpendicular to GO, $Y - mg\sin\theta = mc\ddot{\theta}$

$$\therefore\ Y = mg\sin\theta - \frac{mgc^2\sin\theta}{c^2+k^2}$$

$$= \frac{mgk^2\sin\theta}{c^2+k^2}$$

The resultant reaction R is given by

$$R^2 = X^2 + Y^2$$

$$= \left(\frac{mg}{c^2+k^2}\right)^2\{(3c^2+k^2)^2\cos^2\theta + 4c^4 + 4c^2(3c^2+k^2)\cos\theta + k^4\sin^2\theta\}$$

$$= \left(\frac{mg}{c^2+k^2}\right)^2\left\{\begin{array}{l}k^4\cos^2\theta + 6c^2k^2\cos^2\theta + 9c^4\cos^2\theta + 4c^4 + 12c^4\cos\theta \\ \quad + 4c^2k^2\cos\theta + k^4\sin^2\theta\end{array}\right\}$$

$$= \left(\frac{mg}{c^2+k^2}\right)^2\left\{\begin{array}{l}k^4 + 6c^2k^2\cos^2\theta + 9c^4\cos^2\theta + 4c^4 + 12c^4\cos\theta \\ \quad + 4c^2k^2\cos\theta\end{array}\right\}$$

All the terms are either independent of θ or depend on $\cos\theta$, and all the signs are +. Hence the expression will be a maximum when $\cos\theta$ is a maximum, i.e. when $\theta = 0$ and when $Y = 0$ and so

$$R = X = \frac{mg}{c^2+k^2}(5c^2+k^2)$$

Example 4.—Two flywheels, whose masses are M and m and whose radii are a and b, have their radii of gyration in the ratio of their radii. They are free to rotate in the same plane and a slack belt passes round both. The first is rotating with an angular speed Ω and the second is at rest when the belt is suddenly tightened so that there is no slipping at either wheel. Show that the angular velocity of the second wheel is $Ma\Omega/[b(M+m)]$

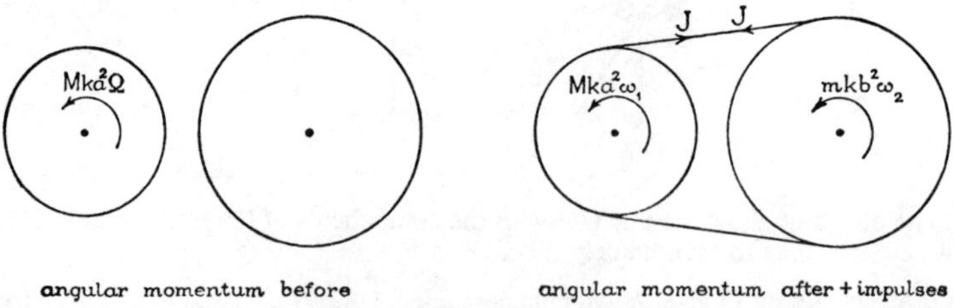

After the belt tightens let the angular velocities be ω_1 and ω_2. Let the impulsive tension in the belt be J. The moments of inertia may be taken as Mka^2 and mkb^2.

For each wheel the moment of the impulse about the axis will be equal to the increase in angular momentum.

$$\therefore\ Mka^2(\omega_1 - \Omega) = -Ja \quad \text{(i)}$$

and

$$mkb^2\omega_2 = Jb \quad \text{(ii)}$$

Since there is no slipping the speed of a point on each wheel must equal the speed of the belt,

$$\therefore\ a\omega_1 = b\omega_2 \quad \dots\dots\dots\dots \quad \text{(iii)}$$

From (i) and (ii) $$\frac{Ma^2(\Omega - \omega_1)}{mb^2\omega_2} = \frac{a}{b}$$

$$\therefore\ Ma(\Omega - \omega_1) = mb\omega_2$$

From (iii) $$Ma\Omega - Mb\omega_2 = mb\omega_2$$

$$\therefore\ \omega_2 = \frac{Ma\Omega}{b(M+m)}$$

Example 5.—A uniform beam of mass M and length $2a$ can rotate freely in a vertical plane about its centre which is fixed. A particle of mass m moving with speed u hits the beam at one end when the beam is horizontal and the particle is moving vertically. If e is the coefficient of restitution, find the angular velocity of the beam and the vertical velocity of the particle immediately after the impact.

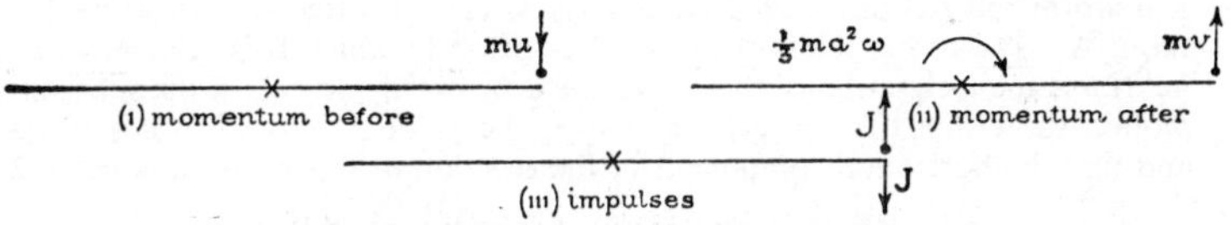

After the impact let the speed of the particle be v and the angular velocity of the rod ω. Let the impulse on the rod at the impact be J.

For the particle the impulse = change in momentum

$$\therefore\ J = m(v + u) \quad \dots\dots\dots\dots \quad \text{(i)}$$

The speed of the end of the rod after the impact will be $a\omega$.

By Newton's Law $$-eu = -v - a\omega \quad \dots\dots\dots\dots \quad \text{(ii)}$$

For the rod the moment of the impulse = the increase in the angular momentum both about the pivot,

$$\therefore\ Ja = \tfrac{1}{3}Ma^2\omega \quad \dots\dots\dots\dots \quad \text{(iii)}$$

From (i) and (iii) $$\tfrac{1}{3}Ma\omega = m(u + v) \quad \dots\dots\dots\dots \quad \text{(iv)}$$

From (ii) and (iv) $$eu - v = \frac{3m}{M}(u + v)$$

$$\therefore\ u(Me - 3m) = v(M + 3m)$$

$$v = \frac{u(Me - 3m)}{M + 3m}$$

$$\therefore\ a\omega = \frac{3mu}{M}\left[1 + \frac{Me - 3m}{M + 3m}\right]$$

$$\omega = \frac{3mu(1 + e)}{(M + 3m)a}$$

EXERCISE 34A

1. A uniform rod suspended from one end just makes a complete revolution. Find the reactions on the pivot in (i) the lowest position, and (ii) when the rod is horizontal, giving in this case the horizontal and vertical components of the reaction.

2. A thin wire circular ring can rotate freely about a fixed horizontal tangent. It falls from the position of unstable equilibrium. Show that when the ring is next vertical the stress on the support is $\frac{11}{3}Mg$, where M is the mass of the ring.

3. A uniform rod of length 1 m hangs from its upper end. What velocity must be given to the lower end in order that the rod may just reach the position of unstable equilibrium? Find in this case the horizontal and vertical components of the reaction at the hinge when the rod is horizontal if its mass is 1 kg.

4. A uniform rod AB is freely hinged at A and is capable of rotation in a vertical plane. It is held with AB horizontal and allowed to fall freely. Find the inclination of the rod to the horizontal when the horizontal reaction at the hinge is maximum.

5. A uniform rod AB of length $2a$ and mass M can turn freely in a vertical plane about A. The rod is released from the position in which B is vertically above A. If θ is the inclination of the rod to the downward vertical, in the subsequent motion show that the vertical reaction at the pivot is zero when $\cos\theta = -\frac{1}{3}$ and that the horizontal component of the reaction in this position is $mg/\sqrt{2}$.

6. A uniform circular lamina can rotate freely about a point P on its circumference which is hinged to a horizontal axis perpendicular to the plane of the lamina. The lamina is released from the position in which the centre is vertically above P. Find in terms of the mass of the lamina M and θ, the angle turned through by the diameter through P, the components of the reaction at P along and perpendicular to this diameter when the diameter has turned through θ.

7. A uniform rod of mass M is free to turn about a fixed smooth pivot at one end. It is held horizontal and released. Prove that when the rod makes an angle θ with the vertical the resultant pressure on the pivot is $\frac{1}{4}Mg\sqrt{(1 + 99\cos^2\theta)}$.

8. A uniform circular disc of mass M and radius a has a particle of mass M rigidly attached to it at a point P on its circumference. The system can rotate freely about a horizontal axis through the centre O of the disc perpendicular to its plane. It is held in the position in which OP is horizontal and released. Find the reaction at the pivot when OP is vertical.

9. A uniform rod AB of mass m and length $2a$ can rotate freely in a vertical plane about the end A. It rests with B vertically below A when it is struck a blow P at right angles to AB at the midpoint of AB. Find P if in the subsequent motion AB just reaches the horizontal position.

10. A uniform rod AB of length $2a$ and mass m is freely pivoted at A and can rotate in a vertical plane. It is held with B vertically above A. Calculate the angular velocity of the rod when AB is horizontal. What will be the impulsive blow on a stop which catches the end B when AB is horizontal and holds it?

11. A uniform rod of length $2a$ and mass m swinging about one end starts from the horizontal position and when vertical is struck a blow at its middle point which reverses and halves its angular velocity. Show that the impulse of the blow is $m\sqrt{(6ga)}$.

12. A uniform gate of length l can swing freely about vertical hinges at one end. When the gate is at rest a particle of mass m and moving with velocity v strikes the gate normally at its midpoint. The moment of inertia of the gate about the hinge is $\frac{1}{8}ml^2$ and the coefficient of restitution is $\frac{1}{2}$. Find the velocity of the particle and the angular velocity of the gate immediately after the impact.

13. A uniform rod AB of length $2a$ and mass M rests on a smooth table and is freely pivoted to the table at A. An inelastic particle of mass m rests against the rod at C where $AC = b$. A horizontal blow of impulse P is given to the rod at D, where $AD = x$, the impulse being normal to the rod. Find the instantaneous angular velocity of the rod.

14. A uniform rod of mass m and length $2a$ is free to rotate in a vertical plane about one end A. It falls from a vertical position and when it is horizontal strikes a fixed inelastic stop at a distance b from A. Show that the impulse of the blow is $\dfrac{2ma}{b}\sqrt{(\frac{2}{3}ag)}$.

15. Two lines of shafting have a common axis of rotation. Their moments of inertia are 2 and 1 kg m² about this axis. The first line is rotating at 300 rev/min and the second is at rest, when they are clutched together. Find the common angular velocity.

16. Two cog wheels of radii a and b are spinning about parallel axes with angular velocities ω and ω_1 in the same sense. If the wheels suddenly become enmeshed show that the speeds of points on their rims become $\dfrac{ab(Ib\omega \sim I'a\omega')}{Ib^2 + I'a^2}$, where I and I' are the moments of inertia of the wheels about their axes.

17. Two flywheels in the same plane are free to rotate about horizontal parallel frictionless axes. A light string has one end attached to the first wheel, is wound round it, passes over to the second wheel, is wound round it, and has its other end attached to the second wheel. A couple of moment G acts on the first wheel so as to make the string start coiling on the first wheel and uncoiling from the second. Prove that the angular acceleration of the second wheel is $\dfrac{Gab}{Ib^2 + Ja^2}$, where a and b are the radii of the wheels and I and J their moments of inertia about their axles.

18. A uniform circular disc is revolving about its centre with angular velocity ω. A point on the rim of the disc is suddenly fixed. Find the new angular velocity of the disc.

EXERCISE 34B

1. A square frame consists of four thin uniform rods each of length $2a$ and mass $\frac{1}{4}m$ rigidly connected. The frame can swing freely in its own plane about a horizontal axis through one corner A. It is released from rest in a position in which AC, a diagonal, is horizontal. Find the resultant reaction on the axis when AC makes an angle θ with the horizontal and show that its magnitude varies between $\frac{2}{5}mg$ and $\frac{11}{5}mg$.

2. A uniform circular hoop of mass M and radius a has a particle of mass m attached to it at a point A on the hoop. The hoop can move freely about a horizontal axis through O, the point diametrically opposite to A. The axis is perpendicular to the plane of the hoop. It is released from rest with OA horizontal. Prove that when OA has turned through an angle θ the angular velocity is $\sqrt{[(g/a) \sin \theta]}$, and that when OA is vertical the action at O is vertical and of magnitude $(2M + 3m)g$.

3. Three equal uniform rods are rigidly connected at their ends to form a triangle ABC of mass M. The triangle swings in a vertical plane about a smooth hinge at C and starts from rest with AB vertical. Show that when AB is inclined at an angle θ to the horizontal the reaction at the hinge has a component perpendicular to AB of $\frac{7}{3}Mg \cos \theta$. If the angular velocity of the triangle is suddenly changed by an impulsive force, find its line of action if there is no impulsive action at C.

4. A uniform rod AB of mass m and length $2a$ has a particle of mass m fixed at B. It is held in a horizontal position with its middle point resting on a fixed small rough peg. If the rod is allowed to move, prove that it will turn through an angle $\tan^{-1}\frac{5}{14}\mu$, where μ is the coefficient of friction, before slipping takes place.

5. A uniform circular lamina centre C is freely pivoted at a point O of the circumference and is held with its plane vertical and OC making 60° with the downward vertical. If the lamina is released, prove that when OC is vertical the reaction at O is $\frac{5}{3}$ of the weight of the lamina.

6. A uniform sphere rolls down a line of greatest slope of a plane inclined at α to the horizontal. Prove that its acceleration is $\frac{5}{7}g \sin \alpha$. Show also that the least coefficient of friction between the plane and the disc must be $\frac{2}{7} \tan \alpha$.

7. Two uniform rods AB and BC each of length $2a$ and of mass m and $3m$ are smoothly pivoted to a fixed point B. The rods are initially in a horizontal straight line ABC and are released from rest so that they collide when A and C are vertically below B. If the rods adhere to each other find the angle the combined rod makes with the vertical when its angular velocity is zero.

8. A uniform rod of length l and mass m is freely hinged at one end and is held in a vertical position with the free end uppermost. The rod is allowed to fall. When it has turned through 60° it is brought suddenly to rest by an impulse applied at its free end perpendicular to the rod. Calculate the impulse.

9. A uniform circular lamina of mass M and radius a is free to move in its own plane, which is vertical, about the extremity of a diameter AB. The lamina hangs in equilibrium with B vertically below A and is struck a blow applied horizontally at B in the plane of the lamina. If the lamina first comes to rest when AB has turned through an angle of 60°, calculate in terms of M and a the impulse of the blow.

10. A uniform rod AB of mass m and length $2l$ is hinged about a horizontal axis perpendicular to the rod through its midpoint and carries at one end A a particle of mass m. If the rod is at rest with A vertically below B, and A is given a velocity v just sufficient to bring the rod to the horizontal position, prove that $v^2 = \frac{3}{2}gl$.

Find what horizontal impulsive force acting at A is necessary to give A this initial velocity if $l = 3$ m and $m = 2$ kg.

11. A uniform rod AOB of length 2 m and mass 10 kg is smoothly pivoted at a point O 50 cm from A. It is released from a horizontal position. When it reaches the vertical it picks up at the end B a stationary mass of 1 kg. Find the angular momentum of the rod just before it picks up this mass and the angle through which it then swings before commencing to return to the vertical.

12. A uniform rod of length $2a$ and mass M is smoothly pivoted at its midpoint, and a mass M is attached to one end A. It is released from rest when horizontal. Show that the angular momentum of the body when A is vertically below the pivot is $Ma\sqrt{(8ag/3)}$. At this instant a horizontal impulse is applied to the rod at its other end B in the plane of the motion. If the rod just reaches the position in which A is vertically above the pivot, find the size of the impulse.

13. A uniform rod AB of mass M and length $2a$ is smoothly pivoted to a fixed point at A and hangs in equilibrium. A particle of mass m moving horizontally with speed v strikes the lower end of the rod B and adheres to it. If the body just rises to the horizontal position, prove that $v^2 = \frac{2}{3}ag\,\dfrac{(M + 3m)(M + 2m)}{m^2}$

14. A uniform hoop of mass m and radius a can move freely about a horizontal axis perpendicular to its plane through a point O of the hoop. When it hangs at rest it is struck a horizontal blow of magnitude $2m\sqrt{(ag)}$ in the plane of the hoop and through its centre. Prove that the initial angular velocity of the hoop is $\sqrt{(g/a)}$. In the subsequent motion prove that the hoop will turn through a right angle before first coming to rest.

15. A uniform rod AB has mass m and length $6a$. Show that its moment of inertia about a point O of its length distant $2a$ from A is $4ma^2$.

The point O of the rod is freely pivoted to a fixed point, and the rod is slightly disturbed from the position in which it was at rest with B vertically above O. Show that the angular momentum when B is vertically below O is $4m\sqrt{(ga^3)}$. If at this instant the rod picks up a stationary particle of mass m at B, find its inclination to the vertical when it next comes to rest.

16. A uniform rod AB of mass M and length $2a$ is freely pivoted at A and is released from rest when horizontal. When the rod is moving through the vertical position it strikes a particle of mass $\frac{1}{3}M$ initially at rest which adheres to its midpoint. Show that the angular velocity is reduced in the ratio $4:5$ by the impact. Find the angle through which the loaded rod moves before coming instantaneously to rest after the impact.

17. A uniform rod of mass m and length $2a$ is freely pivoted at one end and has a mass m attached at the other end. The rod is released from the horizontal position, and when it is vertical its angular velocity is suddenly changed by the action of a horizontal impulsive force; prove that if there is no impulsive reaction at the pivot the impulse is applied at a depth of $16a/9$ below the pivot.

18. A smooth circular hoop is free to swing in a vertical plane about a frictionless pivot at O its highest point. An imperfectly elastic ball whose mass is equal to that of the hoop falls vertically so as to strike the hoop at a point whose angular distance from O is $\alpha(\alpha < \pi/2)$. If the coefficient of restitution is e, show that the ball will rebound horizontally if $3\tan^2\alpha = 2e$.

19. A uniform rod swings in a vertical plane about a smooth pivot at one end. If the rod swings through $\frac{1}{2}\pi$ on each side of the vertical and if, when the rod makes an angle θ with the vertical, the reaction on the pivot makes an angle ϕ with the rod, prove $\tan\theta = 10\tan\phi$.

20. A uniform rod of mass M and length $2a$ is free to rotate about a fixed horizontal axis through its midpoint and perpendicular to its length. When the rod is at rest and horizontal, a particle of mass m which has fallen from rest at a height a above one end of the rod strikes this end and adheres to it. If ω_0 is the initial angular velocity of the rod and ω its angular velocity when the particle is vertically below the axis, show that $\dfrac{\omega^2}{\omega_0^2} = \dfrac{M + 6m}{3m}$.

16

General Motion of a Rigid Body

1. General motion in two dimensions of a rigid body

The position of a lamina which moves in a plane is clearly defined if we know the position of some definite point in the lamina—we shall use as our fixed point in the lamina its centre of gravity—and the angle a fixed line in the lamina makes with a fixed line in space. Thus we require three coordinates: $\bar{x}$ and $\bar{y}$, which give the position of the centre of gravity referred to fixed axes, and θ the angle turned through by a fixed line in the body all in terms of the time t.

2. The motion of the centre of gravity

The motion of the centre of gravity we have already seen on page 230 to be given by the equations

$$M\ddot{\bar{x}} = \Sigma X \text{ and } M\ddot{\bar{y}} = \Sigma Y$$

i.e. the centre of gravity behaves like a particle equal in mass to the mass of the body with all the external forces acting in their proper directions at the centre of gravity. We must now show that the motion of the body can be considered in two distinct parts:

(i) the motion of translation of the centre of gravity and

(ii) the motion of rotation about the centre of gravity.

In other words, we may, so far as rotation is concerned, treat the centre of gravity as a fixed point like the fixed point about which the rotation took place in the previous chapter.

3. The Principle of Linear Momentum

Let (x_1, y_1), (x_2, y_2), (x_3, y_3), etc., be the coordinates of the particles making up the rigid body. Let their masses be m_1, m_2, m_3, etc. Let the particles be subject to external forces whose components parallel to the axes are (X_1, Y_1), (X_2, Y_2), etc., and internal reactions between the particles whose components parallel to these axes are (X'_1, Y'_1), (X'_2, Y'_2), etc.

The equations of motion for the particles are

$$m_1\ddot{x}_1 = X_1 + X'_1 \qquad m_1\ddot{y}_1 = Y_1 + Y'_1 \quad . \quad . \quad . \quad . \quad \text{(i)}$$
$$m_2\ddot{x}_2 = X_2 + X'_2 \qquad m_2\ddot{y}_2 = Y_2 + Y'_2$$

and so on.

Adding these equations for all the particles and remembering that the internal forces occur in equal and balancing pairs, we get, since

$$\Sigma X' = \Sigma Y' = 0$$
$$\Sigma m\ddot{x} = \Sigma X \text{ and } \Sigma m\ddot{y} = \Sigma Y \quad . \quad . \quad . \quad . \quad \text{(ii)}$$

from which it follows that

the rate of change of the linear momentum of the whole system in any direction is equal to the sum of the resolved parts of the external forces in that direction.

Hence if there is a direction in which the sum of the resolved parts of all the external forces is zero, say the x-axis, then $\Sigma X = 0$,

$$\therefore \ \Sigma m\ddot{x} = 0$$

Integrating $\Sigma m\ddot{x} = 0$ we have $\Sigma m\dot{x} =$ constant, i.e. the linear momentum in any direction in which the sum of the resolved parts of all the external forces is zero is constant. This is again the *Principle of Linear Momentum.*

4. The Principle of Angular Momentum

From equations (i) if we multiply each y-equation by the corresponding value of x and vice versa and then subtract the two, we get

$$m_1(x_1\ddot{y}_1 - y_1\ddot{x}_1) = x_1Y_1 - y_1X_1 + x_1Y'_1 - y_1X'_1$$
$$m_2(x_2\ddot{y}_2 - y_2\ddot{x}_2) = x_2Y_2 - y_2X_2 + x_2Y'_2 - y_2X'_2$$

and so on for each particle.

Adding these equations for all the particles of the body we get

$$\Sigma m(x\ddot{y} - y\ddot{x}) = \Sigma(xY - yX) + \Sigma(xY' - yX') \quad . \quad . \quad \text{(iii)}$$

The right-hand side of this equation is the sum of the moments of the external forces and the internal forces about the origin. The internal forces occur in balancing pairs and so

$$\Sigma(xY' - yX') = 0$$

Hence (iii) may be written

$$\Sigma m(x\ddot{y}-y\ddot{x}) = \Sigma(xY-yX) \quad . \; . \; . \; . \; . \quad \text{(iv)}$$

i.e.

$$\frac{d}{dt}\Sigma m(x\dot{y}-y\dot{x}) = \Sigma(xY-yX)$$

The left-hand side is the rate of change of the moment of the momentum of all the particles about the origin, and the right-hand side is the moment of all the external forces about the origin. Hence

the rate of change of the moment of momentum of the system about any fixed point or axis is equal to the sum of the moments of the forces about this axis.

Hence if there is any axis which is fixed and about which the algebraic sum of the moments of the external forces is zero then the moment of momentum about that axis is constant. This is the *Principle of Angular Momentum.*

5. Independence of translation and rotation

The centre of gravity is given by the equations

$$M\bar{x} = \Sigma mx \text{ and } M\bar{y} = \Sigma my$$

where M = total mass. Differentiating these we have

$$M\dot{\bar{x}} = \Sigma m\dot{x} \text{ and } M\dot{\bar{y}} = \Sigma m\dot{y}$$

showing again that the linear momentum is the same as if the whole mass were concentrated at the centre of gravity and moving with it.

Differentiating again $M\ddot{\bar{x}} = \Sigma m\ddot{x}$ and $M\ddot{\bar{y}} = \Sigma m\ddot{y}$

$$\therefore\ M\ddot{\bar{x}} = \Sigma X \text{ and } M\ddot{\bar{y}} = \Sigma Y \text{ from (ii)}$$

This restates what we said at the start of the chapter, namely

the centre of gravity moves as if all the mass were concentrated there and all the external forces acted there in their proper directions.

Going back to equation (iv) $\Sigma m(x\ddot{y}-y\ddot{x}) = \Sigma(xY-yX)$ and changing the origin to the centre of gravity, i.e. putting $x = \bar{x}+x'$ and $y = \bar{y}+y'$ so that (x', y') are the coordinates of the particle relative to the centre of gravity,

$$\Sigma mx' = \Sigma my' = 0$$

By differentiating twice $\quad \Sigma m\ddot{x}' = \Sigma m\ddot{y}' = 0$

We now have

$$\Sigma m\{(\ddot{\bar{x}}+x')(\ddot{\bar{y}}+\ddot{y}')-(\bar{y}+y')(\ddot{\bar{x}}+\ddot{x}')\} = \Sigma\{(\bar{x}+x')Y-(\bar{y}+y')X\}$$

Simplifying and remembering that terms like

$$\Sigma m\bar{x}\ddot{y}' = \bar{x}\Sigma m\ddot{y}' = 0$$

we have

$$M(\bar{x}\ddot{\bar{y}}-\bar{y}\ddot{\bar{x}})+\Sigma m(x'\ddot{y}'-y'\ddot{x}') = \bar{x}\Sigma Y-\bar{y}\Sigma X+\Sigma(x'Y-y'X)$$

but $M\ddot{\bar{x}} = \Sigma X$ and $M\ddot{\bar{y}} = \Sigma Y$

$$\therefore\ \Sigma m(x'\ddot{y}'-y'\ddot{x}') = \Sigma(x'Y-y'X)$$

i.e. $\dfrac{d}{dt}\Sigma m(x'\dot{y}'-y'\dot{x}') = \Sigma(x'Y-y'x)$

This shows that the rate of change of the moment of momentum about the centre of gravity is equal to the sum of the moments of the external forces about the centre of gravity. Thus we may treat the centre of gravity as if it were a fixed point so far as the rotation of the body is concerned.

Hence in dealing with the motion of a rigid body moving in two dimensions we write down

(i) the equations of motion for the centre of gravity by considering all the external forces to act there on the whole mass concentrated there, and

(ii) the equation for the motion about the centre of gravity by taking moments about it as if it were a fixed point.

For this last we need the sum of the moments of the mass-acceleration about the centre of gravity.

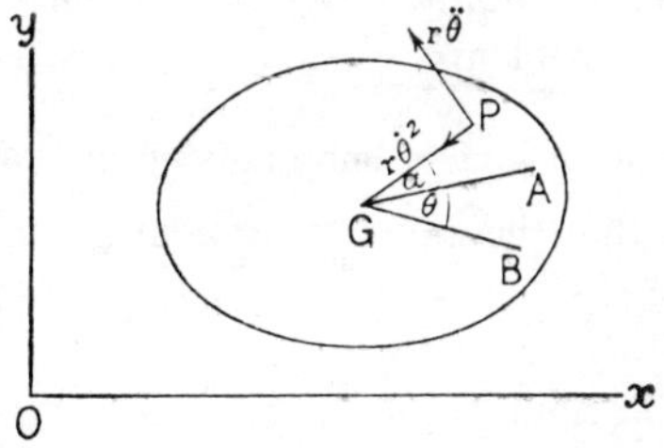

Consider a body moving parallel to the xy-plane and rotating about an axis perpendicular to this plane through its centre of gravity. The figure shows the section of the body in the plane xy. Let P be a typical particle of mass m and let $PG = r$. Let GB be a direction fixed in space and GA a direction fixed in the body.

Let angle $BGA = \theta$, angle $BGP = \phi$, and angle $AGP = \alpha$, then α is fixed and $\phi = \theta + \alpha$. Differentiating twice, $\ddot{\phi} = \ddot{\theta}$. Hence relative to G the acceleration of P has components $r\dot{\phi}^2 = r\dot{\theta}^2$ along PG and $r\ddot{\phi} = r\ddot{\theta}$ perpendicular to GP. The moment of the mass-acceleration relative to G about the axis is $mr^2\ddot{\theta}$.

Therefore relative to G the total moment of the mass-acceleration is $\Sigma mr^2\ddot{\theta}$.

Since $\ddot{\theta}$ is the same for all the particles this is $\ddot{\theta}\Sigma mr^2 = I\ddot{\theta}$, where I is the moment of inertia of the body about G.

6. Force diagram and mass-acceleration diagram

In the solution of problems we shall draw two diagrams

(i) the force diagram showing all the external forces acting on the body and

(ii) the mass-acceleration diagram showing the total mass-acceleration of the system.

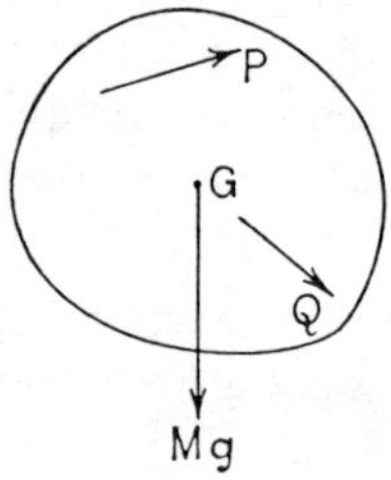

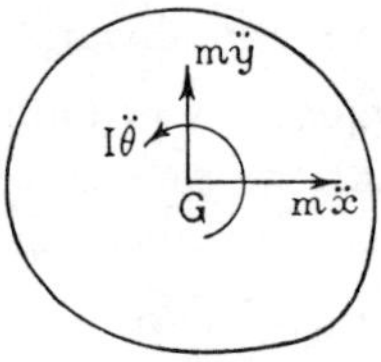

We then get three equations by resolving in two directions and taking moments about a point. Any other equations we use must be implicit in the given problems.

7. Angular momentum of a rigid body moving in two dimensions

We have shown that the angular momentum of the rigid body is

$$\Sigma m(x\dot{y} - y\dot{x}) \quad . \quad . \quad . \quad . \quad . \quad . \quad . \quad \text{(i)}$$

If we change the origin to the centre of gravity $(\bar{x}, \bar{y})$ and if (x', y') are the coordinates of a typical particle relative to the centre of gravity

then $\qquad x = \bar{x} + x'$, and $y = \bar{y} + y'$

and $\qquad \Sigma mx' = \Sigma my' = 0, \quad \therefore \; \Sigma m\dot{x}' = \Sigma m\dot{y}' = 0$

Substituting in (i) we get

$$\Sigma m[(\bar{x}+x')(\dot{\bar{y}}+\dot{y}')-(\bar{y}+y')(\dot{\bar{x}}+\dot{x}')]$$
$$= \Sigma m(x'\dot{y}'-y'\dot{x}')+M(\bar{x}\dot{\bar{y}}-\bar{y}\dot{\bar{x}})+\bar{x}\Sigma m\dot{y}'-\bar{y}\Sigma m\dot{x}'+\dot{\bar{y}}\Sigma mx'-\dot{\bar{x}}\Sigma my'$$

The last four expressions are all zero; the first is the angular momentum about the centre of gravity, which is $I\dot{\theta}$, and the second term is the angular momentum of the whole mass concentrated at the centre of gravity and moving with it.

Hence if M is the mass of a body, V is the velocity of its centre of gravity, ω is the angular velocity about an axis through the centre of gravity, and I is the moment of inertia about this axis, the angular momentum about a fixed point O is

$$MVp+I\omega$$

where p is the perpendicular from O on to the direction of V.

If the external forces have no moment about O this will be constant.

8. Kinetic energy

Using the notation we have employed throughout this chapter, the kinetic energy of our typical particle is $\frac{1}{2}m_1(\dot{x}_1^2+\dot{y}_1^2)$, therefore the kinetic energy of the body is $\frac{1}{2}\Sigma m(\dot{x}^2+\dot{y}^2)$.

Changing the origin to the centre of gravity this is

$$\tfrac{1}{2}\Sigma m[(\dot{\bar{x}}+\dot{x}')^2+(\dot{\bar{y}}+\dot{y}')^2]$$
$$= \tfrac{1}{2}M(\dot{\bar{x}}^2+\dot{\bar{y}}^2)+\dot{\bar{x}}\Sigma m\dot{x}'+\dot{\bar{y}}\Sigma m\dot{y}'+\tfrac{1}{2}\Sigma m(\dot{x}'^2+\dot{y}'^2)$$
$$= \tfrac{1}{2}M(\dot{\bar{x}}^2+\dot{\bar{y}}^2)+\tfrac{1}{2}\Sigma m(\dot{x}'^2+\dot{y}'^2)$$

Therefore the total kinetic energy is equal to the kinetic energy of the whole mass concentrated at the centre of gravity moving with it together with the kinetic energy of the particles in their motion relative to the centre of gravity.

If the velocity of the centre of gravity is V and the body rotates about an axis through the centre of gravity with angular velocity ω then the above reduces to

$$\tfrac{1}{2}MV^2+\tfrac{1}{2}I\omega^2$$

Reverting to equations (i) on page 241, namely

$$m_1\ddot{x}_1 = X_1+X_1' \quad \text{and} \quad m_1\ddot{y}_1 = Y_1+Y_1'$$

and so on, if we multiply each equation by the corresponding velocity component and add up

$$\Sigma m(\dot{x}\ddot{x}+\dot{y}\ddot{y}) = \Sigma(X+X')\dot{x}+(Y+Y')\dot{y}$$

$$\text{i.e.}\quad \frac{d}{dt}[\tfrac{1}{2}\Sigma m(\dot{x}^2+\dot{y}^2)] = \Sigma(X+X')\dot{x}+(Y+Y')\dot{y}$$

The left-hand side is the rate of increase of the kinetic energy and the right-hand side is the rate at which all the forces do work, hence *the increase in the kinetic energy is equal to the total work done.*

9. Worked examples

Example 1.—A uniform sphere of radius a rolls down an inclined plane of inclination α. Find the acceleration of the centre of the sphere.

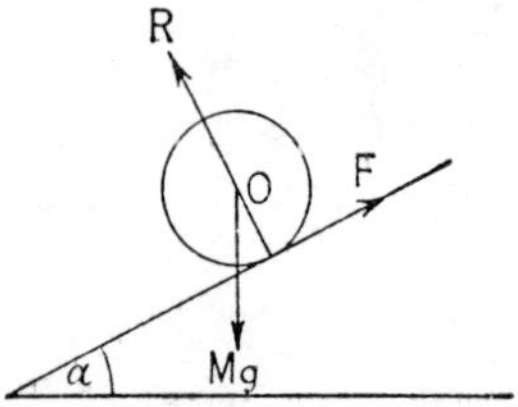

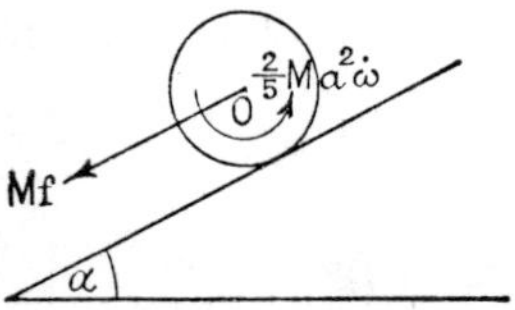

The left-hand figure shows the forces acting on the sphere. R is the normal reaction, F the friction, and Mg the weight. The right-hand figure shows the mass-acceleration. Mf is the mass-acceleration of the centre of gravity and $\frac{2}{5}Ma^2\dot{\omega}$ the mass-acceleration couple.

Resolving normal to the plane,

$$R - Mg\cos\alpha = 0 \quad \text{. . . . . (i)}$$

Resolving along the plane,

$$Mg\sin\alpha - F = Mf \quad \text{. . . . . (ii)}$$

Taking moments about O,

$$Fa = \tfrac{2}{5}Ma^2\dot{\omega} \quad \text{. . . . . (iii)}$$

We have four unknown quantities and so far three equations; the fourth equation we get from the fact that the sphere rolls.

The point of contact is instantaneously at rest and so if u is the linear velocity of O and ω the angular velocity about O,

$$u - a\omega = 0 \text{ or } u = a\omega$$

Differentiating $\quad \dot{u} = f = a\dot{\omega} \quad$ . . . . (iv)

From (ii) and (iii) $\quad Mg\sin\alpha - \frac{2}{5}Ma\dot{\omega} = Mf$

Using (iv) $\quad Mg\sin\alpha = \frac{7}{5}Mf$

$$\tfrac{5}{7}g\sin\alpha = f$$

R and F are now easily obtained from (i) and (ii).

From the equation of energy

Suppose the sphere rolls a distance x down the plane and acquires an angular velocity ω. Let the speed of O in this position be u. As above $u = a\omega$.

The work done on the sphere $= Mgx \sin \alpha$

The kinetic energy acquired by the sphere

$$= \tfrac{1}{2}Mu^2 + \tfrac{1}{2} \times \tfrac{2}{5}Ma^2\omega^2$$

$$\therefore\ Mgx \sin \alpha = \tfrac{1}{2}Mu^2 + \tfrac{1}{2} \times \tfrac{2}{5}Mu^2 \text{ for } u = a\omega$$

$$\therefore\ \tfrac{5}{7}gx \sin \alpha = \tfrac{1}{2}u^2$$

Differentiating with respect to x,

$$\tfrac{5}{7}g \sin \alpha = u\frac{du}{dx}$$

$$\therefore\ \tfrac{5}{7}g \sin \alpha = f$$

Example 2.—The mass of a four-wheeled truck is M. The mass of each pair of wheels is m, and their radii of gyration k. The radius of each wheel is a. The truck is propelled along a horizontal track by a horizontal force P. Neglecting axle friction find the acceleration of the truck.

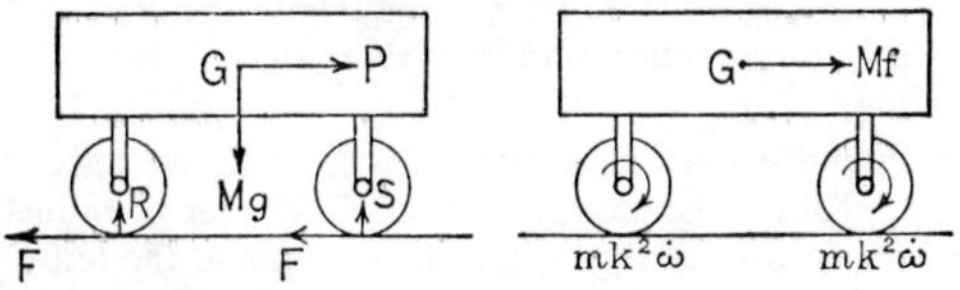

The left-hand figure shows the forces acting and the right-hand figure the mass-acceleration system.

As in example (i), since the wheels roll,

$$f = a\dot{\omega} \qquad \text{(i)}$$

The friction on the two pairs of wheels must be the same, for each must have the same angular acceleration, and moments about the axle for each pair give

$$mk^2\dot{\omega} = Fa \qquad \text{(ii)}$$

Resolving horizontally $\qquad P - 2F = Mf \qquad$ (iii)

From (i) and (ii) substituting for $\dot{\omega}$ and F,

$$P = \frac{2mk^2}{a^2}f + Mf$$

$$f = \frac{a^2P}{Ma^2 + 2mk^2}$$

Or again, if the truck moves a distance x while it acquires a velocity u, and if the angular velocity of the wheels then is ω, from the energy equation

$$Px = \tfrac{1}{2}Mu^2 + \tfrac{1}{2} \times 2mk^2\omega^2$$

$$Px = \tfrac{1}{2}u^2\left[M + \frac{2mk^2}{a^2}\right] \text{ for } u = a\omega$$

Differentiating

$$P = \left(M + \frac{2mk^2}{a^2}\right)u\frac{du}{dx}$$

$$\therefore\ \frac{a^2P}{Ma^2 + 2mk^2} = f$$

Example 3.—A flywheel of moment of inertia I is rotating with angular velocity Ω about a vertical axis. The flywheel contains a pocket at a distance a from the axis into which is dropped a sphere of mass M and moment of inertia i. The sphere has a spin ω about a vertical axis and its centre of gravity has no horizontal velocity. Find the angular velocity of the system after the sphere has come to relative rest in the pocket.

The angular momentum of the system about the vertical axis of the flywheel will remain the same.

Let the angular velocity of the system be ω_1 when the sphere comes to relative rest.

The moment of inertia of the sphere about the axis of the flywheel is $i + Ma^2$,

$$\therefore\ I\omega_1 + (i + Ma^2)\omega_1 = I\Omega + i\omega$$

$$\therefore\ \omega_1 = \frac{I\Omega + i\omega}{I + i + Ma^2}$$

EXERCISE 35A

1. A uniform circular disc rolls down a plane whose inclination to the horizontal is α. Calculate its acceleration and the least value of the coefficient of friction to prevent sliding.

2. One end of a thread wound on a reel is fixed. The reel falls vertically with its axis horizontal. Taking the reel to be a solid cylinder of radius a and mass M, calculate the acceleration of the centre of the reel and the tension in the thread.

3. If you were given a solid and a hollow sphere of the same weight and size, and an inclined plane, how would you find which is the hollow sphere?

4. A uniform circular hoop of radius a and mass m has a particle of mass m attached to it. Initially the particle is at the highest point when it is slightly displaced. If the hoop rolls without slipping, show that its angular velocity when the radius to the particle makes an angle θ with the vertical is given by

$$\omega^2 = \frac{g}{a} \cdot \frac{1 - \cos\theta}{2 + \cos\theta}$$

5. A uniform chain of mass M and length l hangs in equilibrium over a pulley of radius a and moment of inertia I. If it is just started from rest, prove that when the pulley has turned through an agle θ its angular acceleration is given by

$$\frac{2gMa^2\theta}{l(I + Ma^2)}$$

6. A square lamina is suspended by vertical strings tied to two adjacent corners, the side joining these corners being horizontal. If one string is cut prove that the tension in the remaining string is then $\frac{2}{5}mg$, where m is the mass of the lamina.

7. A uniform rod of length $2a$ hangs on a smooth horizontal wire by a small ring attached to it at one end. The rod is held in a horizontal position below the wire and allowed to fall freely. Find the velocities of the ends of the rod when the rod is vertical.

8. A thin straight tube of mass m and length $2a$ can rotate about its centre on a smooth horizontal table. The tube contains a thin rod also of mass m and length $2a$. The system is set in motion with angular velocity ω and the centre of gravity of the rod is moved slightly along the tube. Find the angular velocity of the tube when the rod leaves it.

9. A solid uniform sphere of radius a is rolling with speed v on a horizontal plane in a direction perpendicular to a vertical face of a fixed rectangular step of height h. If the sphere and step are perfectly rough and $h < a$, prove that the sphere will mount the step provided $v^2 > \dfrac{70gha^2}{(7a - 5h)^2}$

10. A rough disc is mounted on a vertical axis through its centre and perpendicular to its plane. A particle of mass m is placed on the disc at a distance r from its centre. The moment of inertia of the disc about the axis is MK^2. The system is initially at rest and a constant couple of moment L is applied to the disc in its plane. Find an expression for the angular acceleration of the disc when it has turned through an angle θ, assuming the particle does not slip. If μ is the coefficient of friction between the particle and the disc, and if slipping takes place when $\theta = \alpha$, show that $rL(1 + 4\alpha^2)^{1/2} = \mu g(MK^2 + mr^2)$.

11. A uniform circular disc of mass M and radius a is free to rotate about a horizontal axis through its centre perpendicular to its plane. A particle P of mass m is placed on the rim of the disc at its highest point. The coefficient of friction between P and the disc is μ. The system is slightly disturbed from rest and θ is the angle turned through by the disc in time t. Find an expression for the angular velocity of the disc at time t assuming slipping has not taken place. Show that slipping will occur when $M \sin\theta = \mu\{(M + 6m)\cos\theta - 4m\}$.

12. A uniform rod of mass m and length $2a$ is free to rotate about a vertical axis through its centre O perpendicular to the rod. Two smooth small rings each of mass $\frac{1}{2}m$ slide on the rod. Initially the rings are on opposite sides of O and each $a/\sqrt{3}$ from O. The rod is given an initial angular velocity of $2\sqrt{(g/a)}$ when the rings are at rest relative to the rod. Show that when the rings are about to slip off the rod its angular velocity is $\sqrt{(g/a)}$ and that the speed of either ring then is $2\sqrt{(\frac{1}{3}ag)}$.

13. A uniform solid cubical block of mass M and edge $2a$ stands on a smooth horizontal table. The block is pierced by a fine straight hole in the vertical plane midway between two parallel vertical faces. The hole passes through A, the middle point of one of the upper edges of the cube, and is inclined at $\alpha(< \pi/4)$ to the horizontal. A smooth particle P of mass m is placed in the hole at A and the system is released from rest. Show that when P issues from the other end of the hole its velocity v relative to the block is given by

$$v^2 \cos\alpha(M + m\sin^2\alpha) = 4ga\sin\alpha(M + m)$$

and that the velocity of the block then is $\dfrac{Mv\cos\alpha}{M + m}$

14. A uniform solid hemisphere of mass m and radius a is held with its plane face against a smooth vertical wall and its lowest point on a smooth horizontal floor. The hemisphere is released. Find the initial pressures on the wall and the floor.

15. A uniform rod of mass M is freely pivoted at one end A and is supported at θ to the horizontal by a string which is perpendicular to the rod and is attached to the other end B of the rod. The string is suddenly cut. Find the reactions of the pivot on the rod along the rod and perpendicular to the rod immediately afterwards.

16. A thin uniform rod of mass m and length $2a$ can rotate freely in a horizontal plane about a vertical axis through its midpoint. A small smooth ring, also of mass m, is threaded on the rod and rests $\frac{1}{2}a$ from the midpoint. The rod is now given an initial angular velocity ω. Find the velocity of the ring along the rod as it leaves the rod and show that the direction of its velocity makes an angle θ with the line of the rod given by $11\sin^2\theta = 3$.

17. A vertical uniform circular disc of radius a rolls without slipping up a line of greatest slope of a plane inclined at α to the horizontal. If the initial angular velocity is ω, show that the disc will roll a distance $\dfrac{3a^2\omega^2}{4g \sin \alpha}$ up the plane before coming instantaneously to rest.

18. A solid uniform spherical ball of radius a rests on top of a fixed sphere of radius r. The ball is disturbed slightly and rolls down the fixed sphere without slipping. Show that the ball leaves the sphere with angular velocity $\sqrt{\left(\dfrac{10g(a + r)}{17a^2}\right)}$

19. A uniform circular disc of radius a and rotating with angular velocity Ω is placed gently on a rough horizontal table whose coefficient of friction is μ. Show that there will be slipping for a time $\dfrac{a\Omega}{3\mu g}$, after which the disc will roll along the plane with angular speed $\frac{1}{3}\Omega$.

EXERCISE 35B

1. A uniform solid cylinder is placed with its axis horizontal on a rough plane making an angle α with the horizontal. If there is no sliding, and the cylinder takes t sec to roll down the plane, find the length of the plane. Find also the time taken for a hollow cylinder to roll from rest through the same distance down the plane.

2. A cylinder of mass M_1 descending a rough plane of inclination α to the horizontal is towing a second cylinder of mass M_2 along a rough horizontal plane. The axes of the cylinders are parallel to the intersection of the planes, and the pull of the rope on each cylinder passes through the axis of the cylinder and is parallel to the plane on which the cylinder rolls. The radius of gyration of each cylinder is p times its radius. Prove that the acceleration of translation of each cylinder is $\dfrac{M_1}{M_1 + M_2} \cdot \dfrac{g \sin \alpha}{1 + p^2}$

3. A uniform circular disc of radius a is free to roll in a vertical plane on a rough horizontal table. To the midpoint of a radius of the disc is attached a particle whose mass is half that of the disc. The system starts to roll when the particle is vertically above the centre. Prove that when the disc has rolled through an angle θ its angular speed is

$$\sqrt{\left(\frac{4g}{a} \cdot \frac{1 - \cos \theta}{17 + 4 \cos \theta}\right)}$$

Find the ratio of the frictional force to the total weight when the radius to the particle is horizontal.

4. A uniform circular disc of mass m and radius a rolls down a plane inclined at α to the horizontal. If the coefficient of friction is μ and a constant breaking couple of moment G is applied to the disc, prove that it will continue to roll down the plane if $\mu > \frac{1}{3} \tan \alpha + \dfrac{2G \sec \alpha}{3mag}$

5. A uniform circular wire ring of radius a is free to rotate about a vertical diameter. A small smooth bead slides on the ring. The masses of ring and bead are the same, and when the bead is at the highest point of the wire it is given an angular speed ω, and the bead is slightly displaced. Find the angular speed of the ring when the bead is at the extremity of a horizontal diameter and show that the bead's speed relative to the ring is then $\sqrt{[\frac{1}{3}(a^2\omega^2 + 6ag)]}$.

6. A uniform rod of length $2l$ rests on a smooth table and passes through a small smooth ring which is freely hinged to the table. Initially the rod is rotating with angular velocity ω about a vertical axis through the ring, and the ring is at a distance a from the midpoint of the rod. Find the angular velocity when the rod is just leaving the ring, and show that the speed of the midpoint of the rod then is $\omega\sqrt{\left(\frac{5}{16}l^2 + \frac{7}{8}a^2 - \frac{3}{16}\frac{a^4}{l^2}\right)}$

7. A smooth particle of mass m is placed at the lowest point of a smooth circular tube of mass M and radius a which can turn freely about a fixed vertical axis coincident with a diameter. The tube is given an initial angular velocity, and the particle is slightly displaced from rest. In the subsequent motion, when the particle reaches its highest point the radius to it makes α with the downward vertical. Find the initial angular velocity of the tube.

8. A smooth straight tube of length a and mass m is free to rotate on a smooth table about one end. A particle, also of mass m, is placed in the tube at its midpoint, and the system is given an initial angular velocity ω. Find the angular velocity of the tube when the particle reaches the end of the tube and show that the speed of the particle relative to the tube then is $\frac{1}{8}a\omega\sqrt{21}$.

9. A non-uniform circular disc, centre C, of radius a has its mass centre G at a distance c from its centre. The disc rolls without slipping along a rough horizontal plane, and the motion takes place in a vertical plane with G slightly displaced from rest from a position vertically above C. Show that when G is level with C the angular velocity of the disc is $\sqrt{\left(\frac{2gc}{a^2 + c^2 + k^2}\right)}$, where k is the radius of gyration of the disc about an axis through G perpendicular to the plane of the disc. Show that when G is at its lowest point the friction at the point of contact of the disc and the plane is zero.

10. A uniform rod is held inclined to the vertical at an angle α with one end on a smooth horizontal table. If the rod is released from rest, find the angular velocity of the rod just before it strikes the plane. Show that the angular acceleration then is $\frac{3g}{4a}$ and that the reaction of the plane on the rod is $\frac{1}{4}mg$, where $2a$ is the length and m is the mass of the rod.

11. A uniform rod of length $2a$ has a small ring of the same mass attached to one end. The ring is free to slide on a smooth horizontal wire. Initially the ring is at instantaneous rest and the rod is vertically below the ring, swinging in a vertical plane through the wire with angular velocity $\sqrt{(g/a)}$. Show that the angular velocity of the rod vanishes instantaneously when the rod is inclined at $\cos^{-1}\frac{7}{12}$ to the vertical, and find the speed of the ring at this instant.

12. A rod of length $2l$ is held vertically with one end resting on a horizontal plane which is rough enough to prevent slipping and is then released. Prove that in the subsequent motion the normal reaction vanishes when the rod makes an angle $\cos^{-1}\frac{2}{3}$ with the vertical and that the angular velocity of the rod then is $\sqrt{(g/l)}$.

13. A rectangular block of mass m rests on a smooth horizontal table. A uniform solid cylinder of mass m' with a rough surface rests on the block; the axis of the cylinder is horizontal and parallel to one edge of the block. The block moves from rest under the application of a horizontal force P applied in a direction perpendicular to the axis of the cylinder and through the centre of gravity of the block. Show that the cylinder will slide on the block if the coefficient of limiting friction between the cylinder and the block is less than $\dfrac{P}{(3m+m')g}$, but that otherwise there will be rolling without slipping.

14. A uniform cube of side a and mass nm is held on a smooth horizontal table. A uniform rod of mass m and length l is hinged at its lower end to the table and rests symmetrically against a smooth face of the cube, its upper end being just below the midpoint of the upper edge of the cube. If the cube is released, prove that its acceleration is $l\ddot{\theta}\cos\theta - l\dot{\theta}^2\sin\theta$, where θ is the inclination of the rod to the vertical. Show that the cube separates from the rod when $3nl\cos^3\theta + 3l\cos\theta = 2a$.

15. A uniform sphere of radius a rests on top of a fixed sphere of radius b and is just displaced. Assuming that the spheres are sufficiently rough to prevent any slipping, show that the moving sphere will leave the fixed sphere when the common normal to the spheres makes an angle with the vertical of $\cos^{-1}\frac{10}{17}$.

16. A movable cylinder of radius a rolls inside a fixed hollow cylinder of radius b so that the axes of symmetry are parallel and horizontal. If the movable cylinder is displaced slightly from the lowest point of the fixed cylinder and makes small oscillations, show that the length of the equivalent simple pendulum is $\frac{3}{2}(b-a)$.

17. A uniform circular hoop is rotating about a horizontal axis through its centre perpendicular to its plane with angular speed Ω when it is placed gently on a horizontal plane whose coefficient of friction is μ. Show that the hoop will slide for a time $\dfrac{a\Omega}{2\mu g}$

10. Impulsive motion

We have seen that the equations of motion of a rigid body are

$$M\ddot{x} = M\dot{u} = X \text{ and } M\ddot{y} = M\dot{v} = Y$$

and

$$Mk^2\ddot{\theta} = Mk^2\dot{\omega} = G$$

where X and Y are the components of the external forces, and G the sum of their moments about the centre of gravity.

Integrating each of these equations and assuming that the initial values of u, v, ω are u_1, v_1, ω_1, we get

$$M(u-u_1) = \int_0^t X\,dt, \qquad M(v-v_1) = \int_0^t Y\,dt$$

$$Mk^2(\omega-\omega_1) = \int_0^t G\,dt$$

If we are concerned with impulsive forces (large forces acting for very short times), $\int_0^t X\,dt$ is the measure of the impulse due to the force X, say I, and $\int_0^t Y dt$ is the measure of the impulse due to the force Y, say J; also $\int G\,dt$ is the measure of the impulsive moment. Hence the above equations may be written

$$M(u-u_1) = I,\ M(v-v_1) = J,\ Mk^2(\omega-\omega_1) = L$$

i.e. the instantaneous changes of linear and angular momentum are the equivalents of the external impulses, and of their moment, respectively.

11. Worked examples

Example 4.—A uniform rod of length $2a$ and mass M is lying on a smooth horizontal table. It is struck a blow of impulse P at a distance x from its centre. The blow is normal to the rod. Find the motion.

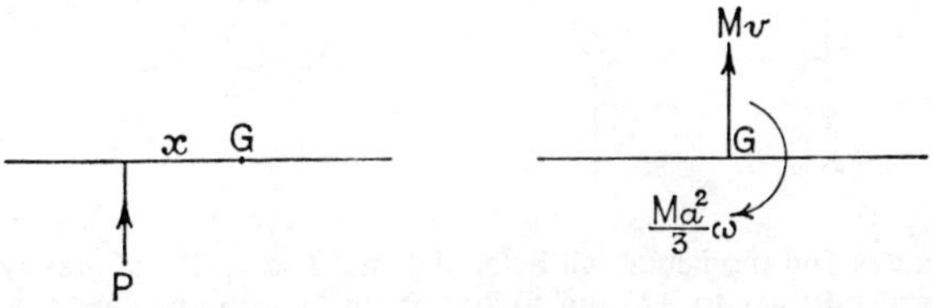

Let v be the velocity of the centre of gravity and ω the angular velocity about G. The diagrams show the impulses and the momenta.

Resolving perpendicular to the rod, $P = Mv$

Taking moments about G, $Px = \frac{1}{3}Ma^2\omega$

$$\therefore\ v = \frac{P}{M} \quad \text{and} \quad \omega = \frac{3Px}{Ma^2}$$

Example 5.—A circular disc is revolving in its own plane about its centre with angular velocity ω when a point on its rim is suddenly fixed. Find the new angular velocity.

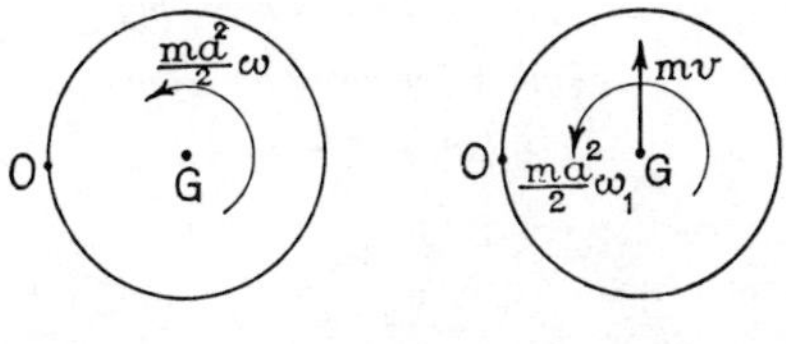

Let O be the point of the disc which is fixed. The impulse acting on the disc to fix O must act through O, and hence the angular momentum about O must be constant.

Let v be the velocity of G and ω_1 the angular velocity about G when O is fixed. The velocity v must be perpendicular to OG. The momentum systems before and after O is fixed are shown in the diagrams.

$$\tfrac{1}{2}ma^2\omega = mva + \tfrac{1}{2}ma^2\omega_1$$

Since O is fixed,

$$v = a\omega_1$$

$$\therefore \tfrac{1}{2}ma^2\omega = ma^2\omega_1 + \tfrac{1}{2}ma^2\omega_1$$

$$\tfrac{1}{3}\omega = \omega_1$$

Example 6.—Two equal uniform rods AB and AC are freely pivoted at A and rest on a smooth horizontal table with angle BAC = 90°. The rod AC is struck a blow normal to it at C. Find the ratio of the initial velocities of the centres of the rods.

Let each rod be of length $2a$ and mass m.

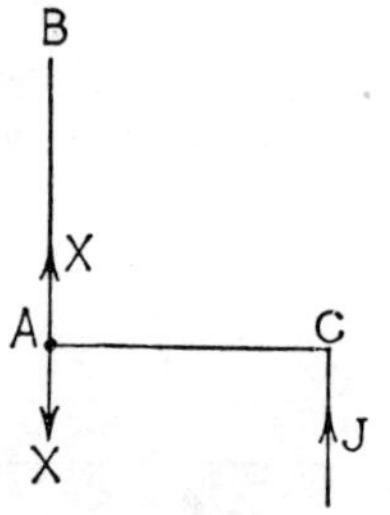

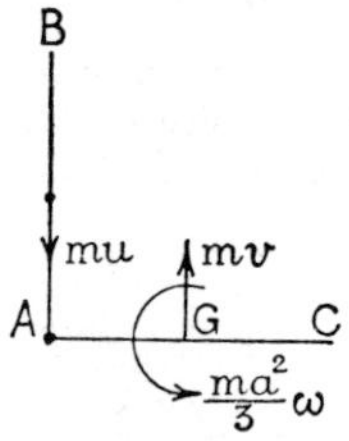

The impulses and momenta will be as shown. The centre of gravity of AC has velocity v perpendicular to AC and rotates about G with angular velocity ω. The centre of gravity of AB will have a velocity along AB, for the only impulse on it is in this direction.

For AC, resolving perpendicular to AC,

$$J - X = mv \quad . \; . \; . \; . \; . \; . \; . \; . \; . \; . \; . \; . \quad \text{(i)}$$

Taking moments about A,

$$2aJ = mva + \tfrac{1}{3}ma^2\omega \quad . \; . \; . \; . \; . \; . \; . \; . \quad \text{(ii)}$$

For AB, resolving along AB,

$$X = -mu \quad . \; . \; . \; . \; . \; . \; . \; . \; . \; . \; . \quad \text{(iii)}$$

The speed of A must be the same when we consider it a point on AB and when a point on AC,

$$\therefore a\omega - v = u \quad . \; . \; . \; . \; . \; . \; . \; . \; . \quad \text{(iv)}$$

From (i) and (iii),

$$J = mv - mu \quad . \; . \; . \; . \; . \; . \; . \quad \text{(v)}$$

From (ii) and (v),

$$\tfrac{1}{2}mv + \tfrac{1}{6}ma\omega = mv - mu$$

$\therefore$ from (iv),

$$\tfrac{1}{2}v + \tfrac{1}{6}(v + u) = v - u$$

$$\tfrac{7}{6}u = \tfrac{1}{3}v$$

$$\frac{u}{v} = \tfrac{2}{7}$$

EXERCISE 36A

1. A uniform rod of length $2a$ lies on a smooth horizontal table. It is struck by a horizontal impulse perpendicular to the rod at a point x from its centre. Find the distance from the centre of the point about which the rod begins to rotate.

2. A uniform rod of mass m and length $2a$ lying on a smooth horizontal table is struck a horizontal blow of impulse P perpendicular to its length at one end. Find the velocities with which the ends of the rod begin to move.

3. Two particles A and B of mass $2m$ and m are connected by a light rod and lie on a smooth horizontal table. The mass A is struck a horizontal blow at $\tan^{-1}\frac{4}{3}$ with AB. Find the ratio of the initial velocities of A and B.

4. A circular disc of radius a lies on a smooth horizontal table. A point P of its rim is made to move along the tangent at P to the disc with speed u. Prove that the angular velocity with which the disc begins to turn is $\dfrac{2u}{3a}$

5. ABCD is a square lamina lying on a smooth horizontal table. The corner A is made to move with speed u along BA. Show that the initial angular velocity of the lamina is $\dfrac{3u}{8a}$

6. Two uniform rods AB and BC of lengths $2a$ and $2b$ and masses m and M are freely jointed at B and lie in a straight line on a smooth horizontal table. An impulse P acts at A normal to AB. Find the impulsive reaction at B and the initial velocity of B.

7. A uniform circular disc is rolling without slipping along a smooth horizontal plane with speed V when its highest point is suddenly fixed. Prove that the disc will make a complete revolution round the point if $V^2 > 24ag$ where a is the radius of the disc.

8. AB and BC are two identical rods freely hinged at B lying in a straight line on a smooth horizontal table. The end A is struck a blow perpendicular to AB. Find the ratio of the initial velocities of A and B.

9. A billiard ball is at rest on a horizontal table when it is struck a horizontal blow in a vertical plane passing through its centre. If the ball begins to roll without slipping, find the height above its centre at which it was struck.

10. A uniform square plate is spinning freely about a diagonal with angular speed ω. Suddenly a corner not on this diagonal is fixed. Prove that the new angular velocity is $\omega/7$.

11. A uniform rod AB hangs vertically from a fixed point A. An equal uniform rod BC is freely hinged to AB at B and it hangs vertically. A horizontal impulse is applied to BC at a point D. If the initial angular velocity of BC is zero prove BD $= \frac{3}{8}$BC.

12. A uniform square plate of side $2a$ and mass m is falling freely with a diagonal vertical when an inelastic string fastened to the middle point of an upper edge suddenly tightens with the string vertical. If before the string tightened the plate was falling with speed u, find its speed immediately after the string tightens, and prove that the impulsive tension in the string is $\frac{4}{7}mu$.

EXERCISE 36B

1. A uniform rod of mass m and length $2a$ rests on a smooth horizontal table. It receives an impulse J perpendicular to it, in the plane of the table at one end. Find the initial speed of the other end of the rod.

2. Two uniform rods AB and BC of equal lengths and masses are smoothly hinged at B and laid in a straight line on a smooth horizontal table. A horizontal impulse perpendicular to its length is applied at the midpoint of AB. Prove that C moves off with $\frac{1}{5}$ of the initial speed of A and in the opposite direction.

3. A string connects a particle of mass m to the end B of a uniform rod AB of mass M. The system is at rest on a smooth horizontal table with the particle at B. If the particle is projected horizontally with velocity V perpendicular to AB, prove that its velocity immediately after the string becomes taut is $\dfrac{4mV}{M+4m}$. Show that the loss of kinetic energy of the system due to the tightening of the string is $\frac{1}{2}\dfrac{mMV^2}{M+4m}$

4. Three particles of equal mass are attached to the ends and the middle of a rigid weightless rod lying on a smooth horizontal table. If an end particle receives an impulse normal to the rod in the plane of the table, prove that the particles will start off with speeds in the ratio 5 : 2 : 1.

5. Two equal heavy particles are connected by a light rod and lie on a smooth horizontal table. If one of the particles receives a blow at right angles to the rod and in the plane of the table, find the speed of the midpoint of the rod and the angular velocity of the rod, if the particle struck has an initial speed V and the length of the rod is l.

6. Two equal uniform rods AB and BC are freely jointed at B and lie on a smooth horizontal table with ABC a straight line. AB receives an impulsive blow at its midpoint at right angles to AB and in the plane of the table. Show that the initial angular velocities of the two rods are equal.

7. Two gear wheels are spinning about parallel axles. Their moments of inertia about the axles are I_1 and I_2, their radii are r_1 and r_2, and their angular velocities in the same sense are ω_1 and ω_2. The axles are moved so that the gears engage. Find the resultant angular velocity of each wheel and show that the impulse acting on the first wheel at the instant of contact has moment about the axle of $\dfrac{I_1I_2(r_1\omega_1 - r_2\omega_2)r_1}{r_1^2I_2 + r_2^2I_1}$

8. A uniform rod of length $2l$ and mass m has a small light ring attached to one end, and the ring is free to slide on a smooth horizontal wire. When the rod is at rest in a vertical position it receives at its lower end a horizontal blow $2mV$ parallel to the wire. Show that the lower end starts off with speed $8V$ and that when the rod is inclined at θ to the vertical

$$(1 + 3\sin^2\theta)l^2\dot{\theta}^2 = 36V^2 - 6gl(1 - \cos\theta)$$

9. A straight uniform rod rests on a smooth horizontal table and a horizontal impulse I, perpendicular to the rod and in the plane of the rod, is applied to the rod at a point P. By finding the initial velocity of any other point Q of the rod show that if the same impulse had been applied at Q perpendicular to the rod and in the plane of the table the initial velocity of P would be the same as that of the point Q.

10. Two equal uniform rods AB and BC are freely jointed at B and are placed on a smooth horizontal table so that the angle ABC is 90°. The rod BA is struck a blow perpendicular to itself in the plane of the table at the end A. Find the ratio of the initial speeds of the centres of gravity of the rods.

11. A light string is wound on a reel of radius a and radius of gyration k about its axis. The free end of the string is tied to a fixed point and the reel is allowed to fall freely. At the instant the string tightens the centre of the reel is descending with speed u and the string is vertical. Prove that the impulsive tension in the string is $\dfrac{muk^2}{a^2 + k^2}$.

12. A circular ring of radius a lies on a smooth horizontal table when a point on the circumference is made to move in the direction of the tangent at the point with speed u. Prove that the initial angular velocity of the ring is $u/2a$.

13. Three equal uniform rods AB, BC, CD, each of mass m and length $2a$ are freely jointed at their ends and lie on a smooth horizontal table in the shape of three sides of a square ABCD. The end A of the rod AB is struck a horizontal blow in the plane of the table parallel to BC of impulse J. Find the initial velocities of the centre of each rod and the angular speeds of the rods.

Revision Examples

Chapter 1

1. An aircraft, travelling in cloud at an airspeed of 240 kn is steered due east by compass. Ten minutes after starting, a break in the cloud reveals that it is directly above a place 6 n miles north and 48 n miles east of the starting point. Find the speed of the wind and its direction, assuming them constant.

2. At 12 noon an aeroplane is 250 n miles NW of its base and is travelling in direction 035° at 400 kn. At the same time an aircraft whose maximum speed is 600 kn leaves the base to intercept the first aeroplane as soon as possible. Find (i) the direction in which the intercepting aircraft should fly and (ii) the time when it intercepts the first aeroplane.

3. Two roads cross at an angle α. A car on one road is approaching the crossing at u km/h and is a km from it. Another car on the other road is b km away from the crossing and is approaching it at v km/h. Show that subsequently the least distance between the cars is $\dfrac{(av \sim bu)\sin\alpha}{\sqrt{(u^2 + v^2 - 2uv\cos\alpha)}}$ km.

4. A river flows with uniform speed V m/s between parallel banks which are a m apart. A boat is propelled at a constant speed $2V$ m/s relative to the water along a straight course from P on one bank to Q on the other and back to P. If the time from P to Q is twice the time from Q to P, show that the total time from P to Q and back again is $\dfrac{2a\sqrt{15}}{5V}$ sec

5. The speed of an aeroplane in still air is u kn and it can carry enough fuel for a flight lasting T hours. The aeroplane has to fly from a point A to a point B and back again without refuelling, where B is a direction $(270 + \alpha)°$ from A. If there is a wind from the west with speed $w(<u)$, show that the speeds of the aircraft on its outward and home journeys are

$$(u^2 - w^2 \sin^2 \alpha)^{1/2} \mp w \cos \alpha \text{ kn}$$

respectively and that the greatest range AB in the given direction is

$$\frac{T}{2}\,\frac{u^2 - w^2}{(u^2 - w^2 \sin^2 \alpha)^{1/2}} \text{ n miles}$$

6. A swimmer in a river finds he can cover a distance d upstream in time t_1 and the same distance downstream in time t_2. Show that the speed of the current is $\dfrac{(t_1 - t_2)d}{2t_1t_2}$. If the width of the river is also d show that he would take a time $\sqrt{(t_1t_2)}$ to swim across it, starting from a point on one bank and crossing in a straight line to the point directly opposite on the other bank.

7. A cyclist is travelling due east with a speed V along a straight level road, and to him the wind appears to blow from the south. When he doubles his speed the wind appears to blow from the south-east. Find the actual velocity of the wind in magnitude and direction. Find also the direction in which the wind appears to blow when the cyclist has trebled his speed.

8. An aeroplane can fly at a speed V in still air and covers a course in the form of an equilateral triangle in time T when there is no wind. It flies the same course in a wind whose speed is kV, where $k < 1$, and whose direction is perpendicular to one side of the triangle. Prove that the time now taken is

$$\frac{T\{(1 - k^2)^{1/2} + (4 - k^2)^{1/2}\}}{3(1 - k^2)}$$

9. To a cruiser C steaming due north at V knots an enemy battleship B distant a n miles away due west appears to be moving SE at $V\sqrt{2}$ knots. Find the true velocity of B and show that when the vessels are nearest together C is NE of B.

If C is within firing range of B when the vessels are not more than b n miles apart, where $b < a$, show that C is liable to be hit any moment within an interval of time lasting $\dfrac{\sqrt{(2b^2 - a^2)}}{V}$ hours.

10. When a motor launch moves northwards at a speed of 20 knots, a pennant on its mast-head points due east. On the return journey when the speed is 20 knots southwards the pennant points 010° 12′. Find the speed and direction of the wind, assuming them to be constant throughout.

Chapter 2

1. The speed of a train travelling between two stations is given in the following table.

Time from start, minutes	0	1	2	3	4	5	6	$6\frac{1}{2}$
Speed, km/h	0	12	24	36	48	48	24	0

Assume that acceleration and retardation are both uniform and draw the speed-time graph using scales of 2 cm = 10 km/h and 2 cm = 1 minute.

(*a*) Find from your graph

(i) the time during which the train is running at full speed (48 km/h);

(ii) the distance in km between the stations.

(*b*) Find also the acceleration and retardation in m/s² giving your answers correct to two decimal places.

2. The figure shows the speed-time graph for a train which starts from rest at one station and comes to rest at another. The time is measured along OX in minutes and the speed along OY in km/h. Find
   (i) the distance in km between the two stations,
   (ii) the acceleration of the train in m/s^2,
   (iii) the time taken to travel the first half of the distance between the stations.

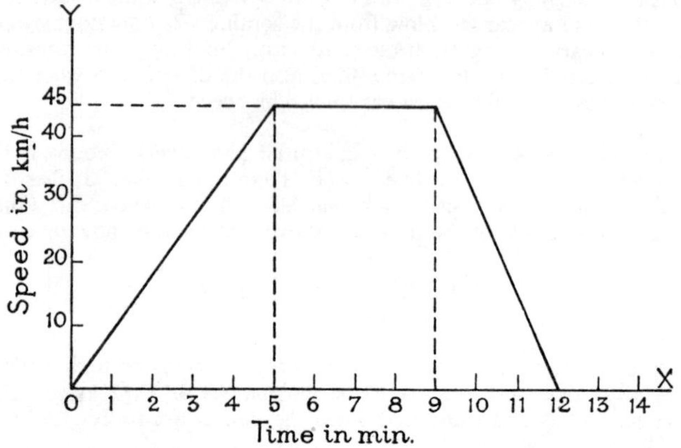

3. A train starts from rest at one station and comes to rest at the next station 2 km away. It travels with uniform acceleration until its speed is 30 km/h; then with uniform speed until the brakes are applied. The retardation is twice the acceleration and the total time of the journey is $5\frac{1}{2}$ min. By means of a velocity-time graph, or otherwise, calculate
   (i) the time in minutes during which the train travelled with uniform speed,
   (ii) the distance in km travelled by the train whilst slowing down,
   (iii) the acceleration of the train in m per sec per min.

4. A smoked glass plate falls vertically past the end of vibrating tuning fork with a style attached producing on the plate a wave-like curve. The distance fallen in 10 consecutive waves is a m and in the next 10 consecutive waves is b m. If the tuning fork makes n complete vibrations per sec, prove that $g = \frac{(b - a)n^2}{100}$ m/s^2.

5. A particle is projected vertically upwards. It passes a point h m vertically above the point of projection after t_1 sec and again after t_2 sec, both measured from the instant of projection. Show that $t_1t_2 = \frac{2h}{g}$ and that the initial velocity was $\frac{1}{2}g(t_1 + t_2)$.

6. A lift starts from rest with constant acceleration f; it then travels with uniform speed and finally slows down to rest with constant retardation f. If the total distance travelled is s and the total time taken is t, show that the time for which it travelled with uniform speed is $\sqrt{\left(t^2 - \frac{4s}{f}\right)}$

7. A particle is projected vertically upwards with a speed of u m/s. After t sec another particle is projected vertically upwards from the same point with the same speed. Prove that the particles will meet at a height $\dfrac{4u^2 - g^2t^2}{8g}$ m above the point of projection.

8. A stone is dropped from a balloon which is rising with uniform acceleration f. After t sec a second stone is dropped. Show that the vertical distance between the stones T sec after the second stone was dropped is $t(f + g)(T + \frac{1}{2}t)$.

9. A cyclist travelling at a speed V begins to free-wheel and observes that after travelling a further distance a in time T his speed is V'. If his retardation is proportional to his speed, prove $\log \dfrac{V}{V'} = \dfrac{T(V - V')}{a}$

10. The velocity v m/s of a particle which travels from rest to rest in a straight line is given by $v = 6t - 3t^2$. Find (i) the maximum speed attained, (ii) the total distance travelled, and (iii) the greatest acceleration.

11. A particle moves in a straight line. The initial speed is u and the retardation is kv^3, where v is the speed at time t. If s is the distance travelled in time t, prove $v = \dfrac{u}{1 + ksu}$ and $t = \frac{1}{2}ks^2 + \dfrac{s}{u}$

Chapter 3

1. A and B are two points on a horizontal plane. At a certain instant the velocity of A is v at an angle α with AB and the velocity of B is u at an angle β with AB. Show that the angular velocity of AB at this instant is $\dfrac{u \sin \beta - v \sin \alpha}{\text{AB}}$

2. Two particles describe a circle of radius a with the same speed u. Show that at any instant their relative angular velocity is u/a.

3. A circle of radius a rolls along a fixed straight line with angular velocity ω. Show that the speeds of the points on its circumference which are at the level of its centre are $a\omega\sqrt{2}$.

4. A particle oscillates with simple harmonic motion on a line 20 cm long with frequency 2000 oscillations per minute. Calculate the greatest velocity and the greatest acceleration of the point correct to three significant figures.

5. A point moves with simple harmonic motion. Its velocity has the values 5 m/s and 4 m/s when its distances from the mean position are 2 m and 3 m. Find the amplitude and the period. Find also the time occupied in passing between the two points if they are on opposite sides of the mean position. Answer to two significant figures.

6. A point moves in a straight line with simple harmonic motion. It has a velocity v_1 when its distance from the centre is x_1 and has a velocity v_2 when its distance from the centre is x_2. Show that the period is $2\pi\sqrt{\left(\dfrac{x_1^2 - x_2^2}{v_2^2 - v_1^2}\right)}$

7. A particle moves in a straight line OAB with simple harmonic motion. It is at rest when at A and B, and OA $= a$ and OB $= b$; its velocity is v when it is at the midpoint of AB. Show that the period is $\dfrac{\pi(b - a)}{v}$

8. The displacement of a particle x is given in terms of the time t by $x = 0{\cdot}1 \cos 3\pi t$; find the initial velocity and the initial acceleration.

9. The amplitude of a simple harmonic motion is a and the maximum speed is V. Find the period, and the acceleration when the displacement is b from the centre.

Chapter 4

1. A regular pentagon ABCDE has its centre at O. Forces 5, 4, 3, 7 N act along OA, OB, OD, OE, respectively. Find the magnitude and direction of their resultant, giving the angle the resultant makes with OE.

2. Three smooth pegs A, B, C, are fixed in a vertical wall. A is uppermost, AB = AC, and BC is horizontal. If the angle BAC = α and a light inextensible string carrying equal weights W at its ends is hung over the pegs, prove that the thrust of the string on the peg B is $2W \sin \alpha/4$, and find the thrust on the peg A.

3. A body of weight 3 kN is supported by two chains which are attached to a small ring in the body and to a beam in the ceiling. If the inclinations of the chains to the vertical are 30° and 45°, find the tension in each chain.

4. A rough plane has inclination $\tan^{-1}\frac{3}{4}$ to the horizontal. A horizontal force of $10g$ N acting on a body of mass 24 kg resting on the plane will just prevent it from sliding down the plane. Find the coefficient of friction between the body and the plane.

5. A uniform chain of length l and weight w per unit length lies along a rough inclined plane of inclination α to the horizontal for the upper half of its length. The other half hangs vertically below the plane and has a mass of weight W attached at the end. Prove that the coefficient of friction between the chain and the plane must be not less than $\dfrac{2W}{wl} \sec \alpha + \tan \alpha + \sec \alpha$.

6. A uniform heavy beam rests in equilibrium with one end on a rough horizontal plane and the other end on a smooth plane inclined at α to the horizontal. If λ is the angle of friction and θ the inclination of the beam to the horizontal, show that the least value of θ is given by $2 \tan \theta = \cot \lambda - \cot \alpha$.

7. The ends of a light inextensible string 24 cm long are tied to two points A and B in a horizontal line 8 cm apart. A small smooth ring slides on the string and is acted on by a force so that the ring rests in equilibrium vertically below A. If the weight of the ring is W show that when the force is least its direction is perpendicular to the bisector of angle ACB. Find its value in this case.

8. AE is a light inextensible string of length $4a$. It has particles of weight w_1, w_2, w_1, attached to it at B, C, D, which are a, $2a$, $3a$, respectively from A. The ends A and E are then attached to fixed points at the same level and the weights hang in equilibrium with AB and DE inclined at α and BC and CD inclined at β, all to the horizontal. Find $\tan \alpha/\tan \beta$ in terms of w_1 and w_2.

9. If ABCDE in question 8 has the form of half a regular octagon, show $w_1/w_2 = 1 + \sqrt{2}$, and in this case find the ratio of the tensions in BC and AB.

10. Three light elastic strings AB, BC, CD, each of natural length a and modulus w are fixed at A and D to points in a horizontal line $2a$ apart. Two particles each of weight w are attached at B and C respectively. Show that in the position of equilibrium AB and CD are inclined to the vertical at an angle θ where $3 \tan \theta = 1 - 2 \sin \theta$.

11. Two smooth spheres each of radius a and weight $2W$ and W are suspended from a fixed point by two light strings, each of length a, the spheres resting in contact with one another. Show that the strings are inclined to the vertical at $19° 6'$ and $40° 54'$.

12. ABC and DEF are two triangles. Their centroids are G and H. Prove

$$\overline{AD} + \overline{BE} + \overline{CF} = 3\overline{GH}$$

Chapter 5

1. A light regular hexagon ABCDEF has masses 2, 4, 8, 6, 10 kg attached at the angular points F, E, D, C, B, respectively. It is suspended freely from the point A, find the angle AF makes with the horizontal in the position of equilibrium.

2. A uniform wire AOB consists of two straight portions AO and OB which are at right angles and of lengths 30 cm and 60 cm. Find the distances of the centre of gravity of the wire from AO and OB. If the wire is freely suspended from A, find the inclination of AO to the vertical.

3. A light rod ABC in which AB = BC has a particle of weight W fixed to the mid-point of AB. The rod is kept at rest in a horizontal position by the action of three like vertical forces applied at A, B, C. The force at B is four times the force at C. Calculate the force at A.

4. ABC is a uniform equilateral triangular lamina. The midpoints of AC and AB are D and E. The portion ADE is removed and BCDE is suspended freely by a string attached to D. Prove that in the position of equilibrium DE makes an acute angle θ with the vertical where $\tan \theta = 5\sqrt{3}/9$.

5. A uniform quadrilateral lamina ABCD of mass 2 kg has the following dimensions: AB = AD = 10 cm, angle CAB = CAD = 30°, and ACB = 45°. It rests in a vertical plane with its edge AB on a horizontal table. Find the least mass which when attached to C will turn the lamina about B.

6. A vertical pole of weight 12 N has its lower end fixed on horizontal ground. A taut rope, which is tied to the pole at the top of the pole 2 m above the ground and to a point on the ground 2 m from the foot of the pole, has a tension of 12 N. Prove that the action of the ground on the foot of the pole consists of a force and a couple. Find the magnitude and direction of the force and the moment of the couple.

7. A uniform rod of length l and weight W is freely hinged at one end to a fixed point. The rod is kept in equilibrium at 60° to the vertical by a couple in the vertical plane of the rod. Calculate the moment of the couple. If the moment of the couple were halved, find the new inclination of the rod and prove that the reaction at the hinge is the same in both cases.

8. A uniform solid consists of a hemisphere and a right circular cone of semi-vertical angle α with coincident plane bases of radii a. Find the distance of the centre of gravity of the solid from the vertex of the cone. Find α when the solid can rest in equilibrium with any point of the curved surface of the hemisphere in contact with a horizontal plane.

9. Three uniform rods PQ, QR, RS, whose weights are proportional to their lengths, a, b, c, respectively are freely jointed at Q and R and rest horizontally on two pegs X and Y. Prove that $XY = \frac{a^2}{2a + b} + \frac{c^2}{2c + b} + b$.

10. Points L, M, N are taken on the sides BC, CA, AB of a triangle ABC and

$$\frac{BL}{LC} = \frac{CM}{MA} = \frac{AN}{NB} = \frac{\mu}{\lambda}$$

Prove that the system of forces represented by AL + BM + CN is equivalent to a couple of moment

$$\frac{2\Delta(\lambda - \mu)}{\lambda + \mu}$$

where Δ is the area of triangle ABC.

11. Two uniform rods AB and BC of the same weight per unit length and of lengths a m and b m respectively are rigidly connected at B so that angle ABC is 90°. If the bent rod is freely suspended from B, prove that the inclination θ of AB to the horizontal is given by $\tan \theta = a^2/b^2$.

Chapter 6

1. A four-wheeled truck of weight W rests on a rough horizontal plane. It is pushed by a horizontal force applied at the back of the truck at a height h above the plane. The centre of gravity of the truck is midway between the axles which are $2a$ apart. The front wheels are locked, and the back wheels are free. The coefficient of friction between the front wheels and the ground is μ. Show that the least force which will move the truck is $\frac{\mu a W}{2a - \mu h}$ provided $\mu h < a$.

2. A uniform circular lamina of radius r and weight W rests with its plane vertical on two fixed rough planes each at α to the horizontal, their line of intersection being perpendicular to the plane of the lamina. Prove that the least couple required to rotate the lamina in its own plane about its centre is $\frac{Wr\mu \sec \alpha}{1 + \mu^2}$ where μ is the coefficient of friction at both points of contact.

3. Two uniform ladders OA, OB, of the same length l and the same weight W, are smoothly jointed at O and stand with A and B in contact with a rough horizontal plane. The coefficient of friction at A and B is μ. If a man of weight W can stand anywhere on the ladders when A and B are $2a$ apart, prove that μ must be not less than $\frac{2a}{3\sqrt{(l^2 - a^2)}}$

4. Two equal uniform rods AB, BC, each of weight W are freely jointed at B. The end A is attached to a smooth pivot and the rods are in equilibrium under the action of a force P acting at C in a horizontal direction. Prove that the angles of inclination of AB and BC to the vertical are $\tan^{-1} \frac{2P}{3W}$ and $\tan^{-1} \frac{2P}{W}$. Find the reactions at A and B on the rod AB.

5. Two equal uniform rods AB, BC, each of weight W are freely jointed at B, and a string connects A to the midpoint of BC. When the string is taut the angle BAC is θ. The rods are placed in a vertical plane with A and C on a smooth horizontal plane. Prove that the tension in the string is

$$\tfrac{1}{4}W\sqrt{(1 + 9 \cot^2 \theta)}$$

6. Four equal uniform rods are freely jointed to form a rhombus ABCD of side a. The rhombus lies on a smooth horizontal plane. An elastic string of natural length a and modulus λ connects A and C. A horizontal force P acts at B along BD and an equal horizontal force acts at D along DB. Prove that if in the position of equilibrium angle ABC is 2θ then $P \tan \theta = \lambda(2 \sin \theta - 1)$.

7. A small ring of weight w fixed to one end A of a uniform rod AB of weight W can slide on a rough rod CAX which is fixed and horizontal. A string CB, equal in length to AB, connects B to a fixed point C in CAX. Prove that AB can be in equilibrium inclined at θ to the vertical if $W \tan \theta \leqslant \mu(3W + 4w)$, where μ is the coefficient of friction between the rod CAX and the ring.

8. Forces 1, 2, 3, 1, 2, P, act along AB, BC, CD, DE, EF, FA, the sides of a regular hexagon ABCDEF. Find P if the system is equivalent to (i) a couple, and (ii) a single force at B.

9. Four uniform heavy rods AB, BC, CD, DA, each of weight W and length $2a$ are smoothly jointed at their ends to form a framework ABCD. The framework is suspended from A and is held in the shape of a square by a light inextensible string whose ends are attached to the midpoints of AD and CD. Find the horizontal and vertical components of the force exerted by AB on BC and show that the tension in the string is $4W$.

10. A uniform thin rod of length l rests inside a sphere of diameter $l\sqrt{2}$ in a vertical plane through the centre of the sphere. If λ is the angle of friction between the sphere and each end of the rod, show that when the rod rests in limiting equilibrium the angle it makes with the horizontal is 2λ.

11. A uniform cylinder of weight w and radius c rests on a rough table with its axis horizontal. A uniform rod ABC of weight W and length $2a$ is smoothly pivoted to the table at A and rests against the curved surface of the cylinder at B. The rod is inclined at α to the horizontal and is in the plane through the centre of gravity of the cylinder normal to its axis. If the components normal to the cylinder of the reactions between it and the rod and the table are R and N, prove $N = w + R$. If the coefficient of friction at both these contacts is μ, show that for equilibrium to be possible $\mu \geqslant \tan \alpha/2$.

12. A rough cylinder of radius r is fixed on a rough horizontal plane, and a uniform rod of length l rests in limiting equilibrium in a vertical plane perpendicular to the axis of the cylinder with one end on the plane and with the rod touching the cylinder. If 2ϕ is the inclination of the rod to the horizontal, prove $2l \sin^2 \phi \cos 2\phi = r \sin 2\lambda$, where λ is the angle of friction at each point of contact of the rod.

13. A and B are two small pegs in a line inclined at θ to the horizontal. B is a vertical height x above A. A thin uniform rod rests on B and passes under A and is in limiting equilibrium when the centre of the rod is a vertical height y above B. If μ_1 and μ_2 are the coefficients of limiting friction between the rod and the pegs A and B respectively, prove $\tan \theta = \mu_2 + (\mu_1 + \mu_2)y/x$.

14. Forces represented by AB, CB, CD, AD, act along the sides of a quadrilateral ABCD. Prove that their resultant is represented by 4 PQ, where P and Q are the midpoints of AC and BD respectively.

15. ABCD is a plane quadrilateral. Forces αAB, βCB, γCD, and δAD act along AB, CB, CD, and AD respectively. If they are in equilibrium prove $\alpha\gamma = \beta\delta$.

16. Forces P, Q, R act along the sides of a triangle ABC, P from B to C, Q from C to A, R from A to B. If their resultant cuts BC produced at D, show that $CD = \dfrac{abR}{cQ - bR}$, where a, b, c, are the lengths of BC, CA, AB.

17. Three equal rods AB, BC, CD, each of length a, are freely jointed at B and C. Light rings fastened to the rods at A and D slide on a rough horizontal wire. The coefficient of friction between the rods and the wire is μ and the system is in equilibrium with BC horizontal. Show that the inclination of AB to the vertical must be less than $\tan^{-1}(3\mu/2)$.

18. Three equal rough cylinders are in contact with one another with their axes parallel and horizontal. The lower two rest on a rough horizontal plane, and the third rests symmetrically on them. If the upper cylinder is about to slip between the other two, prove that the coefficient of friction between them is $2-\sqrt{3}$ if all the surfaces are equally rough.

19. The lower end of a uniform rod rests on a rough horizontal plane, and the upper end is acted upon by a horizontal force so that the rod maintains an inclination α to the horizontal. Show that the coefficient of friction between the rod and the ground must exceed $\frac{1}{2}\cot\alpha$.

Chapter 7

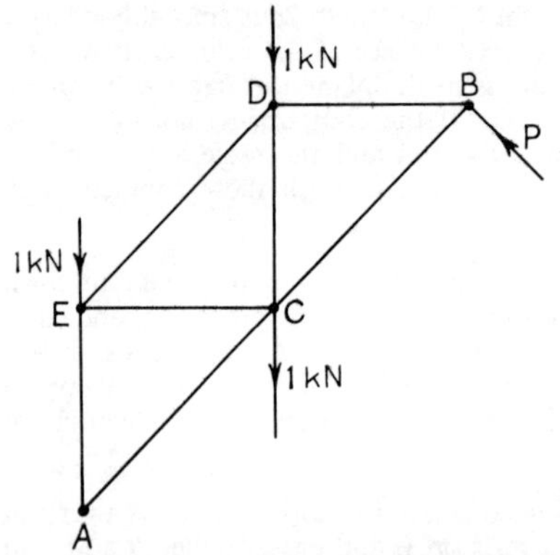

1. In the framework of smooth light rods shown, all the angles are 90° or 45°. It is smoothly pivoted at A; it is supported at B by a force P perpendicular to AB and is loaded as shown. Find the size of P. Show that the force in the rod CD is zero and find the forces in the other rods, saying which are struts and which are ties.

2. A framework is made up of light rods of equal length as shown. A and B rest on smooth supports and there are loads of $W - w$ at D and w at C. Find the thrust in CD and show it is independent of w.

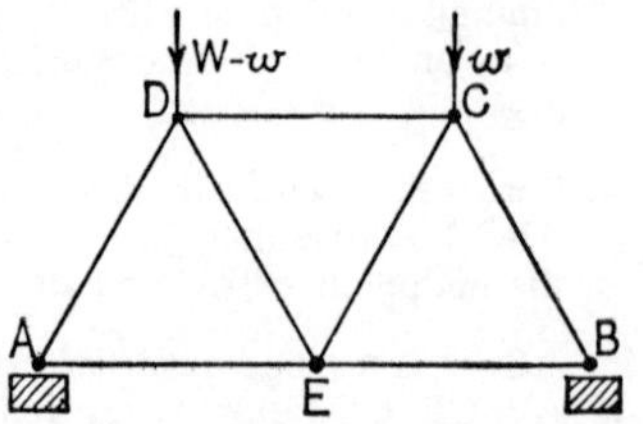

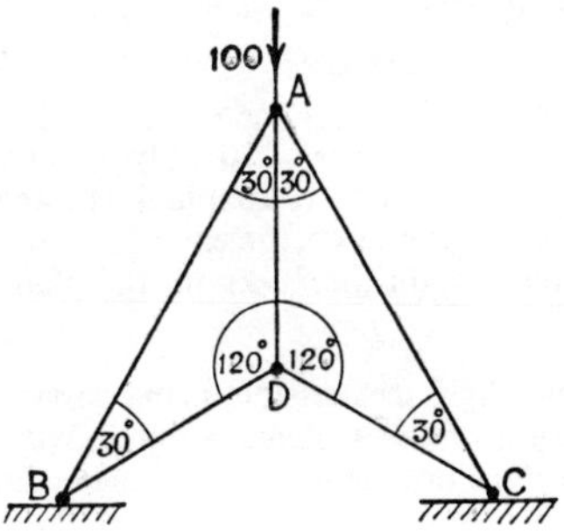

3. The framework shown opposite rests on smooth horizontal supports at B and C and carries a load of 100 N at A. Find the forces in the bars AB, AD, BD.

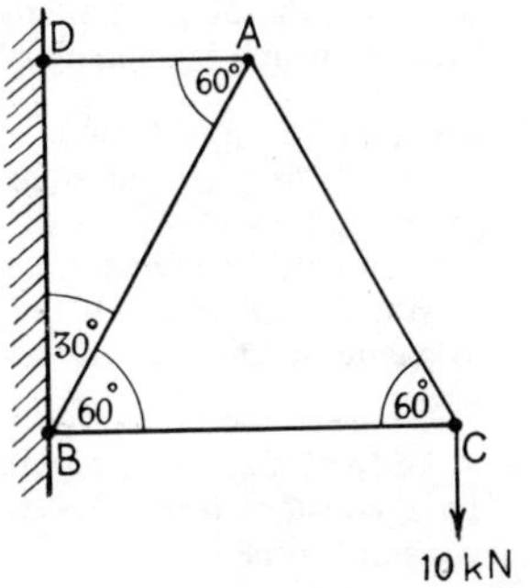

4. In the framework shown opposite the bar AD is horizontal and B is smoothly hinged to a vertical wall. By sketching the force diagram calculate the forces in the bars and the reaction at B.

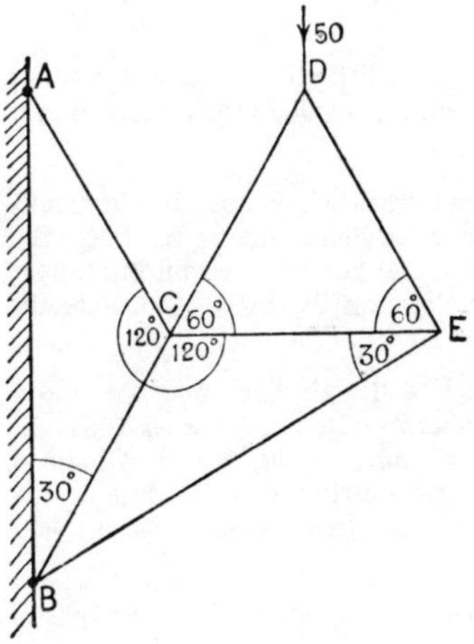

5. The framework of 6 light bars freely jointed to each other is hinged to a vertical wall at A and B. Find graphically the forces in the bars when the framework is loaded with 50 N at D.

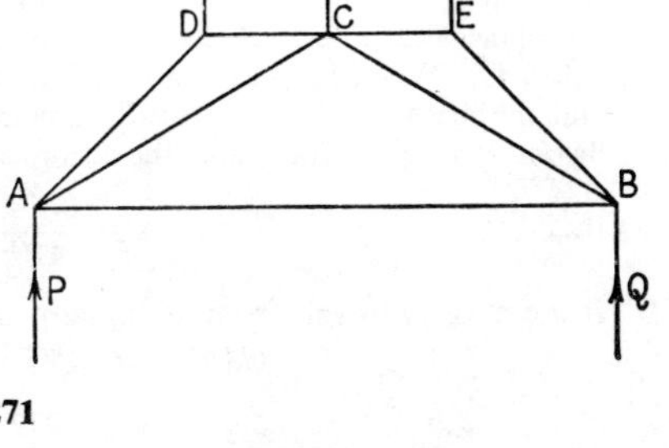

6. The diagram opposite shows a symmetrical framework of 7 light rods freely jointed. It is loaded at D, C, E, as shown and is supported at A and B. Angles CAB and DAB are 30° and 45° respectively. Find P and Q and the forces in AB, AD, AC, and DC.

Chapter 8

1. A horizontal force of 84 N is just sufficient to move a mass of 20 kg lying on a horizontal plank. Find the coefficient of limiting friction. If the plank is tilted until its inclination to the horizontal is 30° and if the coefficient of friction remains the same, calculate the force required to pull the 20 kg up the plank with an acceleration of 1 m/s².

2. Two masses of 30 g and 50 g are connected by a light inextensible string passing over a smooth peg. Calculate the acceleration of the system. After moving for 2 sec the 50-g mass strikes a table and does not rebound. Calculate the time for which it remains on the table.

3. A mass of 5 kg rests on a smooth horizontal table and is connected by a light string which passes at right angles to the edge of the table and over a small smooth pulley to a mass of 2 kg hanging freely. If the system is released from rest with the string taut, calculate the acceleration of the 5-kg mass and the tension in the string. If the table were rough, coefficient of sliding friction $\frac{1}{10}$, calculate in this case the acceleration of the 5-kg mass and the tension in the string.

4. A body of mass 5 kg moving horizontally at 3 m/s is brought to rest in 1 min by a constant force; calculate (*a*) the distance the body moves, and (*b*) the constant force.

5. A packing case of mass 100 kg is moving in a lift. If the vertical reaction between the floor of the lift and the case is 910 N, calculate the acceleration of the lift.

6. A truck whose mass is 5 t is being hauled up a track which rises 1 m vertically for every 70 m of its length by means of a rope parallel to the track. There is a retarding force on the truck due to friction of 800 N. At a certain instant the speed of the truck is 18 km/h and its acceleration 0·1 m/s². Calculate the tension in the rope.

7. A body of mass 2 kg, initially at rest, moves in a straight line under the action of a varying force P N. The graph giving P in terms of the time t sec is a straight line passing through the points $t = 0$, $P = 60$, and $t = 30$, $P = 0$. Obtain a formula giving the acceleration of the body as a function of t. Deduce an expression for the velocity at time t and find the distance covered between $t = 0$ and $t = 30$.

8. Two masses m and M are connected by a light smooth string which passes along a line of greatest slope of an inclined plane and through a small hole in the plane. Initially the mass m is held at a distance l from O measured along the plane and a vertical distance h below O, whilst the mass M hangs freely below O. When motion is allowed to ensue, the mass m moves up the plane and the mass M moves vertically downwards. After the mass m has moved a distance $x(<l)$ up the plane, the string is cut and m just reaches the hole. Prove that

$$x = \frac{(m + M)hl}{M(h + l)}$$

If the mass m travels from its initial position to the hole in time t, show that

$$gh(Ml - mh)t^2 = 2M(h + l)l^2$$

9. A light inextensible string passes over a smooth fixed pulley; it carries at one end a particle of mass 4 kg and at the other end a light pulley over which passes a light string with particles of mass 2 kg and 3 kg at its ends. Neglecting the weight of the pulleys find the accelerations of the three masses when the system is free to move.

The particle of mass 2 kg is replaced by one of m kg. Find m if this particle does not move when the system is released from rest.

10. Prove that the shortest time from rest to rest in which a chain which can just bear a steady load of mass P t can lift a mass of W t through a vertical distance of h m is $\left[\frac{2hP}{g(P-W)}\right]$ sec

11. A light string passes through a smooth fixed ring and carries at one end a mass of $M + m + m_1$ and at the other end a smooth ring of mass M through which passes another light string carrying masses of m and m_1 at its ends. Prove that the ring of mass M rises with acceleration

$$\frac{(m - m_1)^2 g}{2(m + m_1)M + m^2 + 6mm_1 + m_1{}^2}$$

12. A car of mass M kg moves from rest on a horizontal road against a constant resistance of R N. The pull exerted by the engine decreases uniformly with the distance from the starting point, being initially P N and falling to R N after the car has travelled a m. Show that if the car travels x m ($x < a$) from rest in t sec.

$$\left(\frac{dx}{dt}\right)^2 = k^2(2ax - x^2)$$

where $k^2 = (P - R)/(Ma)$. Verify that $x = a(1 - \cos kt)$ is a solution of this equation and show that it satisfies the initial conditions.

13. Two particles of mass m and $2m$ are connected by a light string which passes over a smooth pulley of mass M. This pulley, which does not rotate, is suspended from the roof of a lift by another light string, and the lift moves vertically upwards with acceleration f. Find the tensions in the strings.

14. At speeds over 10 m/s the total tractive force at the rims of the wheels of a 10-t electric car is given by $P(v - 5) = 20{,}000$, where P is the force in N and v is the speed in m/s. Show that the car can accelerate from 36 to 54 km/h in about 240 m.

Chapter 9

1. A golf ball at rest on the ground is struck so that it starts to move with a velocity whose horizontal and upward vertical components are $3v$ and v m/s. In its flight the ball rises to a maximum height of 10 m. Assuming the ground to be horizontal, calculate (i) the value of v, (ii) the horizontal distance travelled by the ball before reaching the ground, and (iii) the direction in which the ball is moving the first time it is 9·1 m above the ground.

2. A stone is thrown with a velocity of 58·8 m/s at an elevation of 30° to the horizontal from the top of a tower 34·3 m high. Calculate (i) the time the stone takes to reach the horizontal plane through the foot of the tower, (ii) the horizontal displacement of the stone from the foot of the tower when it strikes this plane, and (iii) the magnitude of the velocity with which it strikes the plane.

3. A ball projected from a point O on level ground strikes the ground again 32 m from O after a time $2\frac{6}{7}$ sec. Find the horizontal and vertical components of its initial velocity. The ball just passes over a tree standing 6·4 m from O. Find the height of the tree and the speed of the ball at the moment when it is passing the tree.

4. A particle is fired with speed $\sqrt{(\frac{4}{3}gh)}$ from the top of a cliff of height h and strikes the sea at a horizontal distance $2h$ from the gun. Find the two possible angles of projection.

5. Three parallel vertical walls are of height h, H, h, and are equal distances a apart. A ball is thrown in a vertical plane perpendicular to the walls so that it just clears all three tops. Show that the ball will strike the ground at a distance $a\sqrt{\left(\frac{H}{H-h}\right)}$ from the middle wall.

6. A particle is projected from a given point so as just to pass over a wall of height h which is such that the top of the wall is a distance r from the point of projection. Show that the least speed of projection is $\sqrt{[g(h+r)]}$ and that with this speed the particle will have reached its greatest height before grazing the wall.

7. A projectile takes t sec to reach a point on its path and another T sec to reach the horizontal plane through the point of projection. Show that the height of the point reached in t sec above the horizontal plane is $\frac{1}{2}gtT$.

8. A projectile fired from a point on a horizontal plane just clears an object at a horizontal distance a from the point of projection and strikes the plane at a distance R from the point of projection. If the height of the object is h show that the angle of projection is $\tan^{-1}\dfrac{Rh}{a(R-a)}$

9. Three particles A, B, C, are projected simultaneously from the same point and in the same vertical plane. A is projected horizontally with speed u, B is projected with speed V at an angle θ to the horizontal, and C is projected vertically with speed v. Prove that if A, B, C are always collinear

$$\frac{\cos\theta}{u}+\frac{\sin\theta}{v}=\frac{1}{V}$$

10. A particle is projected from a point on an inclined plane with a velocity which is the resultant of u along a line of greatest slope and v vertically upward. Prove that its range on the plane is $2uv/g$.

11. A ball is projected from a point P on a plane inclined at an angle α to the horizontal, the direction of projection lying in a vertical plane through a line of greatest slope and making an angle β ($\beta<\frac{1}{2}\pi-\alpha$) with the upward direction of that line. The coefficient of restitution between ball and plane is e. Show that the condition that the ball returns to P on the nth impact with the plane is $(1-e^n)\tan\alpha\tan\beta=1-e$.

12. From a point in a plane inclined to the horizontal at an acute angle β a particle is projected in a vertical plane perpendicular to the inclined plane. The initial velocity of the particle is V in a direction making an angle α with the upward direction of the line of greatest slope of the plane. Show that the particle will strike the plane perpendicularly if $2\tan\alpha=\cot\beta$. Show also that if this condition is satisfied, the greatest range of the particle on the plane for different values of β is $\dfrac{V^2}{g\sqrt{3}}$

Chapter 10

1. A fire engine can just send a jet of water from a pipe of cross-section 0·005 m^2 to a height of 40 m. Calculate (i) the speed of the water as it leaves the pipe, and (ii) the least power required to project the water.

2. The resistance to the motion of a train whose mass, including that of the engine, is 200 t is 60 N/t. If the greatest power of the engine is 300 kW, find the greatest possible speed of the train on the level, and the acceleration when its speed is 54 km/h, assuming the resistance is still 60 N/t.

3. A car of mass 800 kg can travel on a horizontal road at 108 km/h and can ascend an incline of 1 in 7 at 36 km/h. If the resistances due to friction, etc, are constant and if the power of the car remains the same, find this power.

4. A train of mass 200 t travels from rest to rest between two stations 9 km apart. There is a frictional resistance of 50 N/t throughout the journey. The engine exerts a constant pull of 40 kN until a speed of 54 km/h is reached. This speed is maintained until the brakes, which produce an extra retarding force of $\frac{1}{14}$ of the weight of the train, are applied to bring the train to rest. Find the total time of the journey.

5. The mass of a cyclist and his machine is 80 kg. He cycles along a level road and then ascends an incline which rises 1 m for every 98 m of its length. The frictional resistances to motion are the same on the level and on the incline. If his greatest possible speed on the level is 10 m/s and his greatest possible constant speed on the incline is 6 m/s, prove that the greatest power which he can exert is 120 watts.

6. A pile of mass M is driven into the ground a distance d by a mass m which falls upon it from a height h. Assuming the collision is inelastic and the resistance R of the ground is uniform, show that

$$R = \frac{m^2gh}{(M + m)d} + (M + m)g$$

7. In a Weston differential pulley the larger pulley has a radius R and the smaller r. Show that the velocity ratio is $\frac{2R}{R - r}$. If $r = 7$ and $R = 8$, find the theoretical effort required to lift a load of 200 kg and the percentage efficiency for this load if the actual effort required is 280 N.

8. A 50-kg bag of flour is hoisted vertically through 8 m in one minute by a man working a pulley system. In this time the man working the machine does 4900 J of work. What is the efficiency of the pulley system? At what power is the man working? If the velocity ratio is 5, what average force is the man exerting?

9. A couple of 1·3 N m applied to a screw which has 4 mm pitch, produces a thrust of 10 kN. Calculate, to the nearest whole number, the percentage efficiency of the screw.

10. A differential pulley, the two parts of which have 24 and 25 teeth, is used to raise a weight of 4500 N. What effort must be used if the efficiency is 60 per cent?

11. A force of 20 N will stretch a spring 1 cm. Find the work done to stretch it another 2 cm.

12. A mass of 1 kg is attached to one end of an elastic string 1 m long. The other end of the string is attached to a fixed point O. When the 1-kg mass hangs in equilibrium the string is stretched 25 cm. The mass is held 1 m vertically above O and allowed to fall freely. Find the greatest distance below O it reaches.

Chapter 11

1. Two spheres, whose masses are 20 g and 30 g, are moving towards each other in their line of centres with speeds of 8 and 10 m/s respectively. The coefficient of restitution is $\frac{3}{4}$. Find their speeds after impact.

2. Two masses of $3m$ and m are connected by a string which passes over a smooth pulley. The system is at rest with the mass $3m$ on the ground. A mass also m falls from a height h and hits the mass m and adheres to it and sets the whole system in motion. Find the height above the ground reached by the mass $3m$.

3. A horizontal circular tray has a vertical rim round its edge and A, B, C, D, are points in order on the rim. A small smooth sphere is projected horizontally from A in a direction making an angle α with the radius at A and after impacts at B, C, and D, returns to A. If e is the coefficient of restitution between the sphere and the rim, show that $\tan^2 \alpha = e^3$.

4. A light inextensible string of length $8a$ has a particle of mass 2 kg fixed to one end, a particle of mass 1 kg fixed to the other end, and a particle of mass 2 kg fixed to the middle point. The string passes over a smooth horizontal table along a line perpendicular to the two parallel edges at a distance $4a$ apart. The table is at a height $3a$ above a horizontal inelastic floor. If the system of particles is released from rest when the particles at the ends of the string are each at a height a above the floor, prove that after a time $\sqrt{(10a/g)}$ the suspended mass of 2 kg hits the floor and remains at rest for a time $6\sqrt{\left(\frac{2a}{5g}\right)}$, after which it is jerked into motion with speed $\frac{3}{5}\sqrt{(\frac{2}{5}ag)}$.

5. A small metal ball falls vertically and strikes a smooth fixed plane inclined at angle θ to the horizontal ($\theta < 45°$). The ball rebounds horizontally. Show that a fraction $(1 - e)$ of the kinetic energy is lost during the impact and that, if m is the mass and u the speed of the ball before impact, the impulse on the plane is $mu \sec \theta$.

6. A particle of mass m is attached to a point O by a light inextensible string of length l. The particle is held at a point P above the level of O and at a distance l from O, where OP is inclined at an angle of 60° to the upward vertical. If the particle is allowed to fall from rest at P, prove that at the instant when P is again at a distance l from O the string exerts an impulsive force of magnitude $m\sqrt{(\frac{1}{2}gl)}$ on the particle, and that immediately afterwards the tension in the string is $2mg$.

7. Two particles of masses m and $2m$ are connected by a light inextensible string and lie on a smooth horizontal table. The particles are projected towards each other with speeds $4u$ and u respectively. Show that the particles will move in opposite directions after the collision if the coefficient of restitution exceeds $\frac{1}{5}$. Prove also that the impulsive tension in the string when it becomes taut is $10emu/3$.

8. A train consisting of an engine and two wagons is at rest with the buffers in contact. The mass of the engine is M and the resistance to its motion is R. The mass of each wagon is m and the resistance to its motion is r. The engine exerts a constant tractive force F and moves through a distance a before the first wagon is jerked into motion, and through a further distance a before the second wagon is jerked into motion. Prove that the speed V with which the second wagon begins to move is given by

$$(M + 2m)^2V^2 = 2a(2M + m)(F - R) - 2a(M + m)r$$

9. A particle is projected from a point A on smooth horizontal ground so as to hit a vertical wall at right angles. The coefficient of restitution between the particle and the wall is e and that between the particle and the ground is e_1. If the particle after one bounce from the ground returns to A, prove that $e + 2ee_1 = 1$. Hence show that $e \geqslant \frac{1}{3}$.

10. A hammer-head of mass m_1 moving horizontally with speed v strikes a nail of mass m_2 and there is no rebound, the hammer remaining in contact with the nail in the subsequent motion. Immediately after the impact the nail begins to penetrate horizontally a block of mass m_3 which is free to move on a smooth horizontal plane, penetration being resisted by a constant force R. Prove that the nail will penetrate the block a distance $\dfrac{m_1{}^2m_3v^2}{2R(m_1 + m_2)(m_1 + m_2 + m_3)}$

11. A target of mass M sliding freely along a smooth horizontal plane is struck by a stream of bullets moving in the opposite direction with speed u so that a constant mass σ enters the target per unit time. If initially the speed of the target is v, show that after time t it is $\dfrac{Mv - \sigma ut}{M + \sigma t}$

12. A small body of mass m is projected under gravity from a point O with a velocity whose horizontal and vertical components are u and v. When at the highest point of its trajectory it explodes into two parts each of mass $\frac{1}{2}m$, one part falling vertically downwards with initial velocity V. Show that the distance between the points where the two parts strike the horizontal plane through O is

$$\frac{2u}{g}\{V + \sqrt{(v^2 + V^2)}\}$$

13. Particles A, B, C, of masses m, $2m$, $3m$, are placed on a smooth horizontal table. A light string connects A with B and C. The strings are taut and AB is perpendicular to AC. An impulse is applied to A in the direction of the bisector of angle BAC, so that the strings remain taut. Prove that A starts to move in a direction making an angle $\tan^{-1}1\frac{3}{4}$ with BA.

Chapter 12

1. Two particles of equal mass are connected by a string passing through a hole in a smooth table. One particle is on the table, the other vertically below the hole. How many revolutions per minute would the particle on the table have to make in a circle of radius 20 cm in order to keep the particle below the table at rest?

2. A cyclist travels on a level track of radius 100 m, and the coefficient of friction between the tyres and the ground is $\frac{5}{16}$. Find the greatest speed at which he can travel.

3. A particle of mass m describes a horizontal circle of radius 24 cm inside a smooth hemisphere of radius 25 cm which is fixed with the rim horizontal. A fine thread attached to the particle passes through a smooth hole in the bottom of the hemisphere and supports another mass m hanging at rest. Show that the speed of the first particle is nearly 4 metres per sec.

4. A particle is describing a vertical circle of radius r attached to the end of a light string. If the mass of the particle is m and u is the speed at the lowest point, show that when the string has turned through an angle θ from the downward vertical the tension in the string is $\dfrac{mu^2}{r} - mg(2 - 3\cos\theta)$

If for such a particle the ratio of the maximum to the minimum tension of the string is 3 : 1, show that the ratio of the maximum to the minimum speed is $\sqrt{2} : 1$.

5. A particle of mass m hangs in equilibrium at one end of a light inextensible string of length a whose other end is attached to a fixed point O. If the particle is projected horizontally with speed $2\sqrt{(ag)}$, show that after the string becomes slack the particle will pass through a point on the vertical line through O at a height $9a/16$ above O.

6. A particle is attached by two equal light inextensible strings to two points A and B, distant a apart in the same vertical line, and rotates in a horizontal circle with angular velocity ω. Show that for both strings to remain stretched ω must exceed $\sqrt{(2g/a)}$ and that if the tensions in the strings are in the ratio 2 : 1 then $\omega = \sqrt{(6g/a)}$.

7. A small ring of weight mg is threaded on a light smooth string of length $2a$. The ends of the string are attached to two fixed points A and B. A is a distance a vertically above B. Prove that if the ring describes with uniform angular velocity ω a horizontal circle with centre B then $3a\omega^2 = 8g$.

8. A particle is attached to one end of a taut inextensible string of length r which is describing a vertical circle. The particle passes the lowest point of the circle with speed v and, when the particle is next moving vertically, the string strikes a small peg at a horizontal distance $\frac{1}{3}r$ from the centre of the circle. Prove that the string will become slack before the particle is next moving horizontally unless $v^2 > 4gr$.

9. A horizontal bar OA of length a is made to rotate with uniform angular velocity ω about a vertical axis through the end O. Show that a particle attached by a light string of length l to the end A can describe a horizontal circle of radius $a + l\sin\theta$ if $\omega^2(l\cos\theta + a\cot\theta) = g$, and find the tension in the string.

10. At any point outside the Earth and at a distance x from its centre the acceleration f due to gravity is inversely proportional to x^2. If R is the radius of the Earth and g is the value of f at its surface, express f in terms of R, g, and x.

Find the speed in km/h with which an artificial satellite would describe a circle of radius 7200 km concentric with the Earth and the time in minutes it would take to describe the circle once.
[R = 6400 km and g = 9·8 m/s².]

11. A particle attached to a fixed point by an inextensible string is making complete revolutions in a vertical plane. Prove that the sum of the tensions in positions at the ends of a diameter is the same for all diameters.

12. A particle, attached by a light string which is inelastic and of length l to a fixed point, is describing a horizontal circle of radius nl with uniform speed. The particle is now suddenly reduced to rest and then let go. If in the subsequent motion its speed when the string becomes vertical is equal to half its speed in the original motion, prove that $7n = 4\sqrt{3}$.

13. Two particles of mass m and m_1 (with $m_1 > m$) are connected by a fine thread passing over a small pulley at the top of a smooth fixed solid whose vertical section is a quadrant of a circle. The motion begins when the radius to the particle m is horizontal. Show that when the radius has turned through an angle θ the pressure between the mass m and the surface is

$$\frac{mg\{(3m + m_1)\sin\theta - 2m_1\theta\}}{m + m_1}$$

14. Particles are flung off in all directions in a vertical plane from the circumference of a wheel of radius a rolling in a vertical plane without slipping on a horizontal plane with constant speed V greater than $\sqrt{(ag)}$. Show that the greatest height above the ground reached by any particle is $\dfrac{(V^2 + ag)^2}{2gV^2}$

Chapter 13

1. On a certain day the depth of the water in a harbour at high tide at 7.30 a.m. was 18 m. At the following low tide at 1.45 p.m. the depth was 8 m. Assuming the motion of the surface of the water to be simple harmonic, find (i) the depth of water at noon, (ii) the time at which the depth was 15 m, and (iii) the maximum rate at which the depth decreased.

2. A particle P of mass m is tied to the middle point of an elastic string of modulus mg and natural length $2l$. Each of the ends A and B is attached to a fixed point on a smooth horizontal table, the two points being $4l$ apart. Initially P is released from rest in the line AB and AP $= 3l$. Show that while it moves to a position in which AP $= 2l + x$ the loss of elastic energy in the string is $mg\left(\dfrac{l^2 - x^2}{l}\right)$. Deduce expressions for the speed and acceleration of P and for the time taken to reach the midpoint of AB from the position of rest.

3. A particle of mass m hangs in equilibrium at one end of an elastic string of natural length a and modulus mg. The other end of the string is attached to a fixed point O. Show that in the equilibrium position the particle is at a point C at a distance $2a$ below O. If the particle is projected downwards with velocity $\sqrt{(ag)}$, prove that it will move with simple harmonic motion about C and that the greatest and least lengths of the string during the motion are $3a$ and a.

4. A uniform cork in the form of a circular cylinder has a height h and a mass w per unit volume. It floats with its axis vertical in a large expanse of water whose mass per unit volume is p. Show that in equilibrium a length hw/p is immersed. When the cork is depressed vertically a further small distance x, show that the force of buoyancy is increased by an amount proportional to x. Hence show that if the cork were free to move up and down it would perform a simple harmonic motion of period $2\pi\sqrt{\left(\dfrac{wh}{pg}\right)}$

5. A particle is attached to one end of an elastic string of natural length l, the other end of which is attached to a fixed point O. When the particle hangs in equilibrium the extension of the string is $l/3$. The particle is held at a point vertically below O and distant $l/2$ from O, and is then released. Show that in the subsequent motion the ratio of the time during which the string is stretched to the time during which it is slack is $2\pi/(3\sqrt{3})$.

6. At the place where the value of g is $9{\cdot}8$ m/s² a pendulum beats seconds. How many seconds will it gain in 24 hours at a place where $g = 9{\cdot}85$ m/s²?

7. A seconds pendulum is correct at a place where $g = 9{\cdot}85$ m/s². By what percentage must its length be altered in order that it shall keep correct time at a place where $g = 9{\cdot}8$ m/s²?

8. A particle is attached to one end of an elastic string of length l. The other end of the string is attached to a fixed point O. When the particle hangs freely in equilibrium the extension of the string is d. The particle is held at rest at a point vertically below O such that the extension of the string is $3d$ and is then released. Find the maximum height to which it rises above its starting point and show that the time which elapses before it first returns to its initial position is $\frac{2}{3}(2\pi + 3\sqrt{3})\sqrt{(d/g)}$. (Assume $d < 4l/3$.)

9. A particle is attached to one end of an elastic string of natural length a whose other end is attached to a fixed point O. The particle is let fall from rest at O. If the greatest depth of the particle below O is $a \cot^2 \frac{1}{2}\theta$, show that the modulus of elasticity is $\frac{1}{2}mg \tan^2 \theta$. Show also that the particle reaches its greatest depth below O in a time $\sqrt{\left(\frac{2a}{g}\right)} . [1 + (\pi - \theta) \cot \theta]$.

10. A light spiral spring is fixed at its lower end with its axis vertical. A mass m, which in equilibrium would compress the spring a distance d, falls on the spring from a vertical height h. Prove that it will remain on the spring for a time

$$\sqrt{\left(\frac{d}{g}\right)} . \left\{\pi + 2 \tan^{-1} \sqrt{\left(\frac{d}{2h}\right)}\right\}$$

after which it will be shot off.

Chapter 14

1. In a certain type of chemical reaction the amount x of a substance which has been transformed at time t is such that the rate of increase of x is proportional to the amount of the substance which has not yet been transformed. If the original amount of substance is 100 gm, and 60 gm have been transformed after 2 min, find how much will be transformed after 6 min.

2. A railway truck moves on a level line under a resistance proportional to the square of its speed, the tractive force being zero. The speed is initially 40 m/s and after 10 sec it is 20 m/s. Show that the speed v m/s after t sec from the start is given by $\frac{1}{v} - \frac{1}{40} = \frac{t}{400}$. Show also that the distance covered from $t = 0$ to $t = 10$ is $400 \log_e 2$ m.

3. The temperature θ of a cooling liquid is known to decrease at a rate proportional to $(\theta - \alpha)$, where α is the constant temperature of the surrounding medium. Show that $(\theta - \alpha)$ is proportional to e^{-kt}, where t is the time and k is a positive constant. If $\alpha = 15°$ and θ falls from 60° to 45° in 4 min, find the temperature after a further 4 min, and the time the temperature takes to fall from 45° to 30°.

4. In a solution containing originally a molecules of one substance and a molecules of a second substance the number of molecules x of the second substance changed after time t is given by $\frac{dx}{dt} = k(a - x)^2$, where k is a constant. Find t in terms of x. If one-third of the second substance is changed in 5·7 min, find the time taken to change $\frac{9}{10}$ of it.

5. A particle of mass m travels in a straight line under the influence of a retarding force $m(av^2 + c)$, where a and c are constant. If V_o is the initial speed prove that the particle will come to rest after travelling a distance

$$\frac{1}{2a}\log\left(1 + \frac{aV_0{}^2}{c}\right)$$

6. An engine working at a constant rate moves a mass M against a constant resistance R. If the maximum speed is V, prove that the time taken to reach a speed $\frac{1}{2}V$ from rest is $\frac{MV}{R}(\log 2 - \frac{1}{2})$.

7. A particle of mass m falls from rest against an air resistance of mkv, where k is constant and v is the velocity. Show that the time taken to acquire a velocity of $\frac{g}{2k}$ is $\frac{1}{k}\log_e 2$.

8. When the speed of a particle falling freely is v the resistance to its motion is kmv^2. If the particle started from rest, prove that when it has fallen a distance x

$$v^2 = \frac{g}{k}\{1 - e^{-2kx}\}$$

If x increases from d to $2d$ while the speed increases from V to $\frac{5}{4}V$, find in terms of V the maximum speed of the particle.

9. A vehicle moves from rest on level ground in such a way that its velocity v m/s when it has travelled x m is given by $x = \frac{2v^2}{60 - v}$. Show that its acceleration is given by $\frac{(60 - v)^2}{2(120 - v)}$ and show that the time taken to reach a speed of 15 m/s from rest is $2(\log \frac{4}{3} + \frac{1}{3})$ sec.

10. A body of mass m moves in a straight line under a constant propelling force and a resistance mkv, where v is the velocity. When $t = 0$, $v = 0$, and the velocity tends to a limiting value V. Show that the average velocity during the interval from the start to the instant when $v = \frac{1}{2}V$ is $V\left\{1 - \frac{1}{2\log 2}\right\}$

11. A cyclist freewheeling on a straight level road experiences a retardation proportional to the square of his speed. His speed is reduced from 10 m/s to 5 m/s in a distance of 50 m. Find his average speed during this period.

12. A particle moves on a horizontal table against a resistance proportional to the (speed)^3. If it travels a distance s in a time t whilst the velocity changes from v to u, prove that $\frac{2t}{s} = \frac{1}{v} + \frac{1}{u}$

13. The ends of a uniform chain 24 m long weighing 96 N are attached to two points in the same horizontal line and the middle point of the chain is 8 m below this line. Prove that the parameter of the catenary is 5 m. Find the tension in the chain at its midpoint and its ends. Show that the distance apar of the points of suspension is about 16 m.

14. The ends of a uniform flexible chain of length 56 m are attached to two points A and B where B is 14 m higher than A. The tension at B is $\frac{39}{25}$ of that at A. The lowest point C of the chain lies between A and B. Find the lengths of the chain CA and CB and show that the parameter of the catenary is 15 m.

15. A uniform chain of length 40 m and mass 40 kg hangs from two points at the same level and sags 5 m. Find the tension in the chain at the points of support.

16. A uniform chain of length $2l$ and weight $2wl$ hangs in equilibrium with its ends attached to two points B and C at the same level. The sag is k and the tangents to the chain at B and C are each inclined at α to the horizontal. Show $k = l \tan \frac{1}{2}\alpha$. If the tension at B is double that at A, the lowest point of the chain, prove $k = l/\sqrt{3}$.

Chapter 15

1. A particle of mass m is fixed to the corner B of a uniform square plate ABCD of mass M and side $2a$. The plate can move freely about A in a vertical plane and is held with AB horizontal and above CD. If the plate is released prove that when AB is vertical the angular velocity of the square is $\sqrt{\left[\dfrac{3mg}{a(2M+3m)}\right]}$

2. A solid circular cylinder of radius a is free to turn about its axis which is horizontal. A light string is coiled round the cylinder and carries at its free end a particle of mass M. If the moment of inertia about its axis of the cylinder is I, prove that if the system moves freely the tension in the vertical part of the string is $\dfrac{IMg}{I + Ma^2}$

3. A circular disc of mass m and radius a can turn freely in its own plane about a horizontal axis through its centre. A light string passes over the disc and carries masses m and $\frac{1}{2}m$ at its ends. If there is no slipping between the string and the disc, prove that the acceleration of the masses is $\frac{1}{4}g$.

4. The moment of inertia of a flywheel about its axis is I. There is a constant frictional couple of moment L acting on it. A constant couple G acts on the flywheel for a time t and then ceases to act. Show that if $G > L$ the angle through which the flywheel turns between starting from rest and slowing down to rest is $\dfrac{\frac{1}{2}G(G - L)t^2}{IL}$

5. A uniform circular disc of radius a can move freely about a smooth pivot at a point A of its circumference. When its plane is vertical and the diameter AB is horizontal the point B is given a velocity V vertically downwards. Find the angular velocity of the disc when its centre is vertically below A. Find V if the disc just makes a complete revolution.

6. For a flywheel of mass 1 t the radius of gyration about its axis is 1 m. The flywheel is rotating at 4 radians per sec. Find its kinetic energy and the time it will take to come to rest under the action of a frictional couple of 40 N m.

7. A uniform rod AB of mass m and length $2a$ can move freely about A which is fixed. To B a particle of mass m is attached and the system is released from rest with AB horizontal. Prove that when the rod is vertical the velocity of B is $3\sqrt{(\frac{1}{2}ag)}$.

8. A uniform circular lamina, centre C and mass M, is freely pivoted at a point O on its circumference and is held with its plane vertical at OC at 60° to the downward vertical through O. If the lamina is released, prove that when OC is vertical the reaction on the pivot at O is $\frac{5}{3}Mg$.

9. A uniform rod is free to turn in a vertical plane about a horizontal axis through one end. It is just displaced from the position of unstable equilibrium. Find the angle turned through by the rod when there is no vertical reaction at the axis.

10. A uniform rod of mass m and length $2a$ is freely pivoted at one end and rotates in a vertical plane. It is released from rest when inclined at α to the downward vertical. When the rod is vertical the moving end strikes an inelastic stop and is brought to rest. Prove that the impulse on the stop is $2m\sqrt{(\frac{1}{3}ga)}\sin\frac{1}{2}\alpha$.

11. The end of a uniform rod of mass M and length a is fixed normally to the surface of a uniform sphere of mass m and radius r. If the pendulum is suspended from the free end of the rod, find the length of the equivalent simple pendulum.

12. A uniform circular hoop, centre C, radius a, and mass m, can rotate freely about a horizontal axis perpendicular to its plane through a point O of the hoop. A particle of mass m is fixed to the hoop at A so that angle OCA is 90°. Prove that the length of the equivalent simple pendulum is $4a/\sqrt{5}$.

13. A square lamina of mass M is rotating in its own plane with angular speed ω about its centre when one corner of the lamina strikes a particle of mass m which is at rest. If the particle does not rebound prove that the new angular velocity of the lamina is $\dfrac{M\omega}{M+3m}$

14. A uniform rod of mass M and length $2a$ rotates freely about a vertical axis through one end. A small ring, also of mass M, is initially close to the axis, and the rod is given an angular velocity Ω. Show that the ring will leave the rod with a speed of $a\Omega$ relative to the rod and that at this instant the angular velocity of the rod is $\frac{1}{4}\Omega$.

Chapter 16

1. A cylinder of mass M_1 is descending a rough plane of inclination α to the horizontal towing a second cylinder of mass M_2 on a rough horizontal plane. The axes are parallel to the line of intersection of the planes. The pull of the rope on each cylinder passes through the axis and is parallel to the plane on which the cylinder rolls. The radius of gyration of each cylinder is k times its radius. Prove the acceleration of each cylinder is $\dfrac{M_1 g\sin\alpha}{(M_1+M_2)(1+k^2)}$

2. A uniform rod AB of length $2a$ is held with A on a smooth horizontal table and AB inclined at 30° to the horizontal. Prove that the angular velocity of the rod when it strikes the table after release from rest is $\frac{1}{2}\sqrt{(3g/a)}$.

3. A uniform cylinder of mass M and radius a is rolling along a rough horizontal table under the action of a couple of moment G in a plane perpendicular to its axis. Show that the friction force is $2G/(3a)$.

4. A lorry of total mass M has 4 wheels, each of mass m, radius a, and radius of gyration k about the axles. The lorry runs down a rough plane of inclination α. Neglecting axle friction show that its acceleration is $\dfrac{Ma^2 g\sin\alpha}{Ma^2+4mk^2}$

5. A solid sphere of radius a rests on a rough horizontal table. A horizontal force F acts at the highest point of the sphere. If the sphere rolls prove that the frictional force is $\dfrac{F(a^2 - k^2)}{a^2 + k^2}$ where k is the radius of gyration of the sphere about a diameter.

6. A uniform rod of length $2a$ is freely pivoted at one end and is held at 60° to the downward vertical. The rod is released and rotates in a vertical plane. When the rod is vertical the pivoted end is released. Prove that when next the rod is vertical the centre of the rod will have fallen a distance $\frac{2}{3}\pi^2 a$.

7. A uniform circular disc of radius a is projected on a rough horizontal plane with its plane vertical so that it has an underhand spin ω and its centre has a speed V. Show that if the disc returns to the point of projection $\omega > 2V/a$.

8. A tube of mass M and length $2a$ lies on a smooth horizontal table and contains a particle of mass m close to its centre of mass. Initially the tube is given an angular velocity ω. If I is the moment of inertia of the tube about its mass centre, show that when the particle leaves the angular velocity of the tube is

$$\frac{(M + m)I\omega}{(M + m)I + Mma^2}$$

9. A uniform rod of mass M has masses each m attached to its ends and is at rest on a smooth horizontal table when one of the masses is struck a blow of impulse J perpendicular to the rod. Show that the initial speed of the mass struck is $\dfrac{4J(M + 3m)}{(M + 2m)(M + 6m)}$

10. Three equal uniform rods AB, BC, CD, each of mass m, are freely jointed at B and C and are at rest on a smooth horizontal table with ABCD a straight line. A receives a horizontal impulse J normal to AB. Prove that the initial speed of D is $\dfrac{2J}{15m}$

11. A uniform rod of mass $2m$ and length $2a$ rests on a horizontal table. A smooth sphere of mass m sliding along the table with speed V perpendicular to the rod strikes the rod at a distance $\frac{1}{2}a$ from one end. If the coefficient of restitution between the sphere and the rod is $\frac{1}{2}$ show that after the impact the speed of the sphere is $V/5$.

12. A circular disc of radius 1·4 m and mass m kg has a particle also of mass m kg attached to it 0·7 m from the centre. The disc is placed upright on a rough horizontal plane. When the particle is at its lowest point the speed of the centre of the disc is 2·8 m/s. Show that the disc oscillates over a range of $1{\cdot}4\pi$ m.

13. A thin uniform rod slides under the action of gravity with one end on a smooth horizontal floor and the other end against a smooth vertical wall. The motion is in a vertical plane. If the rod is just displaced from the vertical position, show that when it has turned through an angle θ the acceleration of its mid-point is $\frac{3}{4}g\{5 - 8\cos\theta + 3\cos^2\theta\}^{1/2}$. Prove that the rod will leave the wall when this acceleration is $\frac{3}{4}g$.

Answers

Exercise 1A (p. 5)

1. 4·71, 305° 50′
2. 4·10, 026° 32′
3. 6·26, 28° 9′
4. 6·84, 5·32
5. 55° 46′, 41° 24′
6. 7·53, 21° 59′ with first vector
7. 11·0, 138° 54′
8. 5·70, 147° 26′
9. 3·65, 088° 20′
10. 0·205 at 180°, 11·1

Exercise 1B (p. 6).

1. 8·72, 003° 25′
2. 6·08x, 34° 43′ with first vector
3. 3 : 2
4. 1·15, 0·577
5. 9·28, 297° 57′
6. 23° 9′ or 116° 51′ with first vector
7. 4·03, 194° 11′
8. 70° 58′
9. 2 AC
10. 3 AD

Exercise 2A (p. 6).

1. 34·8 kn from 253°
2. 097°, 246 kn
3. 25 sec, 50 m; 340° 32′, 26·5 sec
5. 311° 30′
6. 49·6 kn from 267°
7. 343° 32′, 44·4 min
8. $1800\left(\frac{1}{t_1}+\frac{1}{t_2}\right)$, $1800\left(\frac{1}{t_1}-\frac{1}{t_2}\right)$
9. 080°; 232 n mile E of A; 38·5 kn from 293°; 094°
10. $\sin\alpha = (w\sin\theta)/u$

Exercise 2B (p. 7).

1. 28 kn, 137°
2. 290°, 263 kn
3. 220 kn, 184°
4. 90°, 153° 26′
5. 167°, 11 min
6. 296°
8. From 260°
9. 32 kn, 186 kn

Exercise 3A (p. 13).

1. (i) 137 m; (ii) 3·0, 2·1, −2·0, −4·0 m/s²
2. (i) 28·4 m; (ii) 19·5 m; (iii) 133 m; (iv) 0·85 m/s²; (v) −2·28 m/s²
3. (i) 405 m; (ii) 573 m; (iii) 0, −0·087, −0·151, −0·175 m/s²
4. (i) 304 m; (ii) 156 m
5. $s = 12t - 3t^2$; (i) 12, 0, −36 m from start; (ii) 12, 24, 60 m
6. (i) 50 m/s; (ii) $166\frac{2}{3}$ m
7. At time 3 sec; as at first
8. 1·5 km approx.
9. (i) $t = 2$; (ii) $t = 4$
10. 389 rev.

Exercise 3B (p. **14**).

1. 78 m

2. (i) 0·14 km; (ii) 14·4 km/h; (iii) 0·34 m/s^2; (iv) 20 sec

3. (i) 0·314 m/s^2; (ii) 0·25 km; (iii) 0·5 km; (iv) 5 min

4. (i) $2\frac{1}{2}$ min; (ii) 0·25 km; (iii) 4·2 m/s/min

5. (i) 45 km/h/min, 12 km/h/min; (ii) 4·08 km

7. (i) 12 m, 2 m/s up; (ii) 12 m, 2 m/s down; (iii) after $2\frac{1}{2}$ sec, $12\frac{1}{2}$ m up; (iv) after 5 sec, 10 m/s down

8. (i) $-0·032$ m/s^2, 6·4 m; (ii) $-0·000\,032$ m/s^2, 7·84 m; (iii) 0, 0, 8 m

9. (i) 0, 5, 8·66, 10, 8·66, 5, 0, -5, $-8·66$, -10, $-8·66$, -5, 0 m/s; (ii) 3 min; (iii) $1\frac{1}{2}$ min to $4\frac{1}{2}$ min; (iv) At starting point

10. (i) 0·175, 0·151, 0·087, 0, $-0·087$, $-0·151$, $-0·175$, $-0·151$, $-0·087$, 0, 0·087, 0·151, 0·175 m/s^2; (ii) 290, 570, 290, -290, -570, -290 m, 0, 290, 860, 1150, 860, 290, 0 m

Exercise 4A (p. **15**).

1. 125 m/s, 5 m/s^2

2. 24 m/s, -40 m/s

3. -30 m/s, 10 m/s^2

4. -20 cm/s^2, -30 m

5. 0·417 m/s^2, 30 sec

6. -54 km/h, 0

7. 1 min, $13\frac{3}{4}$ km

8. 100 m/s, 12 sec

9. 4650 cm/s, 18 sec

10. 1·36 m/s, 2552 m

11. (i) After $2\frac{1}{2}$ sec, 30·625 m up; (ii) After 5 sec, 24·5 m/s; (iii) After 6 sec

12. 3 min 12 sec

13. 75 sec, 36 m/s

14. ± 3 m/s^2, 30 m/s

15. $s = vt - \frac{1}{2}ft^2$

16. 0·5 m/s^2, 63 km/h, 153 km/h

17. 18 cm/s^2

18. $\frac{1}{2}(u+v)$, $\sqrt{\{\frac{1}{2}(u^2+v^2)\}}$

Exercise 4B (p. **16**).

1. 80 cm/s, 1 m

2. 56 m/s, 2·28 km

3. 72 km/h, 2 m/s^2

4. -10^6 m/s^2, 0·001 sec

5. -5 cm/s^2, 0

6. 88 m/s, 22 m/s

7. -2760 cm/s, 3 sec

8. 120 m/min, $-\frac{1}{3}$ m/min^2

9. 0, 0·0025 sec

10. 5 min, 195 m

11. 50 m

12. 1 sec, $1\frac{1}{2}$ sec; $1\frac{1}{8}$ sec

13. 2·5 cm/s^2, 4·75 cm/s

14. 1 sec

15. After 2 sec, 19·6 m up

16. 3 sec

Exercise 5A (p. **20**).

1. (i) 8 m/s, 32 m/s^2; (ii) 20π m/s, $200\pi^2$ m/s^2

2. (i) 2 rad/s, $60/\pi$ rev/min, 40 m/s^2; (ii) 10 rad/s, $300/\pi$ rev/min, 200 m/s^2

3. 2 m/s, $60/\pi$ rev/min

4. Ratio $-4:1$

5. (i) $20\pi^2/9$ m/s^2; (ii) $45/2\pi^2$ m

6. 0·75 m/s^2

7. (i) 2·0, 1·6; (ii) 6600

8. $9·8\sqrt{5}$ m/s^2, to a point 5 m below centre

9. 25 rad/s^2

10. 43·2 km/h

Exercise 5B (p. 21).

1. (i) 100 cm/s, 500 cm/s^2; (ii) $2\pi/9$ cm/s, $\pi^2/405$ cm/s^2
2. (i) $300/\pi$ rev/min; (ii) $30/\pi$ rev/min **3.** 10 rad/s, 9 cm
5. Multiply by 4 **6.** 0·034 m/s^2
7. 250 m **8.** $4\sqrt{2}$ m/s^2, 45° behind radius
9. 8 m/s^2 along radius; -8 m/s^2 along tangent
10. 0, $-10\pi^2$ cm/s^2 along tangent; -10π cm/s, $10\pi^2$ cm/s^2 along radius; 0, $10\pi^2$ cm/s^2 along tangent; 10π cm/s, $10\pi^2$ along radius

Exercise 6A (p. 23).

1. 50 cm **2.** (i) 6π m/s; (ii) $3\sqrt{3}\pi$ m/s
3. $\pi/6$ sec, $\frac{5}{12}$ m **4.** $\pi/6$, $\sqrt{3}\pi/6$, $\pi/3$ m/s
5. 3 sec, $3\sqrt{3}/\pi$ m **6.** $\frac{1}{4}\pi^2$ m/s^2
7. $\dfrac{1}{2\pi^2}$ m **8.** 20π cm/s^2
9. $3{\cdot}75\pi$ m/s $18{\cdot}75\pi^2$ m/s^2 **10.** 3·36 sec, 0·96 sec
11. 2 hr 31 min **12.** $\sqrt{3}v/2$, $v/\sqrt{2}$

Exercise 6B (p. 24).

1. $10\pi^2$ cm/s^2 **2.** $\pm 5\sqrt{2}$ cm **3.** 6π cm/s
4. $\frac{1}{6}$ sec **5.** $\sqrt{20}$ cm, 2π sec **6.** 10 cm, 6 sec
7. 10 cosec 36° $\simeq$ 17·0 cm **8.** $\dfrac{20\sqrt{3}}{\pi}$ cm **9.** $5\sqrt{10} \simeq 15{\cdot}8$ cm
10. 2·59 m, 2·80 sec **11.** 5 cm **12.** Straight line, circle

Exercise 7A (p. 35).

1. 8·602 N, 35° 32′ **2.** 5·36 N, 33° 24′ with 5 N
3. 10 N **4.** 8·72 N at 36° 52′ with 4 N
5. 3·81 N **6.** 8·066 N at 18° 4′ to 12 N
7. 60° **8.** 10 N at 36° 52′ with AB
9. 12·05 N, 81° 45′ **10.** 6 N, 120°
11. 25·53 N, 333° 11′ **12.** 28·63 N, 139° 39′
13. $\sqrt{39}$ N, 76° 6′ **14.** 2 N along GA

15. 110, 55 N **16.** 85, 49 N **17.** $10g$, $24g$ N
18. 85, 49 N **19.** $7\frac{1}{2}$ N, $P = 6$ N **20.** 12 N, 5 N
21. $\frac{1}{2}W$, $\frac{1}{2}W\sqrt{3}$ N **22.** 5 N **23.** $2W\cos\theta$, W
24. 35 cm **27.** $8\frac{2}{3}$ N, $6\frac{2}{3}$ N **28.** 25 N, 2 m

Exercise 7B (p. 37).

1. 13 N, 27° 48′ to 8 N
2. 8 N at 60° to each
3. 36° 52′, 53° 8′
4. 84° 16′
5. 120°
6. 6·2 N at 23° 58′ to 10 N
7. 39 N, 74 N
8. 59 N, 83 N
10. 67° 35′
11. $\frac{W(a^2 + b^2)}{2a^2}$
13. 53° 25′
15. $\sqrt{5}$ N at 333° 26′ with AB
17. $\sqrt{3}$ N parallel to DA
18. 12 N along AD
19. 78 N, $2\frac{1}{2}$ m, 68 N
20. 2·5g N, 4 kg

Exercise 8A (p. 41).

1. 9 N, $8\frac{1}{3}$ N
3. 45·13 N, 83·43 N
4. 1·84, 4·02, 4·02 N
5. $40\sqrt{2}$ N
6. $\frac{3}{8}$, 9 N
7. $\tan\alpha$, $2W\sin\alpha$
8. $\frac{1}{2}$
9. $2r\lambda$
10. $5\sqrt{3} - 8 = \cdot 66$
11. $\frac{W(1 + \mu\sqrt{3})}{\mu + \sqrt{3}}$
12. $W\sin\lambda$, λ to the plane

Exercise 8B (p. 42).

1. 10 N
2. $\tan^{-1}\mu$
3. 3·34 N
4. $\frac{1}{15}\sqrt{3}$, 6 N
6. 11 N
7. $W\tan(\alpha + \lambda)$
8. $\frac{1}{3}\tan\alpha$
9. $\frac{3\sqrt{2} - 1}{3\sqrt{2} + \sqrt{3}} = \cdot 54$
10. 3 kg

Exercise 9A (p. 51).

1. 8 N, 6 cm; 2 N, 24 cm
2. 6, 2 N; 12, 4 N
3. 2 m from 10 kg, 42 cm (approx.)
6. 25 cm
7. $\sqrt{(M_1M_2)}$
8. 120 kg
9. 50 N
10. $(x + y)$ kg
11. $8\frac{1}{3}$, $26\frac{2}{3}$ kg, 5 m towards A
12. $28\frac{1}{3}$, $31\frac{2}{3}$ N, 120 N
13. 12 m, $\sin^{-1}\frac{1}{6} = 9° 36′$
14. $\frac{W\sqrt{(a^2 - b^2)}}{b}$
17. $\frac{W + w}{w}a$
18. 72·42 N, 51·21 N
19. $\frac{2W + 3w}{2W + w}$
20. 10 N
21. 18, 12 N; ±40 cm

Exercise 9B (p. 53).

1. 10 N, 6 cm; 4 N, 15 cm
2. 4, 2 N; 8, 2 N
3. 8, 12 N, $13\frac{1}{3}$ N
4. 1 m, 10 kg
5. 20, 40 N, 5 N
7. 70, 110 N
9. $7\frac{1}{2}$ N
10. 20, 17·32 N
12. $\sin^{-1}\frac{G}{Wa}$
13. 2 m, 30°
14. 26·2 N
15. 25° 52′
17. W

Exercise 10A (p. 63).

1. 4·2 cm, 3 cm
3. $3\frac{1}{2}$ cm, 5 cm
4. $\frac{a\sqrt{61}}{6} = 1{\cdot}3a$
5. $\frac{\sqrt{3}a}{5}$ from the centre
6. $4\frac{1}{2}$ cm, 2 cm, 29° 21′
7. $1\frac{7}{9}$ cm, $2\frac{1}{3}$ cm, 37° 18′
8. 0·5 cm from the centre
9. 5, $1\frac{1}{2}$ cm
10. 4 cm, 18° 26′
11. 1·1 m, 1·26 m
12. 3 cm from BC on the perpendicular bisector
13. 3 cm
15. 2·57 cm
16. 1·32:1
18. $\tan^{-1}\frac{4r}{h}$
20. $\sqrt{3}r$
21. $\frac{3\sqrt{2a}}{\pi + 4}$
22. $\frac{3}{4}$ of the length from the given end

Exercise 10B (p. 64).

1. 23 cm, 15 cm
2. 27 cm, 35 cm
3. $\frac{1}{2}a$, $\frac{3}{5}\sqrt{3}a$
4. $a/6$
5. $\frac{43r}{32}$
7. $\frac{h}{3}\cdot\frac{a + 2b}{a + b}$
9. 2·46 cm
10. 13, 9 cm
12. $(3 - \sqrt{6})a$
14. 4 cm
15. $\frac{3}{4}\frac{(2a - h)^2}{3a - h}$

Exercise 11A (p. 72).

1. 45°, $\frac{1}{2}W\sqrt{5}$, 26° 34′ to vertical
2. 30°, $\frac{1}{4}W\sqrt{13}$ at 13° 54′ to vertical
3. W, W, 30° to the horizontal
4. 86·6 N, $\frac{\sqrt{3}}{6}$
5. 15 N, 9 N
6. 17·5 N, 121·2 N, 7/48
7. 50, 175, 43·3 N
8. 36° 52′
10. $6\sqrt{5}$ 13·4 N
13. $\frac{\sqrt{3}}{5}$

Exercise 11B (p. 74.).

1. $10\sqrt{3}$ N, $10\sqrt{39}$ N, 16° 6′ to vertical
2. 10 N, $10\sqrt{13}$ N at 13° 54′ to the vertical
3. 25 N, 7 N horizontally
4. 16 N, $\frac{48}{61}$
5. 50° 12′
7. 2λ
8. h/a
9. $\frac{W \sin\alpha}{\sin(\alpha + \beta)}, \frac{W \sin\beta}{\sin(\alpha + \beta)}$
10. $W\left(\frac{a}{c}\right)^{\frac{1}{3}}$

Exercise 12A (p. 80).

1. $2\sqrt{3}$ N, 30°, 6 cm
2. 6 N parallel to AD, 80 cm
3. $2\sqrt{2}$ N parallel to AC, $\frac{5}{2}a$
4. −6 N
5. $(p + q + r)$ along AB, $\sqrt{2}(p + q)$ along BO, $\sqrt{2}(q + r)$ along OA
6. 2 N parallel to EF, and $2\sqrt{3}$ cm from it on the side away from D

Exercise 12B (p. **81**).

1. $\sqrt{3}$ N, 30°, 2 cm from A
2. $\sqrt{3}$ N, perp. to BC, E in BA produced so that AE = $8a$
3. $2\sqrt{2}$ N, at 45° to AB cutting BA produced at X where AX = $9a/2$
4. 20, $10\sqrt{2}$, $10\sqrt{2}$ N
5. $\dfrac{21\sqrt{3}}{2}$ N m, 6 N, parallel to the 5 N
6. $7\frac{1}{2}$, 10, $12\frac{1}{2}$ N
7. 12, 16 N
8. $9{\cdot}22F$ at 12° 32′ to DA, $\frac{1}{2}a$

Exercise 13A (p. **85**).

1. $6\sqrt{3}$, $3\sqrt{3}$, 15 N
2. 28 N, 12 m
3. 40° 54′, 132·3 N, 1 m
4. 800 N, 1000 N, at 36° 52′ to the horizontal
5. Perpendicular to the line joining step and centre, $\dfrac{W}{r}\sqrt{(2hr - h^2)}$
8. $\frac{1}{4}W\sec\theta$, $\frac{1}{4}W\tan\theta$, $\frac{3}{4}W$
9. $\dfrac{Wa\sin\theta\sin 2\theta}{2h}$
13. $\sin^{-1}\dfrac{a}{2(a+l)}$
24. $\tan\theta > \dfrac{1}{\mu} - 2$

Exercise 13B (p. **87**).

1. $4\sqrt{3}$, $4\sqrt{3}$, 6 N
2. $\dfrac{W}{1+\cos\alpha}$
3. 40, $20\sqrt{3}$, 10, $10\sqrt{13}$ N
4. Each $20\sqrt{5}$ at 63° 26′ to AD or DA
5. 36° 52′
6. $\tan^{-1}\dfrac{1}{2\mu}$, μW, $W\sqrt{(1+\mu^2)}$
7. $\tan^{-1}\dfrac{1-\mu\mu_1}{2\mu}$, $\dfrac{\mu W}{1+\mu\mu_1}$, $\dfrac{W}{1+\mu\mu_1}$
12. μW

Exercise 14A (p. **93**).

2. $\dfrac{W\sqrt{93}}{6}$ at 68° 57′ to the horizontal, $\dfrac{W\sqrt{21}}{6}$ at 40° 54′ to the horizontal
3. $\dfrac{W\sqrt{3}}{6}$
4. $\dfrac{(W+P)\sqrt{3}}{8}$, $\dfrac{3P-W}{8}$
6. $2W$, $\frac{1}{2}W$ horizontal
7. $\dfrac{W+w}{W+3w}\tan\alpha$
8. 90°
10. $\dfrac{Wa}{\sqrt{(b^2-2ab)}}$
11. $\frac{1}{3}$AB $\cos^2\theta\sin\theta$
12. W horizontal
15. $\frac{1}{2}W\cos\alpha$, $\frac{1}{2}W\sin\alpha\cos\alpha$, $\dfrac{\tan\alpha}{1+2\tan^2\alpha}$
16. $\dfrac{W_1 l\cos 2\alpha}{a\cot^2\alpha}$, $W_1\left(1-\dfrac{l\cos 2\alpha}{a\cot\alpha}\right)$, $\dfrac{W_1 l\cos 2\alpha}{a\cot\alpha}$, $W+\dfrac{W_1 l\cos 2\alpha}{a\cot\alpha}$

Exercise 14B (p. 94).

1. $10\sqrt{3}$ N, $10\sqrt{3}$ N horizontal

2. $\sqrt{3}$ N horizontal at A, $\sqrt{39}$ N at $\tan^{-1} 2\sqrt{3}$ with the horizontal at B and C

3. $4W$ **4.** 2 N **5.** $18\sqrt{3}$ N, $12\sqrt{7}$ N

16. $\dfrac{W(\cos^2\alpha - 3\sin^2\alpha)}{2\cos 2\alpha}$, $-\frac{1}{2}W\tan 2\alpha$, $-\frac{1}{2}W\tan 2\alpha$, $\dfrac{W(3\cos^2\alpha - \sin^2\alpha)}{2\cos 2\alpha}$

17. $\frac{1}{5}m - 2$

Exercise 15A (p. 105).

Tensions are denoted − thrusts +

1. 1·79 N at 45° to BA **2.** 2·65 N at 139° 6′ to AB, 2 m

3. 11·8 N at 167° 16′ to BA, 10·8 cm from B

4. 5·77, 2·11, 7·07 N **5.** −141, −200, +273 N

6. +10, +6·2, +3·8, −6·2 N **7.** −273, 193, −273 N

8. 5, 8·66, 17·32 N

9. *ab* − 5, *bc* − 17·32, *ad* + 17·32, *cd* + 20, *bd* + 10 N

10. AB + 4·47, AD − 4·47, BD + 5, BC + 3·66, CD − 3·66. $P = 6{\cdot}34$, $Q = 3{\cdot}66$

11. 80 cm **12.** 50 N, 130 N

13. 800, 1100 N, 200 N at A **14.** 110 N, 98 N

15. 10·5 N, 25·5 N

16. Reactions 2·24 and 2 kN, *ad* − 2, *bd* + 1·4, *dc* − 1·4, *bc* + 1 kN

17. Reactions 6·3 and 6 kN, AC + 2·83, BC − 6, AD + 4, CD − 2·85, CE − 2, DE + 1·41, DF + 1, EF − 1·41 kN

18. Reactions 2, 2 kN, *ab* + 2·82, *af* − 2, *bc* + 2·12, *cf* − 1, *bf* + ·71 kN

19. *ae* + 1·4, *ad* + 4, *ed* + 1, *ab* − 4·46 kN

20. Reaction 2·54 kN, *be* + ·7, *bc* + 2, *ed* − 2, *ae* − 3·5, *ce* − 3, *cd* + 2·8 kN

21. *ab* + 1·3, *af* − 1·9, *bc* + 1·7, *be* − ·5, *fe* − 1·3, *cd* + 1·7, *ed* − 2·4, *ce* 0, *bf* + ·3 kN

22. AB + 8, BC − $4\sqrt{3}$, CD − $4\sqrt{3}$, CA 0, DE − $5\sqrt{3}$, DA − $2\sqrt{3}$, AE + 10 kN

23. AE − 19·3, AB + 19·3, DE − 38·64, CD + 38·64, EC + 27·3, EB − 10, BC + 23·7 kN

24. AB = AF = −5, BC = $-1\frac{1}{2}$, CD = $-\frac{1}{2}$, BF = 4·33, CF + 0·5, DF − 0·87, EF = −1, DE 0 kN

25. Reactions 1·75, 1·25 kN, AB + 2·02, BC + 0·87, CD + 1·44, ED − 0·72, AE − 1·01, BE + 0·29, EC − 0·29 kN

26. BC − 2, CD − 0·58, CE − 1, AC + 2·31, ED + 1·15, AE + 0·58 kN

27. AC − $\dfrac{2\sqrt{3}}{3}$, CD − 1, BD + $\sqrt{2}$, CB + $\dfrac{\sqrt{3}}{3}$ kN

Exercise 15B (p. 109).

Tensions are denoted − thrusts +

1. 4·36 N, 23° 26′
2. 7·07 N, 45°, 20 cm
3. 50, 150 N, AB − 50$\sqrt{3}$, BC − 100$\sqrt{3}$, AC + 100 N
4. 3$\sqrt{3}$ N m
5. AB − 7·07, AC + 5 N
6. 4·58 N
7. 0·36, 0·64 kN, AB + 0·6, BC + 0·8, AC − 0·48 kN
8. 3 m from A on BA produced
9. 7, 8 kN
10. 1, 1 kN
11. AB = BC = +0·5, AD = DC = −0·5, BD = +0·5 kN
12. Reactions 3, 4 kN, AB + 3$\sqrt{2}$, AC − 3, BC − $\sqrt{2}$, BD + 4, CD 0, CE − 4, DE + 4$\sqrt{2}$ kN
13. Q = 2·52 kN, P = 2·31 kN, AB + 1·15, AE + 1·73, BE − 1·15, EC + 1·15, BC − 1·15, CD − 1·15, ED + 0·58 kN
14. AC − 1, AE + 1·7, CE + 0·58, CD − 1·15, DE + 1·15 kN
15. AB + 1·33, AF − 1·9, BC + 1·67, FE − 1·33, CD + 1·47, ED − 2·36, CE 0, BF + 0·33, BE − 0·47 kN
16. AC + 2·16, AD − 1·08, CE + 1·25, CD − 0·5, DE +.1, DB − 1·95, EB + 2·25 kN
17. 40° 54′, AB − 113, AC − 227, BC + 76 N
18. 31°, AB − 3·1, AD − 5·1, BC − 1·7, CD − 1, BD + 3·6 N
19. P = 2·37 kN, Q = 3·37 kN, AB + 2·37, AD + 3·35, CD − 2·73, BC − 3·35, AC + 1 kN
20. AD + 1·2, AC − 0·84, DC − 0·96 kN
21. 76·4 N, 41° from horizontal. At Q 57·8 N; PQ + 57·8, QR − 57·8, RP + 28·9 N.
22. 7·91 N, 18·4° from horizontal. At R 10·6 N; PQ + 5, QR + 8·66, RP + 3·17 N.
23. 5 kN vertically upward. PQ, QR, − 8·66; PS, SR, QS, + 10 kN.
24. 72·1 N, 16° from horizontal. At Q, 70·3 N; PQ + 60; QR, SQ, − 20; RS, SP, + 20 N.
25. 4 kN, 30° from horizontal. At Q, 3·46 kN; PQ, TP, + 2·31; QR, TQ, − 2·31; RS − 1·15, ST + 0·58, TR + 1·15 kN.

Exercise 16A (p. 122).

1. (i) 28 kN; (ii) 308 kN
2. 375 N; (i) 1·1 m/s²; (ii) 0·1 m/s²
3. 10 m
4. (i) 1 cm/s²; (ii) 2250 m
5. 16 kN
6. 3200 N, $\frac{1}{400}$ sec, 71,000 kN/m²
7. 6·7 m/s
8. (i) 2 sec; (ii) 6·125 m
9. $(n + 1) : (n - 1)$
10. $\frac{1}{2}g(1 - \mu)$, $\frac{1}{2}mg(1 + \mu)$
11. 0·245 N, 23 cm/s²
12. $1025/(n - 1)$ g
13. (i) $\frac{1}{4}g$; (ii) $\frac{1}{12}g$
14. $\dfrac{(M - 2m)g}{M + 4m}$ downward, $\dfrac{3Mmg}{M + 4m}$. Sign of acceleration changed, as expression indicates
15. μg backward; $\mu mg/M$ forward; $\dfrac{Mu}{(M + m)\mu g}$; $\dfrac{mu}{M + m}$; $\dfrac{Mu^2}{2\mu g(M + m)}$

Exercise 16B (p. 123).

1. (i) $0{\cdot}34$ m/s^2; (ii) $0{\cdot}27$ m/s^2 **2.** (i) 15 m; (ii) $\frac{1}{3}g \simeq 3{\cdot}3$ m/s^2
3. (i) $78{\cdot}4$ N; (ii) $0{\cdot}49$ m/s^2 **4.** 140 cm/s
5. (i) $3{\cdot}4$ m/s; (ii) $6{\cdot}75$ m; (iii) $2{\cdot}4$ m/s
6. (i) $nm\left(\frac{xg}{100} + f\right)$ N; (ii) $nm\,\frac{x + y}{100}\,g$ N
7. $\left(\frac{x - z}{100}\right)g$ **8.** $1{\cdot}6\,g$ N; $0{\cdot}6\,g$; $-0{\cdot}2\,g$
9. $19{\cdot}6$ m/s **10.** 22m
12. $\frac{1}{2}g(1 - \sin\alpha - \mu\cos\alpha)$; $\frac{1}{2}mg(1 + \sin\alpha + \mu\cos\alpha)$
13. $\frac{(m_1 - m_2 - \mu m)g}{m_1 + m_2 + m}$; $\frac{m_1 g(\overline{1 + \mu} . m + 2m_2)}{m_1 + m_2 + m}$; $\frac{m_2 g(\overline{1 - \mu} . m + 2m_1)}{m_1 + m_2 + m}$
15. $v = 1 - \frac{1}{20}t$; $v = t - \frac{1}{40}t^2$; 133 m

Exercise 17 (p. 127).

1. LMT^{-1}, $L^{-1}MT^{-2}$, $L^{-3}M$, T^{-2}, MT^{-2}, L^2MT^{-2}
2. (i) L^2MT^{-2}, LT^{-1}, L^2M (ii) LT^{-2}, L, LT^{-1}
(iii) L^2MT^{-2}, LMT^{-1}, L^2MT^{-3} (iv) LT^{-2}, LT^{-2}, T^{-1}
3. (i) $L^3M^{-1}T^{-2}$; (ii) $-p = q = \frac{1}{2}$ **4.** 1,040,000 dynes/sq. cm
5. (i) 10^5 dynes; (ii) 7230 pdl; (iii) 10^5
6. $8{\cdot}31$ **7.** 1340
8. 519 **9.** $0{\cdot}422$ **10.** $0{\cdot}205$

Exercise 18A (p. 135).

1. $44{\cdot}1$ m; $235{\cdot}2$ m; 6 sec
2. $78{\cdot}4$ m; $39{\cdot}2$ m; $40{\cdot}4$ m/s at 14° above horizontal (approx.)
3. 98 m/s; 15° or 75° **5.** 1 sec or 4 sec; 6 sec
6. (i) $\tan^{-1} 1$ or $\tan^{-1} 3$; (iii) $\tan^{-1} 2$; (iii) impossible
7. (i) 9500 m; (ii) 100 m **8.** 28 m/s
9. $\frac{2uv}{g}, \frac{v^2}{2g}$ **10.** (i) $36{\cdot}75$ m; (ii) One half
11. (i) 4 sec; (ii) 85 m (approx.); (iii) $34{\cdot}3$ m/s
12. (i) 20 m/s, $19{\cdot}6$ m/s; (ii) $21{\cdot}225$ m; (iii) 80 m
13. 20 m/s, $24{\cdot}5$ m/s; $15{\cdot}6$ m, $25{\cdot}4$ m/s (approx.)

Exercise 18B (p. 137).

1. (i) $19{\cdot}6$ m/s; (ii) $\sqrt{3} \simeq 1{\cdot}73$ sec; (iii) $9{\cdot}8\sqrt{3} \simeq 17$ m
2. (i) $2\frac{1}{2}$ sec; (ii) 53 m (approx.); (iii) $42{\cdot}5$ m/s (approx.)
3. 50° (approx.)
4. 350 m/s at 8° 8′ to horizontal
5. $29{\cdot}4$ m/s, $22{\cdot}05$ m/s **6.** (i) $34{\cdot}3$ m; (ii) 2 sec **7.** 15 m/s
8. $\tan^{-1} 2 \simeq 63°\ 26'$, or $\tan^{-1} 3 \simeq 71°\ 34'$
9. (i) 173 m/s, 1530 m; (ii) 141 m/s, 2040 m (all approx.)
13. $26{\cdot}6$ m/s, $63{\cdot}5$ m (approx.) **14.** $45° + \frac{1}{2}\alpha$

Exercise 19 (p. **141**).

1. 15°, 45°; 58·8 m, 58·8($\sqrt{3}$ + 1) m **2.** 30°, 21 m/s **6.** 60° **12.** 28 m/s; 69 min

Exercise 20A (p. **150**).

1. 3·6
2. (i) 37·5 kJ; (ii) 3·2 kJ
3. (i) 22 J; (ii) 7·35 J
4. 2·5 g N; 245 W
5. 10 kN; 36 km/h
6. (i) 6 t; (ii) 12 kJ; (iii) 188 kJ; (iv) 75 kW
7. 225 N/t
8. 18 m/s approx.
9. $(T_1 - T_0)\pi nd/60$ W
10. 2 kW
12. 6·25 m/s; 856 W
13. 6 m/s; 12 sec; 24 m/s
14. 2 m/s; 4 m/s^2
15. 6 m/s; 7/2 m/s
16. $mga(2 \sec\theta - \tan\theta)$ + const.; $\theta = 30°$
17. $\lambda x^2/a - mgx$ + const; $mga/2\lambda$
18. $\dfrac{kx^2}{2a^2} - mgx$ + const; mga^2/k

Exercise 20B (p. **152**).

1. 746 W
2. 4700 J
3. 340 J; 72 J
4. 24 kN
5. 9·8 kW
6. approx. 5·2 m/s; approx. 7·6 m/s
7. 900 kW; 0·5 m/s^2
8. 400 kW; 0·01 m/s^2
9. 2 kN; 10 kW
10. approx. 12 kW
11. 230 kW approx.
12. 74 kW approx; 0·0096 m^2
13. 125 sec
14. $2k\pi^2n^2 + 2F\pi n$ J
15. 3 kW; 49,000 J
16. $-2mga \sin\frac{1}{2}\theta$ + const; $\theta = 180°$
17. $mga(2\sin^2\theta - \sin\theta)$ + const; vertical position unstable; inclined position stable

Exercise 21 (p. 156).

1. $2\sqrt{2}$; 88 %
2. 450 N
4. 27 N approx.
5. 4·8, 79%; 5·3, 88%; 5·6, 94%; $P = \frac{1}{6}W + 0{\cdot}43$
6. (i) 59%; (ii) 45%
7. $P \simeq 0{\cdot}067W + 3{\cdot}5$; V.R. = 15; (i) 28%; (ii) 65%; (iii) 52·5 N
8. $400\pi \simeq 1260$; 150 kN
9. (i) 26 N; (ii) 2·4 N
10. $\frac{1}{7}$ m; 214 J; $\frac{6}{7}$ m; 250 N
11. (*a*) 8; P, $2P$, $4P$; (*b*) 7; P, $2P$, $4P$; (*c*) 4; P, $2P$
12. (i) 1 : 3; (ii) $\frac{1}{3}$ rev. clockwise; (iii) $1\frac{1}{3}$ rev. clockwise; (iv) $\frac{3}{4}$ rev. clockwise

Exercise 22A (p. 164).

1. (i) 8 m/s; (ii) 20,000 N s; (iii) 60,000 J
2. 13·6 m
3. 16 kN
4. 1·4 m
5. Approx. 0·94 m
6. (i) 3·5 m/s; (ii) $\frac{5}{7}$ sec; (iii) 1·3 m/s; (iv) 5·63 m (approx.)
7. 324 N (approx.)
8. 640 N
9. (i) 24·5 J, 14 J; (ii) 33°
11. $\sqrt{\left\{\dfrac{2(M-m)ga}{M+m}\right\}}$

Exercise 22B (p. 165).

1. (i) 4·5 m/s; (ii) 30,000 N s; (iii) 50%
2. (i) 221 N; (ii) 2%
4. 14 cm
5. $\dfrac{24{\cdot}5W^2}{(136 - 9{\cdot}8W)(W + 0{\cdot}4)}$ m
6. (i) 20 sec; (ii) 39·6 sec
7. 19° 24′
8. 890 m/s
9. $\frac{24}{29}m\sqrt{(2gh)}$, $\frac{12}{29}m\sqrt{(2gh)}$
10. (i) $\frac{1}{4}$, $\frac{1}{2}$, $\frac{3}{4}$ sec; (ii) 2 N s; $\frac{4}{3}$, $\frac{2}{3}$ N s; 1, $\frac{2}{3}$, $\frac{1}{3}$ N s; (iii) 4, $1\frac{1}{3}$, $\frac{2}{3}$ J
11. 7° 36′
12. $I\sin^2\alpha/m$, $I\sin\alpha\cos\alpha/m$

Exercise 23A (p. 169).

1. 6, 18 m/s
2. −1, 11 m/s
3. 8, 33 m/s
4. −20, 29 cm/s
5. $e = 0{\cdot}9$
6. (i) 2 t; (ii) 0·2; (iii) 26·8 kJ
7. $\frac{1}{5}V(3-2e)$, $\frac{3}{5}V(1+e)$; $\frac{1}{5}V(1+e)(2-e)$, $\frac{2}{5}V(1+e)^2$
8. $2/(1+e)$; $u[\frac{1}{2}(1+e)]^{n-1}$
9. 2·4 m/s; 0·8, 2·4 m/s; 1·05 m

Exercise 23B (p. 170).

2. 6, 11·4 m/s
3. 100 N s, 975 J
4. 9·6, − 12·4 m/s
5. 1 : 15
6. $e = 0{\cdot}2$
8. $\frac{2}{9}$
9. $-V/12$, $V/8$, $13V/72$
10. $4V/3$, $4V/\{3(n+1)\}$

Exercise 24A (p. 174).

1. $e = \frac{1}{2}$, $2\frac{1}{7}$ sec, 4·17 m
2. 45°, 2·5 m
3. Ratio $= e$
4. 84° 19′ with line of centres
6. 13 m/s, 67° 23′ with AB; 17·9 m/s, 26° 34′ with AB produced
7. $u/\sqrt{3}$, perpendicular to original directions
8. For A, $\frac{1}{2}(1-e)u\cos\alpha$, $u\sin\alpha$. For B, $\frac{1}{2}(1+e)u\cos\alpha$, 0
10. First two, $\dfrac{u}{\sqrt{3}}(1+e)$. Third sphere, $\frac{1}{3}u(2-e)$

Exercise 24B (p. 175).

1. $\frac{20}{7}$ sec, $10\sqrt{3} \simeq 17$ m **3.** 2·52 m/s, 0·84 m/s
4. 78° 41′ with line of centres produced
5. 8·14 m/s, 79° 23′ with BA produced; 11·6 m/s, 25° 28′ with AB produced
6. $2ea$
8. $\frac{1}{2}(u_0 + v_0) + \frac{1}{2}(-e)^n(u_0 - v_0)$; $\frac{1}{2}(u_0 + v_0) - \frac{1}{2}(-e)^n(u_0 - v_0)$
9. Angle with AB $= \tan^{-1} \dfrac{v(M + m)}{u(m - eM)}$
10. First, $u/\sqrt{2}$ at 45° to vertical. Second, $u/2$ horizontally

Exercise 25A (p. 179).

1. 0·47 m **2.** 10 rad/sec **3.** 4·4 m/s
4. 11 m/s; 0·58 **5.** 22° 12′; 0·28 **6.** $\dfrac{2v^2}{\sqrt{3g}}$
7. $\dfrac{5m}{2}(\frac{1}{4}\omega^2 + \frac{1}{3}g)$, $\dfrac{5m}{2}(\frac{1}{4}\omega^2 - \frac{1}{3}g)$. $\omega = \sqrt{\dfrac{4g}{3}}$
8. $v = \sqrt{\dfrac{m_2 gr}{m_1}}$; $\dfrac{m_1 I}{m_1 + m_2}$; $\tan^{-1}\left(\dfrac{(m_1 + m_2)v}{I}\right)$ with inward radius

Exercise 25B (p. 17o).

1. $v = \sqrt{(ga \sin\theta \tan\theta)}$; 60° **2.** 66° 29′
3. (i) 54 km/h; (ii) 0·28 **4.** 1·12 m approx.
5. $\dfrac{2g(2M + m)}{m\omega^2}$ **9.** 34 m/s
10. $10mg$; 2·8 m/s

Exercise 26A (p. 184).

1. $g\sqrt{5}$, at $\tan^{-1}(\frac{1}{2})$ below the inward radius **2.** 1 : 7
5. $mv_0^2/a + mg(3\cos\theta - 2)$
10. $v^2 = \dfrac{2ag}{15}(8 - 8\cos\theta + \sin\theta)$, where a = radius of cylinder
12. $\sqrt{\{2ga(1 + \sin\theta)\}}$; $3 + 2\sqrt{2} : 1$

Exercise 26B (p. 185).

1. 120° **2.** $3mg$ **4.** $\frac{1}{2}mg \operatorname{cosec}\alpha$, $mg(3\sin\theta - 2\sin\alpha)$, 30°
6. $mg(3\cos\theta - 2) - mu^2/a$, $\sqrt{(3ag)}$
8. $m(g\cos\alpha + l\omega^2\sin^2\alpha)$ **10.** $\sqrt{(2gh)}$

Exercise 27A (p. 190).

1. $\frac{1}{2}(g \pm \frac{1}{5}\pi^2) \simeq$ 5·9 or 3·9 N **2.** $mg - 64mx$; 15 cm **3.** $2\pi\sqrt{(0{\cdot}3/g)}$; 20 cm
4. 1·8 sec; 28 cm **5.** $2\pi\sqrt{(m/2\lambda)}$ **6.** $2\pi\sqrt{(3m/2\lambda)}$
7. 24 cm/s; 0·15 sec **8.** $a - \dfrac{g}{4\pi^2 n^2}$ **9.** $\sqrt{(10ga)}$
10. 2 cm; 0·12 sec **11.** $(3 + \sqrt{2})mgx/(2a)$ **12.** $\frac{1}{2}\sqrt{\dfrac{17a\lambda}{m}}$

Exercise 27B (p. **192**).

1. 15 cm; $\pi/42$ sec
3. 10·18 a.m.; 2·51 m/hr
4. $\frac{\pi}{24}$ sec; 25 cm
5. $x = 0{\cdot}125 \sin(8t + 0{\cdot}9273)$; 0·125 m; 1 m/s; 0·08 sec
7. $2\pi\sqrt{(a/g)}$; $a/2$ m
8. $9b/2$ from original position; $\sqrt{\frac{b}{g}}\left(\frac{2\pi}{3} + \sqrt{3}\right)$
9. $\sqrt{\tfrac{1}{2}(a^2 + b^2)} - b$
10. 6·9 m/s
11. $x + \tfrac{1}{2}na - nx^2/(2a)$
12. $\tfrac{1}{2}a(3 + \cos \omega t)$

Exercise 28 (p. **197**).

1. (i) 0·635 sec; (ii) 2·84 sec
2. 0·995 m
3. 24·8 cm
4. 88 sec/day, losing
5. 0·10% decrease
6. 72 sec/day loss
7. 0·09% decrease
8. 0·026% increase
9. 110 sec/hr gain
10. Wt × 1·00087; 0·04% increase
11. 33 cm/s
12. 4° 3′
13. $x = a$
14. $3a$
15. $5a$
16. $\frac{1}{2\pi}\sqrt{\frac{Ng}{(N+1)a}}$

Exercise 29A (p. **206**).

1. $\frac{g}{k}(1 - e^{-kt})$
2. $\frac{1}{20}, \frac{1}{400}, x = 400 \log(1 + \frac{t}{20}), v = 20e^{-x/400}$
4. $16\tfrac{2}{3}$ m
5. 60 m/s
12. $m = m_0 e^{-at}, \frac{p}{p_0} = \left(\frac{m_0}{m}\right)^b, p = p_0 e^{abt}$
13. 35·1 min
15. $2\tfrac{4}{7}$

Exercise 29B (p. **207**).

7. $a^2/\sqrt{k}$
9. $\tfrac{3}{4}V$
10. 30·5°, 20·5 min
11. 3906
12. 48·2 sec
13. $i = \frac{E}{R}(1 - e^{-Rt/L})$
15. 44%, $22\tfrac{1}{2}$

Exercise 30 (p. **213**).

1. $260g \simeq 2550$ N
3. $\frac{w(s^2 + k^2)}{2k}$
7. $80g \simeq 780$ N
10. 7650 kN, 7500 kN
11. 37·5 m, $95g \simeq 930$ N
12. 2·5 m

Exercise 31 (p. **221**).

1. $l/3$
2. $\tfrac{1}{2}Ma^2$
3. $\tfrac{1}{3}Ma^2 \sin^2\theta$
4. $\tfrac{4}{5}M(a^2 + b^2)$
6. $\tfrac{1}{2}M(a^2 + b^2), \tfrac{1}{4}M(a^2 + b^2)$
7. $\tfrac{2}{3}Ma^2$
8. $\tfrac{1}{4}Ma^2$
9. $\tfrac{2}{3}Ma^2$
11. $\tfrac{1}{2}Mh^2, \frac{Mh^2}{18}, \tfrac{1}{6}Mh^2$
12. $\tfrac{5}{24}Ma^2$

Exercise 32A (p. **226**).

1. 8 rad/s^2 **2.** 16 rad/s^2, 40 m/s **3.** 4·8 N m

4. 13 N m **5.** 150, 20π N m **6.** $16\frac{2}{3}$ rev, 0·98 m

7. 160 rad/s, 7·1 N m **8.** $\dfrac{2Ft}{Ma}$ **11.** $\dfrac{ga^2}{a^2 + k^2}$

13. $\dfrac{mb^2g}{I + mb^2}, \dfrac{Img}{I + mb^2}$ **16.** $\sqrt{\left(\dfrac{3g}{2a}\right)}, \sqrt{(6ag)}$ **17.** $\sqrt{\left(\dfrac{6ga}{4a^2 + b^2}\right)}$

18. $4 \times \sqrt{\left(\dfrac{3ag}{11}\right)}$ **19.** $\sqrt{(\frac{3}{2}gl)}$ **20.** $\sqrt{\left(\dfrac{6mg}{a(M + 2m)}\right)}$

Exercise 32B (p. **228**).

1. 60 rad/s, 57·3 **2.** 7·5 N, 47·7 **3.** 2 m/s^2, 118 N

4. $\sqrt{\left(\dfrac{3g}{a}\right)}, \dfrac{1}{a}\sqrt{\left[3\left(ga - \dfrac{\pi L}{m}\right)\right]}$ **6.** 270 grams

7. 6·3 N m, 300 rev **8.** $\dfrac{mg(2M + m)}{2M + 3m}$ **9.** $\dfrac{g(2m - M)}{2m + 3M}$

10. $3\sqrt{(\frac{1}{2}ag)}$ **11.** $\dfrac{g}{7}$ m/s^2, $\dfrac{g}{14}$ m/s^2 **12.** 126 sec

13. 3·35 **15.** 0·61 m/s^2, 3·8 m/s **16.** 3·3 m/s

17. $\frac{1}{5}g$, $\frac{6}{5}mg$, $\frac{8}{5}mg$ **18.** $\dfrac{9mga}{2\pi}$ N m **20.** $\dfrac{mga}{\pi}$

Exercise 33 (p. **231**).

1. $\frac{4}{3}a$ **2.** $2\pi\sqrt{\left(\dfrac{3a}{2g}\right)}$ **3.** 2·2 sec

4. 1·35 sec **5.** $\frac{16}{15}a$ **7.** $2\pi\sqrt{\left(\dfrac{l}{\sqrt{3}g}\right)}$, OG $= \dfrac{l}{2\sqrt{3}}$

9. $\frac{4}{3}a$, $\frac{1}{3}a$ **10.** $\frac{56}{55}a$ **11.** $\dfrac{a^2 + 4h^2}{5h}$

13. 105 cm, 75 cm, 20 cm **14.** 72 ma^2, $9a$ from the axis

Exercise 34A (p. **240**).

1. 4 mg, $\frac{3}{2}$ mg, $\frac{1}{4}mg$ **3.** 7·7 m/s; 14·7 N, 2·45 N

4. 45° **6.** $\frac{1}{3}mg(7\cos\theta - 4)$, $\frac{1}{3}mg\sin\theta$

8. $\frac{10}{3}mg$ **9.** $4m\sqrt{\frac{1}{6}ga}$

10. $\sqrt{\left(\dfrac{3g}{2a}\right)}$, $\frac{1}{3}m\sqrt{(6ag)}$ **12.** $\frac{1}{2}v$, $\dfrac{2v}{l}$

13. $\dfrac{3Px}{4a^2M + 3b^2m}$ **15.** 200 rev/min **18.** $\frac{1}{3}\omega$

Exercise 34B (p. **241**).

1. $\frac{1}{5}mg(4 + 117 \sin^2 \theta)^{1/2}$ **3.** Along AB

7. 41° 24′ **8.** $m\sqrt{(\frac{1}{6}lg)}$ **9.** $m\sqrt{(\frac{3}{8}ag)}$

10. 32 N s **11.** $5(\sqrt{\frac{7}{5}}\,g)$, 63° **12.** $4m\sqrt{(\frac{1}{3}ag)}\,.\left(1 - \frac{1}{\sqrt{2}}\right)$

15. 23° 4′. **16.** 66° 25′

Exercise 35A (p. **252**).

1. $\frac{2}{3}g \sin \alpha$, $\frac{1}{3} \tan \alpha$ **2.** $\frac{2}{3}g$, $\frac{1}{3}Mg$

3. The hollow sphere takes the greater time to travel a given length of plane

7. $\sqrt{(6ag)}$ **8.** $\omega/7$

10. $(MK^2 + mr^2)\ddot{\theta} = L$ **11.** $(M + 2m)a\dot{\theta}^2 = 4mg(1 - \cos \theta)$

14. $\frac{15}{56}mg$, $\frac{403}{448}mg$ **15.** $mg \sin \theta$, $\frac{1}{4}mg \cos \theta$ **16.** $\frac{2\sqrt{2}}{3\sqrt{3}}\,a\omega$

Exercise 35B (p. **254**).

1. $\frac{1}{3}gt^2 \sin \alpha$, $2t/\sqrt{3}$ **3.** $\frac{92}{867}$ **5.** $\omega/3$

6. $\frac{(l^2 + 3a^2)\,\omega}{4l^2}$ **7.** $\frac{1}{\cos \frac{1}{2}\alpha}\sqrt{\left[\frac{g}{a}\left(1 + \frac{2m}{M}\sin^2 \alpha\right)\right]}$ **8.** $\frac{7}{16}\omega$

10. $\sqrt{\left(\frac{3g \cos \alpha}{2a}\right)}$ **11.** $\frac{1}{2}\sqrt{(ag)}$

Exercise 36A (p. **259**).

1. $\frac{a^2}{3x}$ **2.** $\frac{4P}{m}, \frac{2P}{m}$ **3.** $\sqrt{5} : 1$

6. $\frac{MP}{2(M + m)}, \frac{2P}{M + m}$ **8.** $\frac{7}{2}$ **9.** $\frac{2}{5}a$ **12.** $\frac{3}{7}u$

Exercise 36B (p. **260**).

1. $\frac{2J}{m}$ **5.** $\frac{V}{2}, \frac{V}{l}$

7. $\frac{r_2(I_1r_2\omega_1 \sim I_2r_1\omega_2)}{I_1r_2^2 + I_2r_1^2}, \frac{r_1(I_2r_1\omega_2 \sim I_1r_2\omega_1)}{I_1r_2^2 + I_2r_1^2}$ **10.** $\frac{3}{7}$

13. $\frac{17P}{12m}, \frac{P}{3m}, \frac{P}{12m}, \frac{7P}{4ma}, 0, \frac{P}{4ma}$

Answers to Revision Examples

Chapter 1 (p. 262).

1. 60 knots from 233° 8′ **2.** 356° 3′, 12·39 p.m.

7. $V\sqrt{2}$ from 225°. From 116° 34′ **9.** V knots due E

10. 21·25 knots from 189° 48′

Chapter 2 (p. 263).

1. $1\frac{1}{2}$ min, 3·2 km, 0·06 and 0·22 m/s² **2.** 6 km, 0·042, $6\frac{1}{2}$ min

3. $2\frac{1}{2}$ min, $\frac{1}{4}$ km, 4·2 m/s per min app. **10.** 3 m/s, 4 m, 6 m/s²

Chapter 3 (p. 265).

4. 20·9 m/s, 4390 m/s² **5.** 4·2 m, 4·7 sec, 0·204 sec

8. $0, -0{\cdot}9\pi^2$ **9.** $\frac{2\pi a}{V}, -\frac{V^2 b}{a^2}$

Chapter 4 (p. 266).

1. 7·55 N, 325° 42′ **2.** $2W\cos\frac{1}{2}\alpha$

3. $3(\sqrt{3}-1), \frac{3}{2}(\sqrt{6}-\sqrt{2})$ kN **4.** $\frac{16}{63}$

7. $W/\sqrt{10}$ **8.** $\frac{2w_1 + w_2}{w_2}$ **9.** $(\sqrt{2}-1):1$

Chapter 5 (p. 267).

1. 16° 50′ **2.** 20 cm, 5 cm, 38° 39′ **3.** $\frac{7}{12}W$

5. $\frac{5\sqrt{3}+3}{3}$ kg **6.** 22·1 N at 67° 30′ to the horizontal, 17 N m

7. $\frac{1}{4}Wl\sqrt{3}$, 25° 39′ **8.** $\frac{a(3\cot^2\alpha + 8\cot\alpha + 3)}{4(\cot\alpha + 2)}$, 30°

Chapter 6 (p. 268).

4. $\sqrt{(P^2 + 4W^2)}$ at $\tan^{-1}\frac{2W}{P}$ to the horizontal, $\sqrt{(P^2 + W^2)}$ at $\tan^{-1}\frac{W}{P}$ to the horizontal

8. 3, −9 **9.** $\frac{1}{2}W$, 0

Chapter 7 (p. **270**).

1. ED $+\sqrt{2}$, DB $+1$, BC $-\frac{1}{2}\sqrt{2}$, EC -1, AC $+\frac{1}{2}\sqrt{2}$, DC 0, EA $+2$, $P=\frac{1}{2}\sqrt{2}$

2. $W/(2\sqrt{3})$

3. AB + 87, DA = DB = −50 N

4. AD $-\dfrac{20\sqrt{3}}{3}$, AC $-\dfrac{20\sqrt{3}}{3}$, BC $+\dfrac{10\sqrt{3}}{3}$, AB $+\dfrac{20\sqrt{3}}{3}$, $10\sqrt{\frac{7}{3}}$ at 40° 54′ to the horizontal

5. AC $-\dfrac{100\sqrt{3}}{3}$, BC $-\dfrac{50\sqrt{3}}{3}$, CD $+\dfrac{50\sqrt{3}}{3}$. DE $+\dfrac{50\sqrt{3}}{3}$, CE $-\dfrac{100\sqrt{3}}{3}$ BE + 50.

6. 7, 7; −9·2, +5·66, +6, +4

Chapter 8 (p. **272**).

1. $\frac{3}{7}$, 202 N

2. 2·45 m/s², 1 sec

3. 2·8 m/s², 14 N; 2·1 m/s², 15·4 N

4. 90 m, $\frac{1}{4}$ N

5. 0·7 m/s² down

6. 2000 N

7. $f=30-t$, $v=30t-\frac{1}{2}t^2$, 9000 m

9. $g/11$, $-g/11$, $3g/11$ m/s², $2\frac{1}{4}$

13. $\frac{4}{3}m(f+g)$, $\frac{1}{3}(8m+3M)(f+g)$

Chapter 9 (p. **273**).

1. 14 m/s, 120 m, 5° 43′ to the horizontal

2. 7 sec, 356 m, 64·3 m/s

3. 11·2 m/s, 14 m/s, 6·4 m, 14 m/s

4. 45° and 18° 26′

Chapter 10 (p. **275**).

1. 28 m/s, 55 kW

2. 90 km/h, 0·04 m/s²

3. 16·8 kW

4. 11 min

7. 12·5g N, $43\frac{3}{4}$%

8. 80%, 82W, 120 N app.

9. 13%

10. 150 N

11. 0·8 J

12. 2·28 m

Chapter 11 (p. **276**).

1. −10·9, 2·6 m/s

2. $h/5$

Chapter 12 (p. **277**).

1. 67 (approx.)

2. 17·5 m/s

9. $mg\sec\theta$

10. 27,000 km/h, 100 min

Chapter 13 (p. **279**).

1. 9·8 m, 9·48 a.m. $\dfrac{\pi}{4500}$ m/s

2. $\sqrt{\left[\dfrac{2g(l^2-x^2)}{l}\right]}$, acceleration $=-2gx/l$, $\frac{1}{2}\pi\sqrt{(l/2g)}$

6. 220 sec

7. −0·51%

8. $\frac{9}{2}d$

Chapter 14 (p.**280**).

1. 93·6 g **3.** 35, 6·8 min **4.** 102·6 min

8. $\frac{4\sqrt{7}}{7}V$ **11.** 10 log 2 m/s **13.** 20, 52 N

14. 20 m, 36 m **15.** $42\frac{1}{2}g \simeq 420$ N

Chapter 15 (p.**282**).

5. $\sqrt{\left(\frac{3V^2 + 16ag}{12a^2}\right)}$, $4\sqrt{(\frac{1}{3}ag)}$ **6.** 8000 J, 1 min 20 sec

9. 70° 32′ **11.** $\frac{2}{15}\left[\frac{5Ma^2 + 6mr^2 + 15ma^2}{Ma + 2mr + 2ma}\right]$

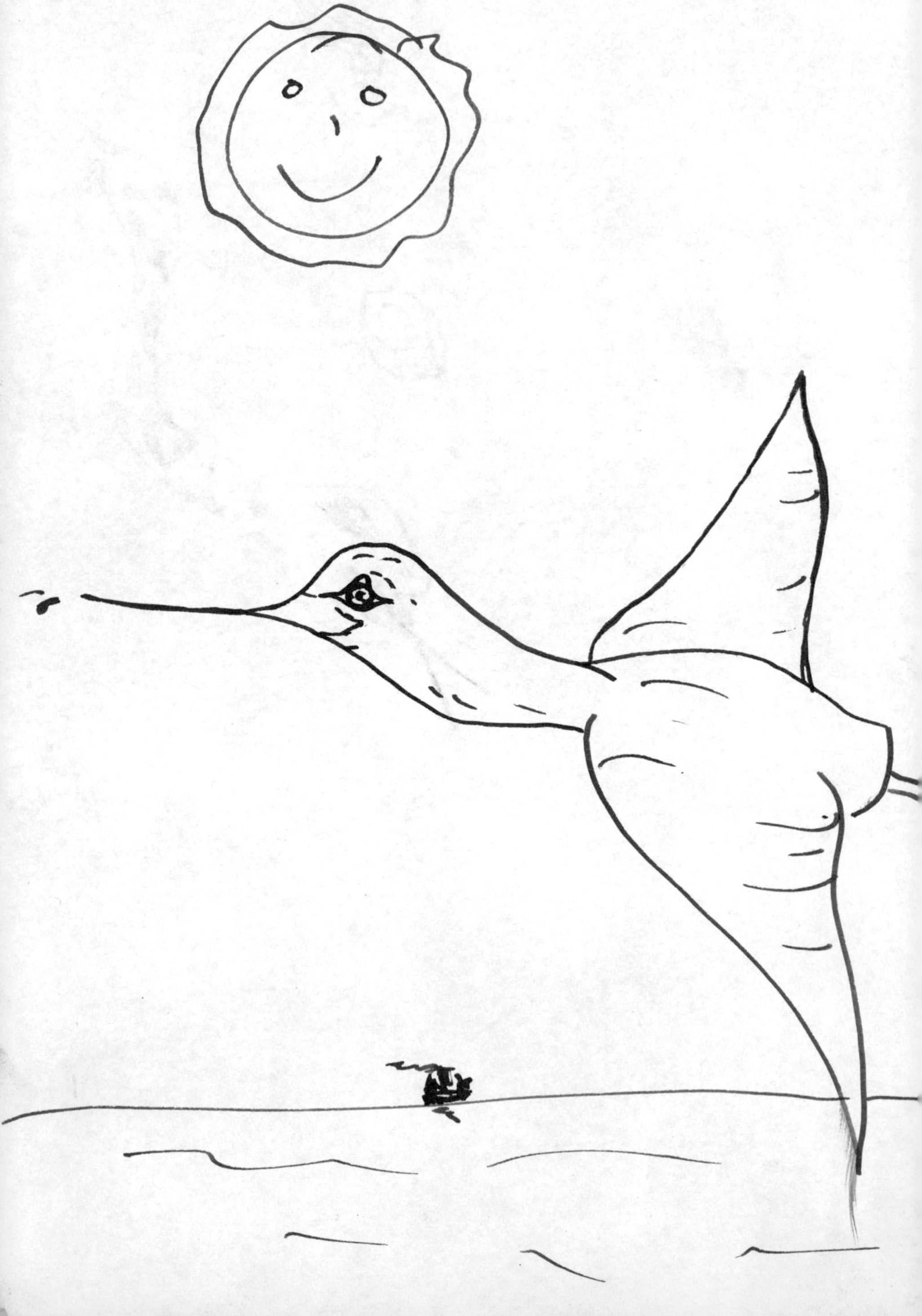

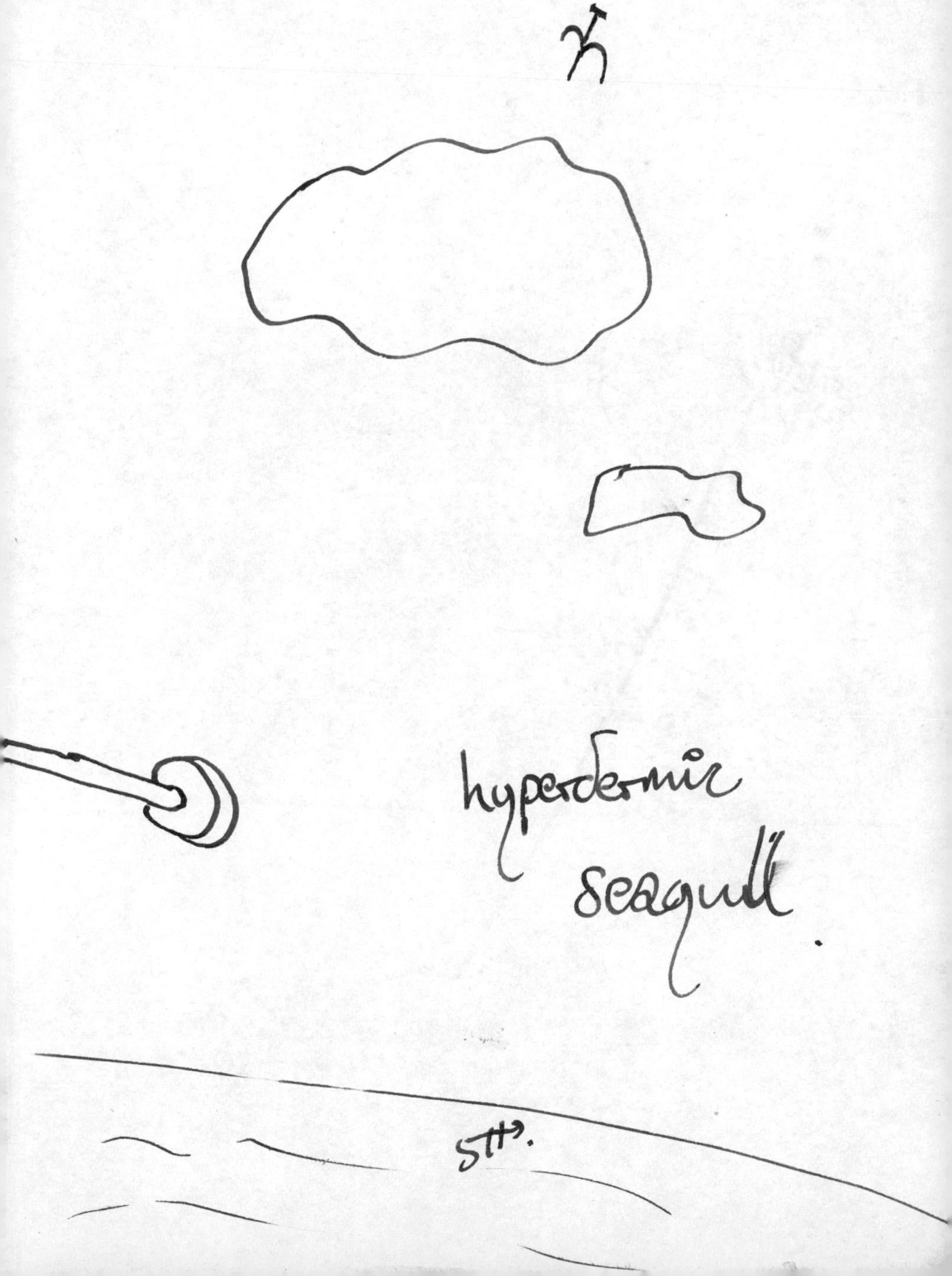
hyperdermic seagull
5th.

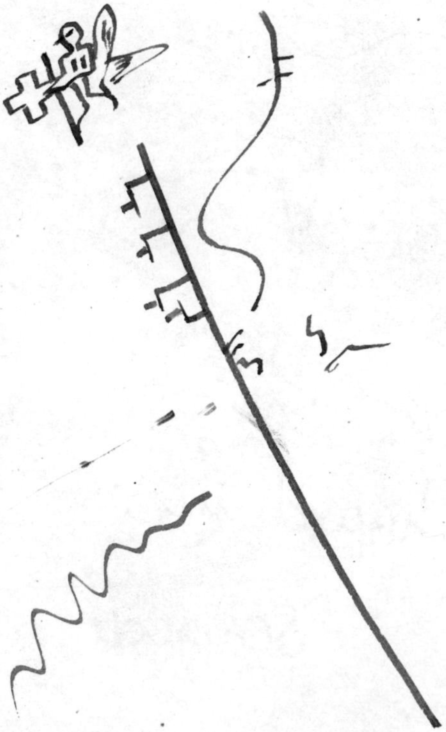